COMMUNICATION TECHNOLOGY UPDATE

8TH EDITION

August E. Grant & Jennifer H. Meadows, Editors

In association with
Technology Futures, Inc.

OXFORD AMSTERDAM BOSTON LONDON NEW YORK PARIS
SAN DIEGO SAN FRANCISCO SINGAPORE SYDNEY TOKYO

Editors
August E. Grant
Jennifer H. Meadows
Technology Futures Inc.
Production Editor   Debra R. Robison
Art Director   Helen Mary V. Marek

Focal Press is an imprint of Elsevier Science.

Library of Congress Cataloging-in-Publication Data
A catalog record for this book is available from the Library of Congress.

ISBN: 0-240-80494-5

British Library Cataloguing-in-Publication Data
A catalogue record for this book is available from the British Library.

The publisher offers special discounts on bulk orders of this book.
For information, please contact:

Manager of Special Sales
Elsevier Science
225 Wildwood Avenue
Woburn, MA 01801-2041
Tel: 781-904-2500
Fax: 781-904-2620

For information on all Focal Press publications available, contact our World Wide Web home page at: http://www.focalpress.com

10 9 8 7 6 5 4 3 2 1

Printed in the United States of America

TABLE OF CONTENTS

Updates can be found on the
Communication Technology Update Home Page
http://www.tfi.com/ctu/

PREFACE

This is the tenth anniversary of the *Communication Technology Update*. Yes, it is also the eighth edition—the first five editions were a year apart, and the last three, two years apart. We've made some additions and changes to reflect this anniversary, including adding a retrospective chapter on the last 10 years of communication technologies (Chapter 25) and moving and extending the chapter providing comparable statistics on all technologies to the front of the book (Chapter 2) to provide context for the discussions that follow.

The first step in preparing this edition was taking a look back at the previous editions. Through all of the changes in technology, cover art, graphics, etc., one constant was apparent—the strong contributions of a host of authors who repeatedly performed the extraordinary feat of synthesizing hundreds of pages of information into chapter form for inclusion in the *Update*.

To acknowledge these authors individually would require a separate chapter. Instead, we want to acknowledge them in groups. The most prominent group is faculty members who have applied substantial expertise in one or more areas of technology to create their chapters. Our special gratitude goes out to those who come back edition after edition, continuing to meet impossible deadlines and producing material that is more up-to-date than can be found in any other book.

The second group consists of students, including both graduate and undergraduate, who took on the challenge of producing updates that had to stand alongside those created by top scholars in the field. It is a testament to these students and their work that we have never received a single negative comment about including student work alongside that of faculty.

We also want to acknowledge the contribution that readers make to the *Update*. Your e-mail and other comments, suggestions, etc. have helped us to improve every edition of the book over previous editions. The fact that the *Update* has become a staple in so many academic programs is the greatest compliment paid to the book.

Focal Press has proven to be a strong partner in the distribution of the book, and we are grateful for their continuing efforts to introduce new readers to the *Update.*

Finally, we want to express our deepest satisfaction and gratitude to Technology Futures, Inc. for this 10-year partnership. As President of TFI, Larry Vanston was the first to commit himself to this publication, and he has continued to provide valuable advice and direction to this project. Although she is no longer involved with the *Update,* Julia Marsh played a major role in developing the book and overseeing its early growth, and we are grateful for her efforts. Production Director Deb Robison and Art Director Helen Mary Marek have worked tirelessly on each of the eight editions to produce the highest-quality product in an incredibly short period of time. When discussing how we condense all of the editing, layout, etc. into a two-month period, most people express amazement at how that quality is achieved with so much speed. The difference is Deb and Helen Mary, and we appreciate their efforts and their spirit.

What's next for the *Update*? We will periodically update the *Communication Technology Update* home page (www.tfi.com/ctu) to supplement the text with updated information and links to a wide variety of information available over the Internet. And, in Spring 2003, we will begin planning our next edition—and looking for input from you on how the *Update* can best serve your needs.

As always, we encourage you to suggest new topics, glossary additions, and possible authors for the next edition of this book by communicating directly with us via e-mail, snail mail, or voice. Thank you!

Augie Grant and Jennifer Meadows
May 13, 2002

Augie Grant
Focus 25 Research & Consulting
25 Crossbow Lakes Court
Columbia, SC 29212-1654
Phone: 803.749.3578
augie@Focus25.com

Jennifer H. Meadows
Department of Communication Design
California State University, Chico
Chico, CA 95929-0504
Phone: 530.898.4775
jmeadows@csuchico.edu

COMMUNICATION TECHNOLOGY UPDATE

8TH EDITION

1

THE UMBRELLA PERSPECTIVE ON COMMUNICATION TECHNOLOGY

August E. Grant, Ph.D.[*]

Communication technologies are the nervous system of contemporary society, transmitting and distributing sensory and control information, and interconnecting a myriad of interdependent units. Because these technologies are vital to commerce, control, and even interpersonal relationships, any change in communication technologies has the potential for profound impacts on virtually every area of society.

One of the hallmarks of the industrial revolution was the introduction of new communication technologies as mechanisms of control that played an important role in almost every area of the production and distribution of manufactured goods (Beniger, 1986). These communication technologies have evolved throughout the past two centuries at an increasingly rapid rate. The evolution of these technologies shows no signs of slowing, so an understanding of this evolution is vital for any individual wishing to attain or retain a position in business, government, or education.

The economic and political challenges faced by the United States and other countries since the beginning of the new millennium clearly illustrate the central role these communication systems play in our society. Just as the prosperity of the 1990s was credited to advances in technology, the economic challenges that followed were linked as well to a major downturn in the technology sector. The

---

[*] Senior Consultant, Focus 25 Research & Consulting, and Visiting Associate Professor, College of Mass Communications and Information Studies, University of South Carolina (Columbia, South Carolina).

aftermath of the September 11 tragedy led many to propose security measures, including control and monitoring of communication technologies that make extensive use of the technologies discussed in this book.

This text provides you with a snapshot of the process of technological evolution. The individual chapter authors have compiled facts and figures from hundreds of sources to provide the latest information on more than two dozen communication technologies. Each discussion explains the roots and evolution, the recent developments, and the current status of the technology as of mid-2002. In discussing each technology, we will deal not only with the hardware, but also with the software, organizational structure, political and economic influences, and individual users.

Although the focus throughout the book is on individual technologies, these individual snapshots comprise a larger mosaic representing the communication networks that bind individuals together and enable us to function as a society. No single technology can be understood without understanding the competing and complementary technologies and the larger social environment within which these technologies exist. As discussed in the following section, all of these factors (and others) have been considered in preparing each chapter through application of the "umbrella perspective." Following this discussion, an overview of the remainder of the book is presented.

DEFINING COMMUNICATION TECHNOLOGY

The most obvious aspect of communication technology is the hardware—the physical equipment related to the technology. The hardware is the most tangible part of a technology system, and new technologies typically spring from developments in hardware. However, understanding communication technology requires more than just studying the hardware. It is just as important to understand the messages communicated through the technology system. These messages will be referred to in this text as the "software." It must be noted that this definition of "software" is much broader than the definition used in computer programming. For example, our definition of computer software would include information manipulated by the computer (such as this text, a spreadsheet, or any other stream of data manipulated or stored by the computer), as well as the instructions used by the computer to manipulate the data.

The hardware and software must also be studied within a larger context. Rogers' (1986) definition of "communication technology" includes some of these contextual factors, defining it as "the hardware equipment, organizational structures, and social values by which individuals collect, process, and exchange information with other individuals" (p. 2). An even broader range of factors is suggested by Ball-Rokeach (1985) in her "media system dependency theory," which suggests that communication media can be understood by analyzing dependency relations within and across levels of analysis, including the individual, organizational, and system levels. Within the system level, Ball-Rokeach (1985) identifies three systems for analysis: the media system, the political system, and the economic system.

These two approaches have been synthesized into the "Umbrella Perspective on Communication Technology" illustrated in Figure 1.1. The bottom level of the umbrella consists of the hardware and software of the technology (as previously defined). The next level is the organizational infrastructure: the group of organizations involved in the production and distribution of the technology. The top level is the system level, including the political, economic, and media systems, as well as other groups of individuals or organizations serving a common set of functions in society. Finally, the "handle" for the umbrella is the individual user, implying that the relationship between the user and a technology must be examined in order to get a "handle" on the technology. The basic premise of the umbrella perspective is that all five areas of the umbrella must be examined in order to understand a technology.

(The use of an "umbrella" to illustrate these five factors is the result of the manner in which they were drawn on a chalkboard during a lecture in 1988. The arrangement of the five attributes resembled an umbrella, and the name stuck. Although other diagrams have since been used to illustrate these five factors, the umbrella remains the most memorable of the lot.)

Figure 1.1
The Umbrella Perspective on
Communication Technology

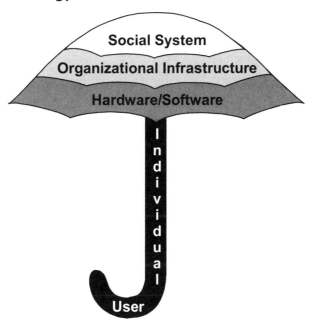

Source: A. E. Grant

Factors within each level of the umbrella may be identified as "enabling," "limiting," "motivating," and "inhibiting." *Enabling factors* are those that make an application possible. For example, the fact that coaxial cable can carry dozens of channels is an enabling factor at the hardware level, and the decision of policy makers to allocate a portion of the spectrum for cellular telephony is an enabling factor at the system level (political system).

Limiting factors are the opposite of enabling factors. Although coaxial cable increased the number of television programs that could be delivered to a home, most analog coaxial networks cannot transmit more than 100 channels of programming. To the viewer, 100 channels might seem to be more than is needed, but to the programmer of a new cable television channel unable to get space on a filled-up cable system, this hardware factor represents a definite limitation. Similarly, the fact that the policy makers discussed above permitted only two companies to offer cellular telephone service in each market is a system-level limitation on that technology.

Motivating factors are those that provide a reason for the adoption of a technology. Technologies are not adopted just because they exist. Rather, individuals, organizations, and social systems must have a reason to take advantage of a technology. The desire of local telephone companies for increased profits, combined with the fact that growth in providing local telephone service is limited, is an organizational factor motivating the telcos to enter the markets for new communication technologies. Individual users who desire information more quickly can be motivated to adopt electronic information technologies.

Inhibiting factors are the opposite of motivating ones, providing a disincentive for adoption or use of a communication technology. An example of an inhibiting factor at the software level might be a new electronic information technology that has the capability to update information more quickly than existing technologies, but provides only "old" content that consumers have already received from other sources. One of the most important inhibiting factors for most new technologies is the cost to individual users. Each potential user must decide whether the cost is worth the service, considering his or her budget and the number of competing technologies.

All four types of factors—enabling, limiting, motivating, and inhibiting—can be identified at the system, organizational, software, and individual user levels. However, hardware can only be enabling or limiting; by itself, hardware does not provide any motivating factors. The motivating factors must always come from the messages transmitted (software) or one of the other levels of the umbrella.

The final dimension of the umbrella perspective relates to the environment within which communication technologies are introduced and operate. These factors can be termed "external" factors, while ones relating to the technology itself are "internal" factors. In order to understand a communication technology or to be able to predict the manner in which a technology will diffuse, both internal and external factors must be studied and compared.

Each communication technology discussed in this book has been analyzed using the umbrella perspective to ensure that all relevant factors have been included in the discussions. As you will see, in most cases, organizational and system-level factors (especially political factors) are more important in the development and adoption of communication technologies than the hardware itself. For example, political forces have, to date, prevented the establishment of a world standard for high-definition television (HDTV) production and transmission. As individual standards are selected in countries and regions, the standard selected is as likely to be the product of political and economic factors as of technical attributes of the system.

Organizational factors can have similar powerful effects. For example, the entry of a single company, IBM, into the personal computer business in the early 1980s resulted in fundamental changes in the entire industry. Finally, the individuals who adopt (or choose not to adopt) a technology, along with their motivations and the manner in which they use the technology, have profound impacts upon the development and success of a technology following its initial introduction.

Each chapter in this book has been written from the umbrella perspective. The individual writers have endeavored to update developments in each area to the extent possible in the brief summaries provided. Obviously, not every technology experienced developments in each of the five areas, so each report is limited to areas in which relatively recent developments have taken place.

OVERVIEW OF BOOK

The next chapter, "Communication Technology Timelines," provides a broad overview of most of the technologies discussed later in the book, allowing you to compare them along a number of dimensions: the year each was first introduced, growth rate, number of current users, etc. This chapter co-anchors the book to highlight commonalties in the evolution of individual technologies, as well as present the "big picture" before we delve into the details.

The technologies discussed in this book have been organized into three sections: electronic mass media, computers and consumer electronics, and telephony and satellite technologies. These three are not necessarily exclusive; for example, direct broadcast satellites (DBS) could be classified as either an electronic mass medium or a satellite technology. The ultimate decision regarding where to put each technology was made by determining which set of current technologies most closely resembled the technology from the user's perspective. Thus, DBS was classified with electronic mass media. This process also locates the discussion of a cable television technology—cable modems—in the "Broadband Networks" chapter in the telephony section.

Each chapter is followed by a brief bibliography. These reference lists represent a broad overview of literally thousands of books and articles that provide details about these technologies. It is hoped that the reader will not only use these references, but will examine the list of source material to determine the best places to find newer information since the publication of this *Update*.

Most of the technologies discussed in this book are continually evolving. As this book was completed, many technological developments were announced but not released, corporate mergers were under discussion, and regulations had been proposed but not passed. Our goal is for the chapters in this book to establish a basic understanding of the structure, functions, and background for each technology, and for the supplementary Internet home page to provide brief synopses of the latest developments for each technology. (The address for the home page is http://www.tfi.com/ctu.)

The final two chapters attempt to draw larger conclusions from the preceding discussions. The first of these two chapters presents a look back at how the communication technology landscape has developed by comparing the current state of the technologies in this book with their status 10 years ago. The final chapter then attempts to place these discussions in a larger context, noting commonalties among the technologies and trends over time. It is impossible for any text such as this one to ever be fully comprehensive, but it is hoped that this text will provide you with a broad overview of the current developments in communication technology.

BIBLIOGRAPHY

Ball-Rokeach, S. J. (1985). The origins of media system dependency: A sociological perspective. *Communication Research, 12* (4), 485-510.

Beniger, J. (1986). *The control revolution.* Cambridge, MA: Harvard University Press.

Rogers, E. M. (1986). *Communication technology: The new media in society*. New York: Free Press.

2

COMMUNICATION TECHNOLOGY TIMELINE

Dan Brown, Ph.D.[*]

Following in the footsteps of earlier works, this chapter traces the timelines of various American communication media from their outsets, (Brown & Bryant, 1989; Brown, 1996; Brown, 1998; Brown, 2000). The goal is to provide consistent data across these media that allow you to better understand each medium, as well as identify patterns across media.

Consideration of the "big picture" is important before addressing individual technologies. Figure 2.1 compares the introduction date of each technology. The timeline provides context for the data that appear in this chapter. The focus on entertainment among Americans is reflected in the ubiquity of television and radio, technologies that exist in nearly every U.S. household. These devices required nearly 50 years to reach current penetration levels. What new technologies will nearly all households use three or four decades from now? The Internet appears to be the most likely choice, and computers must necessarily precede the Internet in penetration as long as computers remain the primary vehicle for Internet access.

Figure 2.1 also shows that the number of new communication technologies introduced in each decade appears to be accelerating. Technology had become more pervasive at the time of introduction of the VCR as compared with the level of technological sophistication in effect when the telephone was introduced. However, many technologies from earlier years do not appear in the graph, usually because they have since become less important or have failed. Examples include 8-track audiotapes, quadraphonic sound, 3D television, CB radios, 8mm film cameras, etc. In their day, each of these technologies would have earned its own chapter in this book, but history has banished them to

---

[*] Associate Dean of Arts & Sciences, East Tennessee State University (Johnson City, Tennessee).

footnotes. Many of the technologies that receive attention in this volume may suffer the same fates in the coming years. The remainder of this chapter attempts to provide one way to help predict which ones will succeed or fail by studying the history of comparable technologies.

Figure 2.1
Communication Technology Timeline

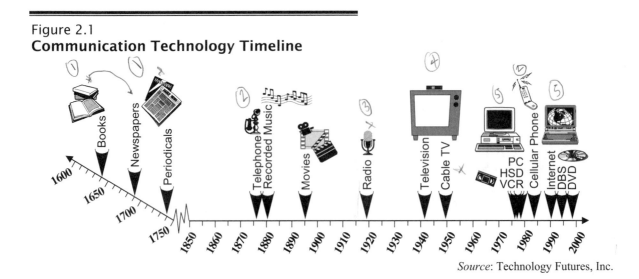

Source: Technology Futures, Inc.

To aid in understanding rates of adoption and use, a premise of this chapter is that non-monetary measures are a more consistent measure of a technology's impact than the dollar value of sales. More meaningful media consumption trends emerge from examining changes in non-monetary media units and penetration (i.e., percentage of marketplace use, such as households) rather than on dollar expenditures. Box office receipts from motion pictures offers a notable exception, with unit attendance figures reported along with the dollar amounts.

Another premise of this chapter is that government sources should provide as much of the data as possible. Researching the growth or sales figures of various media over time quickly reveals conflict in both dollar figures and units shipped or consumed. Such conflicts exist even among government reports. For example, marked differences occurred for some years between the *Statistical Abstract* tables and *Industrial Outlook* tables in reporting motion picture box office receipts. From time to time, unit definitions changed, as with newly published book titles, sometimes resulting in a loss of meaningful trend portrayals. Government sources, although frequently based on private reports, provide some consistency to the reports. Readers should use caution in interpreting data for individual years and instead emphasize the trends over several years.

PRINT MEDIA

The American printing industry is the largest such industry among the printing countries of the world. Although competition and consolidation of ownership reduced the number of American printing firms to 62,000 in 1999, 4.6% fewer than the figure for a decade earlier, the printing industry still enjoyed "unparalleled demand for its products" (U.S. Department of Commerce/International Trade Association, 2000, p. 25-1). Foreign investment in printing is bringing a more global characteristic to the industry, with half of the 20 largest American book publishers having foreign ownership (U.S. Department of Commerce/International Trade Association, 1999, p. 25-5).

NEWSPAPERS

Publick Occurrences, Both Foreign and Domestick was the first American newspaper, appearing in 1690 (Lee, 1917). Figure 2.2 and Table 2.1 show extremely slow growth in American newspaper firms and newspaper circulation until the 1800s. Early growth suffered from relatively low literacy rates and the lack of discretionary income among the bulk of the population. The progress of the industrial revolution brought money for workers and improved mechanized printing processes. Lower newspaper prices and the practice of deriving revenue from advertisers encouraged significant growth beginning in the 1830s. Newspapers made the transition from the realm of the educated and wealthy elite to a mass medium serving a wider range of people from this period through the Civil War era (Huntzicker, 1999).

Figure 2.2
**Newspaper Firms & Daily Newspaper
Circulation**

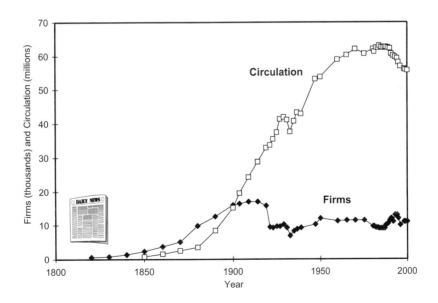

Table 2.1
**Newspaper Firms & Daily Newspaper
Circulation, 1704-2000**

Year	Firms	Circulation*	Year	Firms	Circulation*
1704	1		1933	6,884	37.6
1710	1		1935	8,266	40.9
1720	3		1937	8,826	43.3
1730	7		1939	9,173	43.0
1740	12		1947	10,282	53.3
1750	14		1950	12,115	53.8
1760	18		1960	11,315	58.9
1770	30		1965	11,383	60.4
1780	39		1970	11,383	62.1
1790	92		1975	11,400	60.7
1800	235		1980	9,620	62.2
1810	371		1981	9,676	61.4
1820	512		1982	9,183	62.5
1830	715		1983	9,205	62.6
1840	1,404		1984	9,151	63.1
1850	2,302	0.8	1985	9,134	62.8
1860	3,725	1.5	1986	9,144	62.5
1870	5,091	2.6	1987	9,031	62.8
1880	9,810	3.6	1988	10,088	62.7
1890	12,652	8.4	1989	10,457	62.6
1900	15,904	15.1	1990	11,471	62.0
1904	16,459	19.6	1991	11,689	60.0
1909	17,023	24.2	1992	11,339	60.0
1914	16,944	28.8	1993	12,597	60.0
1919	15,697	33.0	1994	12,513	59.0
1921	9,419	33.7	1995	12,246	57.0
1923	9,248	35.5	1996	10,466	57.0
1925	9,569	37.4	1997	10,042	57.0
1927	9,693	41.4	1998	10,504	56.2
1929	10,176	42.0	1999	10,530	56.0
1931	9,299	41.3	2000	10,696	55.8

* In millions

Note: The data from 1704 through 1900 are from Lee (1973). The data from 1904 through 1947 are from U.S. Bureau of the Census (1976). The number data between 1947 and 1986 are from U.S. Bureau of the Census (1986). The data from 1987 through 1988 are from U.S. Bureau of the Census (1995c). The data for 1988 and 1989 are from U.S. Bureau of the Census (1997). Data from 1989 through 1997 are from U.S. Bureau of the Census (1999). Data after 1997 are from U.S. Bureau of the Census (2001).

The Mexican and Civil Wars stimulated public demand for news by the mid-1800s, and modern journalism practices, such as assigning reporters to cover specific stories and topics, began to emerge. Circulation wars among big city newspapers in the 1880s featured sensational writing about frequently outrageous stories. The number of newspaper firms and newspaper circulation soared. Although the number of firms would level off in the 20th century, circulation continued to rise.

The number of morning newspapers more than doubled after 1950, despite a 16% drop in the number of daily newspapers over that period. Circulation remained higher at the start of the new millennium than it was in 1950, although it inched downward throughout the 1990s. Although circulation actually increased in many developing nations, both American newspaper circulation and the number of American newspaper firms remain lower than the respective figures posted in the early 1990s. Total newspaper circulation in 2000 declined for the third year in a row to 55.8 million. Per capita daily newspaper circulation fell to 0.20 in 2000 after four consecutive years at 0.21.

More than 80% of American daily newspaper circulation comes from morning editions. Between 1995 and 2000, the number of morning daily newspapers rose by 16.8% to 766,000. Evening dailies declined during the same period by 18.4% to 727,000 (U.S. Department of Commerce/International Trade Association, 2000). In 2000, the number of newspaper firms in America grew by 1.6% from the previous year to 10,696, marking the third consecutive year of increases after a four-year run of losses (U.S. Bureau of the Census, 2001).

Shifts from analog production techniques to digital methods strengthened the relationship between newspaper publishers and printers. Flexibility from digital production methods increased newspapers' ability to deliver zoned editions that reduced unprofitable fringe readership in areas that lie far from central newspaper production facilities. By the end of the 20th century, more than two-thirds of American daily newspapers maintained Web sites that offered a variety of content, including classified advertising. The use of the Web to increase exposure to classified advertising is especially notable, as ads account for more than 40% of newspaper advertising revenues. "Despite growth in readership of newspapers on the Internet, an expanding U.S. population continues to prefer to purchase the printed edition rather than viewing the electronic edition" (U.S. Department of Commerce/International Trade Association, 2000, p. 25-6).

PERIODICALS

By the end of 1999, the American periodicals industry included 4,700 firms that employed 120,000 people. Consumer magazines accounted for 63% of industry revenues, and business magazines generated 28%. Consumer magazines reached an aggregated 365 million readers in 1999, nearly 75% of the more than 500 million in annual U.S. periodical circulation. Consumer magazine circulation through subscriptions enjoyed gradual annual increases after 1980, with exceptions in 1992 and 1996. However, sales of single copies "have been in a long-term decline for decades" (U.S. Department of Commerce/International Trade Association, 2000, p. 25-10).

"The first colonial magazines appeared in Philadelphia in 1741, about 50 years after the first newspapers" (Campbell, 2002, p. 310). Few Americans could read in that era, and periodicals were costly to produce and circulate. Magazines were often subsidized and distributed by special interest groups, such as churches (Huntzicker, 1999). *The Saturday Evening Post* began in 1821, and it became the longest running magazine in American history. It was also first to target women as an audience and distribute to a national audience. By 1850, nearly 600 magazines were operating.

By the early 20th century, national magazines became popular with advertisers who wanted to reach wide audiences. No other medium offered such opportunity. However, by the middle of the century, many successful national magazines began dying in the face of advertisers' preferences for the new medium of television and the increasing costs of periodical distribution. Magazines turned to smaller niche audiences and were more effective at reaching these audiences than television. Table 2.2 and Figure 2.3 show the number of American periodical titles by year, revealing that the number of new periodical titles nearly doubled from 1958 to 1960.

Table 2.2
Published Periodical Titles

Year	Titles	Year	Titles	Year	Titles
1904	1,493	1947	4,610	1987	11,593
1909	1,194	1954	3,427	1988	11,229
1914	1,379	1958	4,455	1989	11,556
1919	4,796	1960	8,422	1990	11,092
1921	3,747	1965	8,990	1991	11,239
1923	3,829	1970	9,573	1992	11,143
1925	4,496	1975	9,657	1993	11,863
1927	4,659	1980	10,236	1994	12,136
1929	5,157	1981	10,873	1995	11,179
1931	4,887	1982	10,688	1996	9,843
1933	3,459	1983	10,952	1997	8,530
1935	4,019	1984	10,809	1998	12,448
1937	4,202	1985	11,090	1999	11,751
1939	4,985	1986	11,328	2000	13,019

Note: The data from 1904 through 1958 are from U.S. Bureau of the Census (1976). The data from 1960 through 1985 are from U S. Bureau of the Census (1986). The data from 1987 through 1988 are from U S. Bureau of the Census (1995c). The data for 1988 through 1991 are from U.S. Bureau of the Census (1997). The data from 1991 through 1996 are from U.S. Bureau of the Census (1999). The data after 1996 are from U.S. Bureau of the Census (2001).

Figure 2.3
Published Periodical Titles

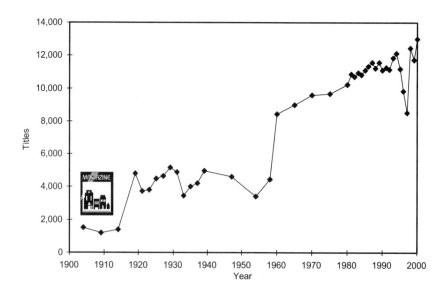

The number of periodical titles increased by 10.8% in 2000 over the number for the previous year. This increase seems particularly strong, given that the number of titles declined in four of the previous five years. In the 10 years beginning in 1990, the average annual gain in the number of periodical titles reached only 20, despite the average of 788 new titles published annually in the 1990s. The difference occurred from the high mortality rate, as evidenced by a loss in total titles in 6 of the 10 years in the decade. "Approximately two-thirds of all new titles fail to survive beyond four or five years" (U.S. Department of Commerce/International Trade Association, 2000, p. 25-9).

The issuing of periodical titles bears a strong positive correlation with the general economic health of the country. Other important factors include personal income, literacy rates, leisure time, and attractiveness of other media forms to advertisers. With the decline in network television viewing throughout the 1990s, magazines became more popular with advertisers, particularly those advertisers seeking consumers under age 24 and over age 45. Both of those groups seem likely to increase in numbers.

Following the lead of newspapers, almost all magazine publishers operate Web sites. These sites offer content, solicit audience interaction, and peddle print wares. In an environment that becomes increasingly cluttered with media choices, Web sites provide opportunities to engage readers and win them as subscribers to print media (U.S. Department of Commerce/International Trade Association, 2000).

BOOKS

Books obviously enjoy a history spanning many centuries. Stephen Daye printed the first book in colonial America, *The Bay Psalm Book*, in 1640 (Campbell, 2002). Books remained relatively expensive and rare until after the industrial revolution. Linotype machines developed in the 1880s allowed for mechanical typesetting. After World War II, the popularity of paperback books helped the industry expand.

Table 2.3 shows new book titles published by year from the late 1800s through 2000. These data show a remarkable, but potentially deceptive, increase in the number of new book titles published annually, beginning in 1997. The U.S. Bureau of the Census reports that the data provided were based on material from R. R. Bowker, which changed its reporting methods beginning with the 1998 report. Ink and Grabois (2000) explained the increase as resulting from a change in the method of counting titles "that results in a more accurate portrayal of the current state of American book publishing" (p. 508). Data for previous years came from databases compiled, in part by hand, by R. R. Bowker. The older counting process tallied only books included in the Library of Congress Cataloging in Publication program. This program included publishing by the largest American publishing companies, but omitted such books as "inexpensive editions, annuals, and much of the output of small presses and self publishers" (p. 509). Ink and Grabois observed that the U.S. ISBN Agency assigns more than 10,000 new publisher ISBN prefixes annually.

Table 2.3
Published Book Titles

Year	Total Titles	Year	Total Titles	Year	Total Titles	Year	Total Titles
1880	2,076	1911	11,123	1942	9,525	1980	42,377
1881	2,991	1912	10,903	1943	8,325	1981	48,793
1882	3,472	1913	12,230	1944	6,970	1982	46,935
1883	3,481	1914	12,010	1945	6,548	1983	53,380
1884	4,088	1915	9,734	1946	7,735	1984	51,058
1885	4,030	1916	10,445	1947	9,182	1985	50,070
1886	4,676	1917	10,060	1948	9,897	1986	52,637
1887	4,437	1918	9,237	1949	10,892	1987	56,057
1888	4,631	1919	8,594	1950	11,022	1988	55,483
1889	4,014	1920	8,422	1951	11,255	1989	53,446
1890	4,559	1921	8,329	1952	11,840	1990	46,738
1891	4,665	1922	8,638	1953	12,050	1991	48,146
1892	4,862	1923	8,863	1954	11,901	1992	49,276
1893	5,134	1924	9,012	1955	12,589	1993	49,756
1894	4,484	1925	9,574	1956	12,538	1994	51,663
1895	5,469	1926	9,925	1957	13,142	1995	62,039
1896	5,703	1927	10,153	1958	13,462	1996	68,175
1897	4,928	1928	10,354	1959	14,876	1997	119,262
1898	4,886	1929	10,187	1960	15,012	1998	120,244
1899	5,321	1930	10,027	1961	18,060	1999	119,357
1900	6,356	1931	10,307	1962	21,904	2000	96,080
1901	8,141	1932	9,035	1963	25,784		
1902	7,833	1933	8,092	1964	28,451		
1903	7,865	1934	8,198	1965	28,595		
1904	8,291	1935	8,766	1966	30,050		
1905	8,112	1936	10,436	1967	28,762		
1906	7,139	1937	10,912	1968	30,387		
1907	9,620	1938	11,067	1969	29,579		
1908	9,254	1939	10,640	1970	36,071		
1909	10,901	1940	11,328	1975	39,372		
1910	13,470	1941	11,112	1979	45,182		

Note: The data for 1880-1919 include pamphlets; 1920-1928, pamphlets included in total only; thereafter, pamphlets excluded entirely. Beginning 1959, the definition of "book" changed, rendering data on prior years not strictly comparable with subsequent years. Beginning 1967, the counting methods were revised, rendering prior years not strictly comparable with subsequent years. The data from 1904 through 1947 are from U.S. Bureau of the Census (1976). The data from 1975 through 1983 are from U.S. Bureau of the Census (1984). The data from 1984 are from U.S. Bureau of the Census (1985). The data from 1985 and 1989 through 1992 are from U.S. Bureau of the Census (1995c). The data from 1986 and 1987 are from U.S. Bureau of the Census (1990). The data from 1988 are from U.S. Bureau of the Census (1992). The data from 1985 and 1989 are from U.S. Bureau of the Census (1995c). The data from 1989 through 1993 are from U.S. Bureau of the Census (1997). The data from 1993 through 1996 are from U.S. Bureau of the Census (1999). The data from 1997 through 2000 are from U.S. Bureau of the Census (2001).

Figure 2.4 shows trends in publishing book titles through 1996. Figure 2.5 shows similar trends from 1997 through 2000, after the change in the methods of counting book titles published annually.

Figure 2.4
Published Book Titles

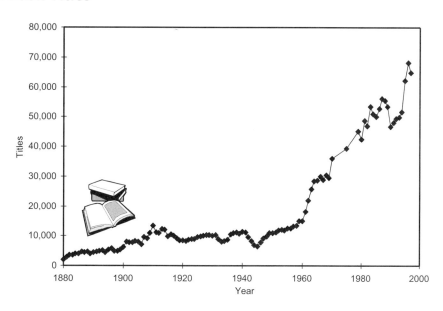

Figure 2.5
Published Book Titles, 1997-2000

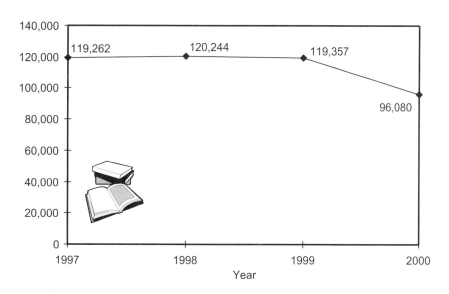

More than 40% of adult Americans (79,218,000) reported reading books at least once as leisure activity during the previous 12 months in 1999, and more than 20% (43,919,000) reported such participation at least twice each week. Annual expenditures for reading material per American consumer between 1992 and 1999 fluctuated somewhat from year to year, but mostly held steady overall, dropping from $162 in 1992 to $159 in 1999 (U.S. Bureau of the Census, 2001). The 1999 figure marked the second consecutive decline from the previous year, but it remained higher than the lowest figure of the interval ($156) in 1996. The highest of these figures ($166) occurred in 1993. The expenditure for reading per person accounted for 8.4% of the total expenditures per person for the combination of entertainment and reading in 1999, down from 10.4% in 1992 and ending a run of two consecutive years of increases in the proportion of entertainment spending per person per year.

The outlook for the growth of reading of printed matter includes both positive and negative indicators. The number of Americans older than 45 years is expected to grow by 10.1 million through 2004 over the same figure in 2000, but growth is also occurring in the number of households with Internet access, as is documented later in this chapter. Relatively new book forms, such as Internet sales and print-on-demand, promise to aid publishers' profitability by making books more accessible to consumers (Publishing, 2002).

TELEPHONE

Alexander Graham Bell became the first to transmit speech electronically, that is, to use the telephone, in 1876. By June 30, 1877, 230 telephones were in use, and the number rose to 1,300 by the end of August, mostly to avoid the need for a skilled interpreter of telegraph messages. The first exchange connected three company offices beginning in Boston on May 17, 1877, reflecting a focus on business rather than residential use during the early decades of the telephone. Hotels became early adopters of telephones as they sought to reduce the costs of employing human messengers, and New York's 100 largest hotels had 21,000 telephones by 1909. After 1894, non-business telephone use became ordinary, in part because business use lowered the cost of telephone service. By 1902, 2,315,000 American telephones were in service (Aronson, 1977). Table 2.4 and Figure 2.6 document the growth to near ubiquity of telephones in American households and the expanding presence of wireless telephones.

Table 2.4
**Telephone Penetration and Cellular Systems
and Subscribers, 1920-2000**

Year	Households with Telephones (%)	Year	Households with Telephones (%)	Cellular Systems	Cellular Subscribers (thousands)
1920	35.0	1957	75.5		
1921	35.3	1958	76.4		
1922	35.6	1959	78.0		
1923	37.3	1960	78.3		
1924	37.8	1961	78.9		
1925	38.7	1962	80.2		
1926	39.2	1963	81.4		
1927	39.7	1964	82.8		
1928	40.8	1965	84.6		
1929	41.6	1966	86.3		
1930	40.9	1967	87.1		
1931	39.2	1968	88.5		
1932	33.5	1969	89.8		
1933	31.3	1970	90.5		
1934	31.4	1975			
1935	31.8	1979			
1936	33.1	1980	93.0		
1937	34.3	1981			
1938	34.6	1982			
1939	35.6	1983			1
1940	36.9	1984	91.8		100
1941	39.3	1985	92.2		350
1942	42.2	1986	92.2		682
1943	45.0	1987	92.5		1,231
1944	45.1	1988	92.9	517	2,069
1945	46.2	1989	93.0	584	3,509
1946	51.4	1990	93.3	751	5,283
1947	54.9	1991	93.6	1,252	7,557
1948	58.2	1992	93.9	1,506	11,033
1949	60.2	1993	94.2	1,529	16,009
1950	61.8	1994	93.9	1,581	24,134
1951	64.0	1995	93.9	1,627	33,786
1952	66.0	1996	93.8	1,740	44,043
1953	68.0	1997	93.9	2,228	55,312
1954	69.6	1998	94.1	3,073	69,209
1955	71.5	1999	94.2	3,518	86,047
1956	73.8	2000	94.1	2,440	109,478

Note: 1950-1982 data applies to principal earners filing reports with FCC; earlier data applies to Bell and independent companies. Beginning in 1959, data includes figures from Alaska and Hawaii. The data for 1986 and 1987 are estimates. The data to 1970 are from U.S. Bureau of the Census (1976). The data from 1970 through 1982 are from U.S. Bureau of the Census (1986). The data after 1982 are from U.S. Department of Commerce (1987). The data from 1986 and 1987 are from U.S. Bureau of the Census (1992, 1993). The data for 1987 through 1989 are from U.S. Bureau of the Census (1997). The telephone households data from 1989 through 1998 are from U.S. Bureau of the Census (1999), except that households with telephones for 1998 are from FCC (2000). Cellular telephone data from 1990 through 1994 are from U.S. Bureau of the Census (1999). Households with telephones for 1999 and cellular subscribers for 1994 through 1999 are from U.S. Bureau of the Census (2001). Households with telephones for 2000 are from FCC (2001). Cellular systems from 1990 and 1994 through 2000 are from U.S. Bureau of the Census (2001). Systems from 1991 and 1992 are from U.S. Bureau of the Census (1998).

Figure 2.6
**Telephone Penetration and Cellular
Telephone Systems and Subscribers**

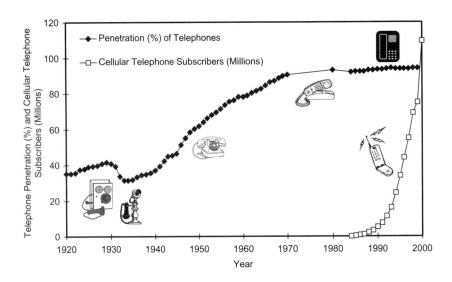

Guglielmo Marconi sent the first wireless data messages in 1895. The growing popularity of telephony led many to experiment with Marconi's radio technology as another means for interpersonal communication. By the 1920s, Detroit police cars had mobile radiophones for voice communication (ITU, 1999). The Bell System offered radiotelephone service in 1946 in St. Louis, the first of 25 cities to receive the service. Bell engineers divided reception areas into cells in 1947, but cellular telephones that switched effectively among cells as callers moved did not arrive until the 1970s. The first call on a portable, handheld cell phone occurred in 1973. However, by 1981, only 24 people in New York City could use their mobile phones at the same time, and only 700 customers could have active contracts. The FCC began offering cellular telephone system licenses by lottery in June 1982 (Murray, 2001). Other countries, such as Japan in 1979 and Saudi Arabia in 1982, operated cellular systems earlier than the United States (ITU, 1999). Table 2.4 and Figure 2.6 show the growth in American cellular systems and subscribers from 1983 through 2000.

The number of cellular telephone systems grew from 751 employing 21,382 people in 1990 to 2,440 systems with 184,449 employees in 2000. During the same period, the average monthly bill declined from $80.90 to $45.20 (U.S. Bureau of the Census, 2001). By the end of 2001, the 130 million U.S. wireless subscribers nearly doubled the number from 1998, and users talked an average of 422 minutes each month (Selingo, 2002). That increase represented a 75% increase from the average of 242 minutes per month two years earlier.

However, some sources reported that cellular phones were approaching the saturation point (Stellin, 2002), noting a slowing of the growth rate. One factor could be the high rate of problems with using cell phones relative to using other communications media. J. D. Power & Associates reported that cellular telephone companies conducted more than three times as many service calls as local telephone service providers over one-year period that ended in the spring of 2001, and the FCC received more than 3,000 complaints about cellular telephone service providers in the third quarter of 2001.

About one-third of the complaints addressed poor service or false advertising. During the same period, only 48 complaints reached the FCC regarding cable television operators (Selingo, 2002). Census data in Table 2.4 show reductions in the growth rate during 1994-1999, with a small increase in the rate from 24% to 27% from 1999 to 2000. The nearly 110 million wireless subscribers shown in Table 2.4 for 2000 falls far short of the 2002 number of 130 million reported in popular media, only 10% of whom use their phones for data, as opposed to voice, transmissions (Hafner, 2002b).

Other competitors exist among the various telephone technologies. The Telecommunications Act of 1996 encouraged competition among communications providers. For example, the Act allowed cable television operators to provide telephone services as part of efforts to promote competition with telephone companies. The FCC (2002) found that, although a few major cable operators offered telephone services, most such companies were waiting for technological development with Internet protocol (IP) telephony to mature to a greater degree than was available.

Hafner (2002b) pointed out that major wireless services providers announced high-speed wireless data services by early 2002, suggesting ensuing appeals designed to attract many new customers. However, these wireless data services involve multiple incompatible technologies, often entail extra fees, sometimes offer access speeds no faster than telephone dial-up modems, and may require purchasing different phones than the ones used for standard wireless voice communications.

MOTION PICTURES

In the 1890s, George Eastman improved on work by and patents purchased from Hannibal Goodwin in 1889 to produce workable motion picture film. The Lumière brothers projected moving pictures in a Paris café in 1895, hosting 2,500 people nightly. William Dickson, an assistant to Thomas Edison, developed the kinetograph (an early motion picture camera) and the kinetoscope (a motion picture viewing system). A New York movie house opened in 1894, offering moviegoers several coin-fed kinetoscopes. Edison's vitascope, which expanded the length of films over those shown via kinetoscopes and allowed larger audiences to simultaneously see the moving images, appeared in public for the first time in 1896. In France in that same year, Georges Méliès started the first motion picture theater. Short movies became part of public entertainment in a variety of American venues by 1900 (Campbell, 2002), and average weekly movie attendance reached 40 million people by 1922.

Average weekly motion picture theater attendance, as shown in Table 2.5 and Figure 2.7, rose annually from the earliest available census reports on the subject in 1922 until 1930. After falling dramatically during the great depression, attendance regained growth in 1934 that continued until 1937. Slight declines in the prewar years were followed by a period of strength and stability throughout the World War II years. After the end of the war, average weekly attendance reached its greatest heights, averaging 90 million movie-goers weekly from 1946 through 1949. With the advent of television in the 1950s, weekly attendance would never again reach these levels.

Table 2.5
**Motion Picture Attendance and Box Office
Receipts, 1922-1999**

Year	Average Weekly Attendance (millions)	Receipts ($ million)	Year	Average Weekly Attendance (millions)	Receipts ($ million)	Hours Per Person Per Year
1922	40		1961	42.0	921	
1923	43		1962	43.0	903	
1924	46	1.7	1963	42.0	904	
1925	46		1964	44.0	913	
1926	50		1965	44.0	927	
1927	57		1966		964	
1928	65		1967		989	
1929	80	720	1968		1,045	
1930	90	732	1969		1,099	
1931	75	719	1970	18.0	1,162	
1932	60	527	1971	14.0	1,214	
1933	60	482	1972	15.0	1,583	
1934	70	518	1973	16.0	1,524	
1935	80	556	1974	18.0	1,909	
1936	88	626	1975	20.0	2,115	
1937	88	676	1976	20.0	2,036	
1938	85	663	1977	20.0	2,372	
1939	85	659	1978	22.0	2,643	
1940	80	735	1979	22.0	2,821	
1941	85	809	1980	20.0	2,749	
1942	85	1,022	1981	21.0	2,966	
1943	85	1,275	1982	23.0	3,453	
1944	85	1,341	1983	23.0	3,766	
1945	85	1,450	1984	23.0	4,030	
1946	90	1,692	1985	20.3	3,749	
1947	90	1,594	1986	19.6	3,780	
1948	90	1,506	1987	20.9	4,250	
1949	70	1,451	1988	20.9	4,460	
1950	60	1,376	1989	21.8	5,030	
1951	54	1,310	1990	22.8	5,020	
1952	51	1,246	1991	21.9	4,800	
1953	46	1,187	1992	22.6	4,870	11
1954	49	1,228	1993	23.9	5,200	12
1955	46	1,326	1994	24.8	5,400	12
1956	47	1,394	1995	24.3	5,500	12
1957	45	1,126	1996	26.3	5,900	12
1958	40	992	1997	26.7	6,366	12
1959	42.0	958	1998	28.5	6,949	13
1960	40.0	951	1999	28.2	7,448	13

Note: The data to 1970 are from U.S. Bureau of the Census (1976). The data from 1970, 1975, and 1979 through 1985 are from U.S. Bureau of the Census (1986). The box office data from 1971 are from U.S. Bureau of the Census (1975). The box office receipts data from 1972 through 1988 are from U.S. Department of Commerce (1988), and the data for 1988 through 1992 are from U.S. Department of Commerce (1994). The 1991 attendance came from U.S. Bureau of the Census (1996). The data for 1993 through 1996 are from U.S. Department of Commerce (1998). The data for 1997 through 1999 are from U.S. Bureau of the Census (2001). All data for hours per person per year are from U.S. Department of Commerce (1999), and these figures after 1997 represent projected estimates.

Figure 2.7
Motion Picture Attendance and Box Office Receipts

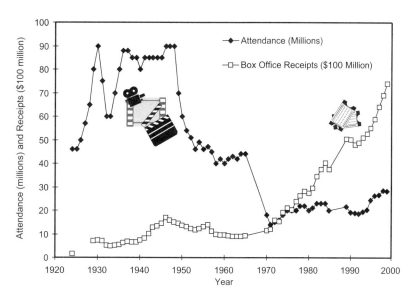

Although a brief period of leveling off occurred in the late 1950s and early 1960s, average weekly attendance continued to plummet until a small recovery began in 1972. This recovery signaled a period of relative stability that lasted into the 1990s. Through the last decade of the century, average weekly attendance enjoyed small but steady gains.

Box office revenues, which declined generally for 20 years after the beginning of television, began a recovery in the late 1960s. That recovery began to skyrocket in the 1970s, and the explosion is still underway. Much of the increase in revenues is due to increases in ticket prices and inflation, rather than from increased popularity of films with audiences. Total motion picture revenue from box office receipts declined during recent years, as studios realized revenues from television and video-cassettes (U.S. Department of Commerce/International Trade Association, 2000). In fact, motion picture studios realized 54.8% of their total revenue in 2000 from home video (FCC, 2002).

RECORDING

Thomas Edison expanded on experiments from the 1850s by Leon Scott de Martinville to produce a talking machine or phonograph in 1877 that played back sound recordings from etchings on tin foil. Edison later replaced the foil with wax. Emile Berliner created the first flat records from metal and shellac that played on his gramophone in the 1880s and provided for the mass production of recordings. The early standard recordings played at 78 revolutions per minute. When shellac became a scarce commodity during World War II, records were manufactured from polyvinyl plastic. In 1948, CBS Records produced the long-playing record that turned at 33-1/3 revolutions per minute (rpm), extending the playing time from three to four minutes to 10 minutes. RCA countered in 1949 with 45 rpm records that were incompatible with machines that played other formats. After a five-year war of formats, record players were manufactured that would play recordings at all of the speeds (Campbell, 2002).

The Germans used plastic magnetic tape for sound recording during World War II. After the Americans confiscated some of the tapes at the end of the war, the technology became a boon for western audio editing and multiple track recordings that played on bulky reel-to-reel machines. By the 1960s, the reels were encased in cassettes, which would prove to be deadly competition in the 1970s for single song records playing at 45 rpm and long-playing albums playing at 33-1/3 rpm. At first, the tape cassettes were popular in 8-track players. As technology improved, high sound quality was obtainable on tapes of smaller width, and 8-tracks gave way to smaller audio compact cassettes. Thomas Stockholm began recording sound digitally in the 1970s, and the introduction of compact disk (CD) recordings in 1983 decimated the sales performance of the earlier analog media types (Campbell, 2002). Table 2.6 and Figure 2.8 trace the rise and fall of the sales of recordings of respective types. The advent of digital recording also enabled new forms of illegal distribution and piracy of music on the Internet.

Figure 2.8
Recorded Music Unit Shipments

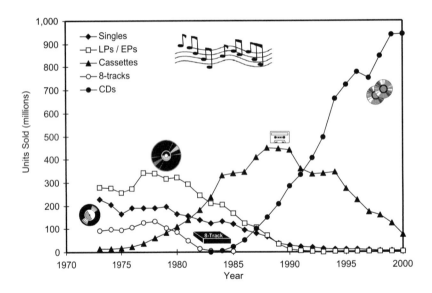

Table 2.6
**Recorded Music Unit Shipments, 1943-2000
(Millions)**

Year	Singles	LPs/EPs	Cassettes	8-tracks	CDs	Total
1973	228.0	280.0	15.0	91.0		614
1974	204.0	276.0	15.3	96.7		592
1975	164.0	257.0	16.2	94.6		531.8
1976	190.0	273.0	21.8	106.1		590.9
1977	190.0	344.0	36.9	127.3		698.2
1978	190.0	341.3	61.3	133.6		726.2
1979	195.5	318.3	82.8	104.7		701.3
1980	164.3	322.8	110.2	86.4		683.7
1981	154.7	295.2	137.0	48.5		635.4
1982	137.2	243.9	182.3	14.3		577.7
1983	125.0	210.0	237.0	6.0	1	579
1984	132.0	205.0	332.0	6.0	6	681
1985	121.0	167.0	339.0	4.0	23	654
1986	93.9	125.2	344.5		53	616.6
1987	82.0	107.0	410.0		102.1	701.1
1988	65.6	72.4	450.1		149.7	737.8
1989	36.6	34.6	446.2		207.2	724.6
1990	27.6	11.7	442.2		286.5	768
1991	22.0	4.8	360.1		333.3	720.2
1992	19.8	2.3	336.4		407.5	766
1993	15.1	1.2	339.5		495.4	851.2
1994	11.7	1.9	345.4		662.1	1,021.1
1995	10.2	2.2	272.6		722.9	1,007.9
1996	10.1	2.9	225.3		778.9	1,017.2
1997	7.5	2.7	172.6		753.1	935.9
1998	5.4	3.4	158.5		847.0	1,014.3
1999	5.3	2.9	123.6		938.9	1,070.7
2000	4.8	2.2	76.0		942.5	1,025.5

Note: "Singles" refer to vinyl singles. The data for all years prior to 1983 are from U.S. Department of Commerce (1986). The data from 1983 through 1985 are from U.S. Bureau of the Census (1986). The data for 1986 through 1989 are from U.S. Bureau of the Census (1986). The data for 1986 through 1989 are from U.S. Bureau of the Census (1995c). The data after 1989 are from U. S. Bureau of the Census (1999). Data for 1999 through 2000 are from U.S. Bureau of the Census (2001).

RADIO

Guglielmo Marconi's wireless messages in 1895 on his father's estate led to his establishing a British company to profit from ship-to-ship and ship-to-shore messaging. He formed an American subsidiary in 1899 that would become the American Marconi Company. Reginald A. Fessenden and Lee De Forest separately transmitted voice by means of wireless radio in 1906, and a radio broadcast from the stage of a performance by Enrico Caruso occurred in 1910. Various American companies and Marconi's British company owned important patents that were necessary to the development of the infant industry, so the American firms formed the Radio Corporation of America (RCA) to buy out the patent rights from Marconi. Debates still rage over the question of the first broadcast station; the contenders include KDKA in Pittsburgh, WHA in Madison (Wisconsin), WWJ in Detroit, and KQW in San Jose.

In 1919, Dr. Frank Conrad of Westinghouse broadcast music from his phonograph in his garage in East Pittsburgh, and Westinghouse's KDKA in Pittsburgh announced the presidential election returns over the airways on November 2, 1920. By January 1, 1922, the Secretary of Commerce had issued 30 broadcast licenses, and the number of licensees swelled to 556 by early 1923. By 1924, RCA owned a station in New York, and Westinghouse expanded to Chicago, Philadelphia, and Boston. In 1922, AT&T withdrew from RCA and started WEAF in New York, the first radio station sponsored by commercials. In 1923, AT&T linked WEAF with WNAC in Boston by the company's telephone lines for a simultaneous program. This began the first network, which grew to 26 stations by 1925. RCA linked its stations with telegraph lines, which failed to match the voice quality in the transmissions of AT&T. However, AT&T wanted out of the new business and sold WEAF in 1926 to the National Broadcasting Company, a subsidiary of RCA (White, 1971).

The 1930 penetration of radio sets in American households reached 40% and approximately doubled in the next 10 years, passing 90% by 1947. Table 2.7 and Figure 2.9 show the rapid rate of increase in the number of radio households from 1922 through the early 1980s, when the rate of increase declined. However, the increases continued until 1993, when leveling began.

Although thousands of radio stations were transmitting their sounds via the Internet by 2000, Channel1031.com became the first station to cease using FM and move to the Internet in September of that year (Raphael, 2000). Many other stations were operating only on the Internet when questions about fees for commercial performers and royalties for music played on the Web arose. Many operations were threatened by the requirements for separate payments for these rights by Web radio. The U.S. Copyright Office also announced in February 2002 that it would take comments through April 2002 on a proposed rule requiring radio stations to keep detailed logs of songs transmitted over the Internet (U.S. Copyright Office, 2002).

Table 2.7
Radio Households & Penetration, 1922-1998

Year	HHs with Sets (000s)	%	Year	HHs with Sets (000s)	%	Year	HHs with Sets (000s)	%
1922	60		1948	37,623		1974	70,800	
1923	400		1949	39,300	93.4	1975	72,600	98.6
1924	1,250		1950	40,700		1976	74,000	
1925	2,750		1951	41,900		1977	75,800	
1926	4,500		1952	42,800		1978	77,800	
1927	6,750		1953	44,800		1979	79,300	
1928	8,000		1954	45,100		1980	79,968	99.0
1929	10,250		1955	45,900	95.9	1981	81,600	99.0
1930	13,750	40.3	1956	46,800	95.7	1982	82,691	99.0
1931	16,700		1957	47,600	95.8	1983	83,078	99.0
1932	18,450		1958	48,500	96.1	1984	84,553	99.0
1933	19,250		1959	49,450	96.1	1985	85,921	99.0
1934	20,400		1960	50,193	95.1	1986		99.0
1935	21,456		1961	50,695	94.7	1987		99.0
1936	22,869		1962	51,305	93.7	1988	91,100	99.0
1937	24,500		1963	52,300	94.6	1989	92,800	99.0
1938	26,667		1964	54,000	96.2	1990	94,400	99.0
1939	27,500		1965	55,200	96.1	1991	95,500	99.0
1940	28,500	80.3	1966	57,200	97.6	1992	96,600	99.0
1941	29,300		1967	57,500	97.1	1993	97,300	99.0
1942	30,600		1968	58,500	96.2	1994	98,000	99.0
1943	30,800		1969	60,600	97.4	1995	98,000	99.0
1944	32,500		1970	62,000	97.8	1996	98,000	99.0
1945	33,100		1971	65,400		1997	98,000	99.0
1946	33,998		1972	67,200		1998		99.0
1947	35,900	91.8	1973	69,400				

Note: Authorization of new radio stations and production of radio sets for commercial use was stopped from April 1942 until October 1945. 1959 is the first year for which Alaska and Hawaii are included in the figures. The data prior to 1970 are from U.S. Bureau of the Census (1976). The households with sets data from 1970-1972 are from U.S. Bureau of the Census (1972). The households with sets data from 1973 and 1974 are from U.S. Bureau of the Census (1975). The households with sets data from 1975 through 1977 are from U.S. Bureau of the Census (1978). The households with sets data from 1978 and 1979 are from U.S. Bureau of the Census (1981). All data from 1988 through 1994 are from U.S. Bureau of the Census (1995c). The data after 1994 are from U.S. Bureau of the Census (1999). Penetration for 1998 is from U.S. Bureau of the Census (2001).

Figure 2.9
Radio and Television Households

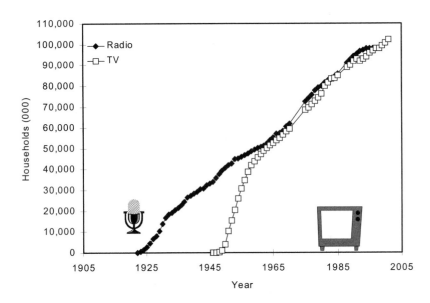

TELEVISION

Paul Nipkow invented a scanning disk device in the 1880s that provided the basis from which television would develop. In 1927, Philo Farnsworth became the first to electronically transmit a picture over the air. Fittingly, he transmitted the image of a dollar sign. In 1930, he received a patent for the first electronic television, one of many patents for which RCA would be forced, after court challenges, to negotiate. By 1932, Vladimir Zworykin discovered a means of converting light rays into electronic signals that could be transmitted and reconstructed at a receiving device. These innovations were combined to create television, which was first demonstrated publicly by RCA at the 1939 World's Fair.

In 1941, the FCC designated 13 channels for use in transmitting black-and-white television, and the commission issued almost 100 television station broadcasting licenses before placing a freeze on new licenses in 1948. The freeze that offered time to settle technical issues ran longer because of the American involvement in the Korean War (Campbell, 2002). As shown in Table 2.8, nearly 4,000 households had television sets by 1950, a 9% penetration rate that would escalate to 87% a decade later. Penetration has remained steady at about 98% since 1980. Figure 2.9 illustrates the meteoric rise in the number of households with television by year from 1946 through the turn of the century.

American television standards set in the 1940s provided for 525 lines of data composing the picture. By the 1980s, Japanese high-definition television (HDTV) increased that resolution to more than 1,100 lines of data in a television picture. This increase enabled much higher quality in televised content to be transmitted with less electromagnetic spectrum space per signal. In 1996, the FCC approved a digital television transmission standard and authorized broadcast television stations a second channel for a 10-year period to allow a transition to HDTV. That transition will eventually make all older analog television sets obsolete because they cannot process HDTV signals (Campbell, 2002).

Table 2.8
Television Households and Penetration, 1946-2001

Year	HHs with Sets (000s)	%	Year	HHs with Sets (000s)	%
1946	8		1975	68,500	97
1947	14		1976	69,600	
1948	172		1977	71,200	
1949	940		1978	72,900	
1950	3,875	9	1979	74,500	
1951	10,320		1980	76,300	98
1952	15,300		1981	79,900	98
1953	20,400		1982	81,500	98
1954	26,000		1983	83,300	98
1955	30,700		1984	83,800	98
1956	34,900		1985	84,900	98
1957	38,900		1986	85,900	98
1958	41,924		1987	87,400	98
1959	43,950		1988	89,000	98
1960	45,750	87	1989	90,000	98
1961	47,200		1990	92,000	98
1962	48,855		1991	93,000	98
1963	50,300		1992	92,000	98
1964	51,600		1993	93,000	98
1965	52,700		1994	94,000	98
1966	53,850		1995	95,000	98
1967	55,130		1996	96,000	98
1968	56,670		1997	97,000	98
1969	58,250		1998	98,000	98
1970	59,550	95	1999	99,400	98
1972			2000	100,802	
1973			2001	102,185	
1974					

Note: 1959 is the first year for which Alaska and Hawaii are included in the figures. The data dealing with households with television to 1971 are from U.S. Bureau of the Census (1976). The data dealing with households with television from 1980 through 1984 are from U.S. Bureau of the Census (1985). The data about penetration for all other pre-1987 years and all data for 1985 and 1986 are from U.S. Bureau of the Census (1986), and data from 1987-1991 are from U.S. Bureau of the Census (1995c). The data from 1991 through 1996 are from FCC (1999). The penetration data from 1996 through 1999 are from U.S. Bureau of the Census (2001). The households with sets data from 1996 through 2001 are from FCC (2001).

The FCC set May 2002 as the deadline by which all American commercial television broadcasters were required to be broadcasting digital television signals. Progress toward digital television broadcasting fell short of FCC requirements that all of the affiliates of the top four networks in the top 10 markets transmit digital signals by May 1, 1999. Within the 10 largest television markets, all except one network affiliate had begun HDTV broadcasts by August 1, 2001. By that date, 83% of American television stations had received construction permits for HDTV facilities or a license to broadcast HDTV signals, and 229 stations were already broadcasting digital television signals (FCC, 2002). About two million households had high-definition television by early 2002 (Nichols, 2002).

A San Francisco market research firm estimated that 25% of American households in 2002 enjoyed home theater, a combination of entertainment equipment including a television set with at least a 27-inch screen, an audio/video receiver with surround-sound capability, four or more speakers, and a VCR or DVD player. The television set is the obvious focal point of such arrangements, and 37% of American households have sets with screens that reach 30 inches or larger (Hafner, 2002a).

CABLE TELEVISION

Cable television began as a means to overcome poor reception for broadcast television signals. John Watson claimed to have developed a master antenna system in 1948, but his records were lost in a fire. Robert J. Tarlton of Lansford (Pennsylvania) and Ed Parsons of Astoria (Oregon) set up working systems in 1949 that used a single antenna to receive programming over the air and distribute it via coaxial cable to multiple users (Baldwin & McVoy, 1983). At first, the FCC chose not to regulate cable, but after the new medium appeared to offer a threat to broadcasters, cable became the focus of heavy government regulation. Under the Reagan administration, attitudes swung toward deregulation, and cable began to flourish. Table 2.9 and Figure 2.10 show the growth of cable systems and subscribers, with penetration remaining below 25% until 1981 but passing the 50% mark before the 1980s ended.

The rate of growth in the number of cable subscribers slowed over the last half of the 1990s. Penetration, after consistently rising every year from cable's outset, declined every year after 1997. The Federal Communications Commission reports annually to Congress regarding the status of competition in the marketplace for video programming. The 2001 report (FCC, 2002) revealed that cable television remained the primary mode of consumer access to video programming, but the franchised cable operators' share of multichannel video programming distributors (MVPDs) slipped from 80% in 2000 to 78% in 2001.

However, the viewing share of non-premium basic cable programming rose from 46% to 48% from mid-2000 to mid-2001. The FCC reported that, despite a slight drop in basic cable penetration, cable television growth occurred from 2000 in the number of homes with access to cable services (104 million, 97.1% of TV homes), viewing shares of both basic (average 45.5% share) and premium services, and channel capacity of cable systems. Although pay-per-view channels grew by more than 20% during 1999-2001, the number of basic cable channels offered fell by 11.6% and premium channels declined by 7%. The National Cable and Telecommunications Association (NCTA) (2001) presented a more optimistic picture of the extent of cable television use in the United States than the FCC report (2002). The NCTA data show 9,947 cable operators serving 72,958,000 subscribers, or 69.2% of American households, all three numbers eclipsing their counterparts from government sources that appear in Table 2.9 and that are illustrated in Figure 2.10.

The FCC report cited the U.S. Bureau of Labor Statistics in concluding that the price of cable services increased by 4.24% during the period, a higher rate than the 3.25% increase in the Consumer

Price Index. The report estimated that cable revenue would reach $53.11 per subscriber per month by the end of 2001. In a period of increasing cable industry revenues (up 15.4% from 2000), cable operators continued to expand offerings of broadband services that support Internet access and telephone communications.

Digital cable subscriptions enjoyed rapid growth from the 8.7 million subscribers at the end of 2000. Within another six months, the estimated count reached 12 million, with projections for 15.1 million digital cable subscribers by the end of 2001 (FCC, 2002).

Figure 2.10
Cable Television Households and Penetration

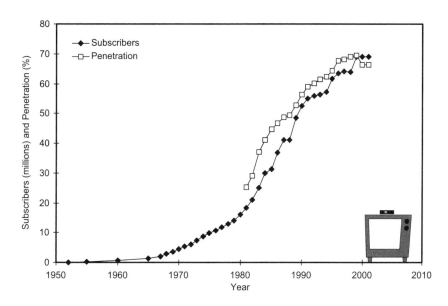

The FCC (2002) noted that video on demand (VOD) services showed signs of growing popularity by the end of 2001, when the services were estimated to have generated revenues exceeding $65 million. The commission cited projections of $420 million in VOD revenue occurring in 2002, which allows subscribers to view movies at times of their own choosing or at frequently scheduled times.

Table 2.9
Cable Television Households and Penetration, 1946-2001

Year	Systems	Subscribers (000s)	Penetration (%)
1952	70	14	
1955	400	150	
1960	640	650	
1965	1,325	1,275	
1967	1,770	2,100	
1968	2,000	2,800	
1969	2,260	3,600	
1970	2,490	4,500	
1971	2,639	5,300	
1972	2,841	6,000	
1973	2,991	7,300	
1974	3,158	8,700	
1975	3,506	9,800	
1976	3,681	10,800	
1977	3,832	11,900	
1978	3,875	13,000	
1979	4,150	14,100	
1980	4,225	16,000	
1981	4,375	18,300	25.3
1982	4,825	21,000	29.0
1983	5,600	25,000	37.2
1984	6,200	30,000	41.2
1985	6,600	31,275	44.6
1986	7,600	36,933	46.8
1987	7,900	41,100	48.7
1988	8,500	41,100	49.4
1989	9,050	48,600	52.8
1990	9,575	52,600	56.4
1991	10,704	54,900	58.9
1992	11,075	55,800	60.2
1993	11,217	56,400	61.4
1994	11,230	57,200	62.4
1995	11,126	58,000	63.4
1996	11,119	65,300	67.8
1997	10,950	66,500	68.2
1998	10,845	66,000	67.2
1999	10,700	67,000	67.5
2000	10,400	68,500	66.4
2001	10,300	69,000	66.3

Note: The systems and subscribers data are from U.S. Bureau of the Census (1986). The penetration data through 1986 are from U.S. Bureau of the Census (1986). The penetration data through 1986 are from U.S. Bureau of the Census (1986) except for 1970, 1980, and 1987 through 1995. Of the latter, 1987 data are from U.S. Bureau of the Census (1988), and data from 1970, 1980, and 1988 through 1995 are from U.S. Bureau of the Census (1995c). The penetration data for 1995-1997 are from FCC (1998). The number of systems from 1995 through 1997 are from U.S. Bureau of the Census (1998). The number of systems from 2000 through 2001 are from *Television & Cable Factbook* (2001). The number of subscribers from 1996-1999 and all other data from 1998-1999 are from U.S. Bureau of the Census (2001). Subscriber and penetration data for 2000 through 2001 are from FCC (2002).

DIRECT BROADCAST SATELLITE AND OTHER CABLE TV COMPETITORS

Competitors for the cable television industry include a variety of technologies. FCC reports (e.g., FCC, 2002) distinguish between home satellite dish (HSD) systems and direct broadcast satellite (DBS) systems. Both are included as multichannel video program distributors (MVPDs), which include cable television, wireless cable systems called multichannel multipoint distribution services (MMDS), and private cable systems called satellite master antenna television (SMATV). Along with these video sources, the FCC includes the Internet, but the agency does not yet find that Internet video has become a competitor to traditional video sources.

Home satellite dishes in the early 1980s spanned six feet or more in diameter. Sales of dishes topped 500,000 only twice (1984 and 1985) between 1980 and 1995 (Brown, 2000), when home satellite system subscribers apparently peaked before a steady decline in numbers. Conversely, smaller dish DBS subscribers have multiplied every year since their numbers were first reported by census data in 1993. SMATV subscriptions generally grew after 1993, although declines occurred in 1998 and no growth occurred in 2001 relative to the previous year. MMDS subscriptions peaked in 1996, although the slide slowed in 2001. The FCC (2000) report forecast that the approval by the commission of two-way and digital MMDS services would eventually result in this part of the spectrum's use primarily for data transmission services. Table 2.10 and Figure 2.11 track trends of these non-cable video delivery types.

The FCC also considers several types of services as potential MVPD operators. These services include home video sales and rentals, the Internet, electric and gas utilities, and local exchange carriers (LECs). The latter category includes telephone service providers that were allowed by the Telecommunications Act of 1996 to provide video services to homes at the same time that the Act allowed cable operators to provide telephone services. The expected competition from LEC organizations for other video services did not occur, and the FCC (2002) noted that existing LEC providers had left the arena of providing video, with the exceptions of reselling of DBS services and efforts by new LECs to offer video services via telephone lines. The same report added a new category of video program distributor in its 2002 report: broadband service providers. These services offer video, voice, and telephone capability to their customers.

Table 2.10
Non-Cable MVPD TV Households, 1946-2001

Year	Home Satellite Dish Subscribers	DBS Subscribers	MMDS Subscribers	SMATV Subscribers
1991	764		180	965
1992	1,023		323	984
1993	1,612	<70	397	1,004
1994	2,178	602	600	850
1995	2,341	1,675	800	950
1996	2,300	3,500	1,200	1,100
1997	2,184	5,047	1,100	1,163
1998	2,028	7,200	1,000	940
1999	1,783	10,078	821	1,450
2000	1,477	12,987	700	1,500
2001	1,000	16,070	700	1,500

Note: Home satellite dishes represent the four- to eight-foot dishes. DBS uses smaller dishes. The data through 1995 are from FCC (1995). Subscriber data for 1996 and 1997 are from FCC (1998). Data from 1997 through 2001 are from FCC (2002).

Figure 2.11
Non-Cable MVPD TV Households

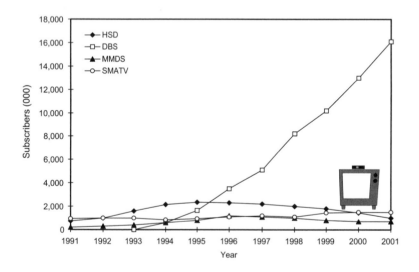

Both cable television and other MVPD companies deliver a variety of communications services to households (FCC, 2000). Beyond the basic video channels, DBS operations provide Internet access, and MMDS and SMATV operators offer Internet access along with local and long distance telephone communications.

As the growth of cable television appeared to be peaking and dropping, the growth of non-cable MVPD increased by more than 15% in mid-2001 over figures from the previous year (FCC, 2002). This growth appeared to occur primarily as increases in subscribers to direct broadcast satellite, which grew at a rate of 2.5 times the rate of growth of cable television subscribers during that year. The Satellite Home Viewer Improvement Act of 1999 granted permission to DBS carriers for carriage of local broadcast stations in their local markets. From mid-2000 to 2001, DBS subscribers increased by 250% of the growth rate of cable subscribers, reaching 18.2% of MVPD subscribers.

The FCC (2002) cited DirecTV as reporting that about half of its subscribers dropped cable TV before connecting with DBS. DirecTV also reported that about 20% of its subscribers lived in circumstances other than single-family homes. The commission estimated that 20% to 23% of cable operator income comes from multiple dwelling units. Multiple dwelling units comprise an important market segment that includes a variety of non-single-family dwelling units, such as apartment buildings, townhouses, condominiums, mobile home parks, hospitals, nursing homes, and hotels. The FCC cited a report that 21.4 million such units existed in the United States in 2001, with the figure anticipated to reach 23.3 million by 2005.

The FCC attempted to reduce barriers to new services (Promotion of, 2000) that might serve residential and multiple dwelling units. The commission required utilities to provide reasonable, nondiscriminatory access to telecommunications and cable providers that would facilitate delivery of services to potential customers. For example, utility poles and conduits in rights-of-way became available to content providers other than the utilities themselves.

Whether consumers live in multiple- or single-family dwellings, they usually have few choices of providers of MVPD or video services. Most cable operators enjoy monopolies through franchise agreements that restrict competitors. The FCC reported effective cable competition in just 419 (1.3%) of 33,000 communities that offer cable television (FCC, 2002). DBS offers a form of competition for cable for most viewers other than people in multiple-family buildings, and advertisements for the two types of services tout the differences between the two.

VIDEOCASSETTE RECORDERS

Although videocassette recorders (VCRs) became available to the public in the late 1970s, competing technical standards slowed the adoption of the new devices. After the longer taping capacity of the VHS format won greater public acceptance over the higher-quality images of Betamax, the popularity of home recording and playback rapidly accelerated, as shown in Table 2.11 and Figure 2.12.

The FCC (2002) report on competition in the video marketplace listed VCR penetration at about 90%, with multiple VCRs in nearly 46 million households. Spending for rented or purchased recorded home video exceeded $19 billion in 2000, an increase of nearly 10% over spending in the previous year. The popularity of home video prompted the commission to view home video as a viable competitor to the MVPD options described earlier. Other types of home video devices threaten the long-term growth of VCRs, but strengthen options for consumers.

About 25 million digital videodisc (DVD) players existed in American homes by the end of 2001, and DVD shares of home video rental and sales doubled in 2000, reaching 6.9% of video rental revenue and 16% of video sales revenue (FCC, 2002). During 2001, VCR sales declined by about 35% at the same time that sales of DVD players rose by nearly 50% (Reuters, 2002). Table 2.11 and Figure 2.13 trace the rapid growth in DVD sales. Personal video recorders (PVRs) debuted during 2000, and about 500,000 units were purchased by the end of 2001 (FCC, 2002). Compared with VCRs, both of the latter devices constituted relative newcomers to home video, and both offered the advantages of digital audio and video. By early 2002, major satellite content providers, such as EchoStar and DirecTV, began building digital video recording devices into satellite television receivers as a means of enhancing their competitiveness with cable television (Lee, 2002).

Whether the VCR weathers the competitive storms of these new media forms may hinge on the adaptability of the technology. In early 2002, four motion picture studios announced the release of recent films on high-definition videotape, called *D-Theater*, which is designed for use with high-definition television sets (Nichols, 2002).

Table 2.11
Households with VCRs and DVDs and VCR Penetration, 1978-2000

Year	VCRs (000s)	VCR Penetration (%)	DVD Players (000s)
1978	402		
1979	478		
1980	804	1.1	
1981	1,330	1.8	
1982	2,030	3.1	
1983	4,020	5.5	
1984	7,143	10.6	
1985	18,000	20.8	
1986		36.0	
1987	43,000	48.7	
1988	51,000	58.0	
1989	58,000	64.6	
1990	63,000	68.6	
1991	67,000	71.9	
1992	69,000	75.0	
1993	72,000	77.1	
1994	74,000	79.0	
1995	77,000	81.0	
1996	79,000	82.2	
1997	82,000	84.2	
1998	83,000	84.6	500
1999	84,000	84.6	2,500
2000			3,300

Note: The data from 1978 are from U.S. Bureau of the Census (1982). The data from 1979 through 1984 are from U.S. Bureau of the Census (1984). The data from 1985 are from U.S. Bureau of Census (1986). The penetration data are from U.S. Bureau of the Census (1986). The VCR sales data for 1985 and 1988 through 1994 are from the U. S. Bureau of the Census (1995c). The VCR sales data from 1987 are from U. S. Bureau of the Census (1990). The VCR penetration data from 1987 are from U. S. Bureau of the Census (1988). The VCR penetration data from 1988 through 1992 are from U.S. Bureau of the Census (1995c). The data from 1993 through 1997 are from U.S. Bureau of the Census (1999). The data from 1998 through 1999 are from U. S. Bureau of the Census (2001). DVD data are from FCC (2002).

Figure 2.12
Households with VCRs and VCR Penetration

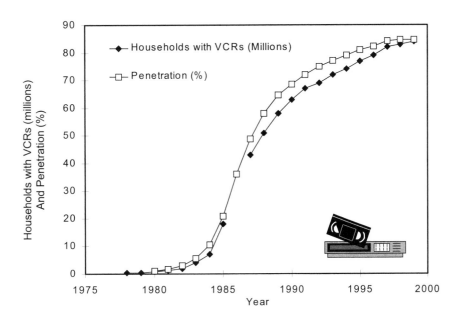

Figure 2.13
Households with DVDs

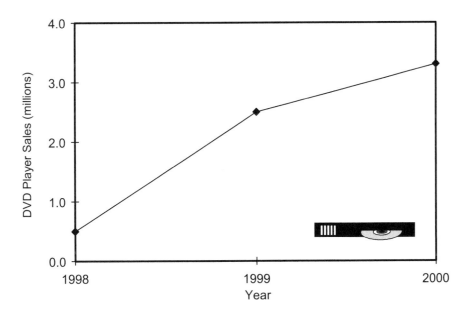

PERSONAL COMPUTERS

The history of computing traces its origins back thousands of years to such practices as using bones as counters (Hofstra University, 2000). Intel introduced the first microprocessor in 1971. The MITS Altair, with an 8080 processor and 256 bytes of RAM, sold for $498 in 1975, and was the first desktop computer for individuals. In 1977, Radio Shack offered the TRS80 home computer, and the Apple II set a new standard for personal computing, selling for $1,298. Other companies began introducing personal computers, and, by 1978, 212,000 personal computers were shipped for sale.

Figure 2.14 and Table 2.12 trace the rapid and steady rise in computer shipments and home penetration. By October 1997, 81,013,000 Americans used personal computers (U.S. Bureau of the Census, 2001). About 51% of those users were males, 84% were white, 6.1% were black, 5% were Hispanic, and 56.6% lived in families with $50,000 or more in annual income. Among computer users, 23.5% used computers either six or seven days each week, and 24% used them one day or less each week.

By 2000, 86,289,000 adults reported having used the Internet either at home or at work within the previous 30 days, 51% of American households had computers, and 41.5% had Internet access (U.S. Bureau of the Census, 2001). The proportion of households with Internet access had risen from 26.2% in 1998. Among the users in 2000, 87.5% were between 18 and 54 years of age, but only 12.5% were 55 years of age or older. Among the users, only 32.9% lived in households with incomes under $50,000 per year.

Figure 2.14
Homes with Personal Computers

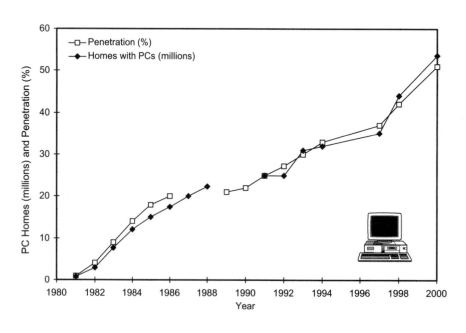

Table 2.12
Personal Computer Shipments and Home Use, 1978-2000

Year	PCs Shipped	Homes with PCs	% of Homes with PCs*
1978	212,000		
1979	246,000		
1980	371,000		
1981	1,110,000	750,000	1%
1982	3,530,000	3,000,000	4%
1983	6,900,000	7,640,000	9%
1984	7,610,000	11,990,000	14%
1985	6,750,000	14,960,000	18%
1986	7,040,000	17,380,000	20%
1987	8,340,000	19,970,000	23%
1988	9,500,000	22,380,000	25%
1989	9,329,970		
1990	9,848,593		
1991	10,903,000	25,000,000	25%
1992	12,544,374	25,000,000	27%
1993	14,775,000	31,000,000	30%
1994	18,605,000	32,010,000	33%
1995	22,582,900		
1996	25,000,000		
1997	30,000,000	35,020,000	37%
1998	35,400,000	44,205,000	42%
1999			
2000		53,700,000	51%

* The percentage of homes with PCs for 1997 is from Newburger (1999). The homes with PCs for 1997 are calculated by multiplying the percentage of homes with PCs times the number of homes with TVs from Table 2.8. The percentage of homes with computers for 1998 is from FCC (2000). The percentage of homes with computers for 2000 is from U.S. Bureau of the Census (2001). Homes with computers for 2000 are from the National Telecommunications & Information Administration (2000).

Note: Shipments from 1978 are from U.S. Bureau of the Census (1983); from 1979 and 1980 are from U.S. Bureau of the Census (1984). The data from 1981 through 1988 are from U.S. Bureau of the Census (1992). Shipments from 1989 and 1990 are from U.S. Bureau of the Census (1993), for 1991 through 1993 are from U.S. Bureau of the Census (1995c), and for 1994 through 1995 are from U.S. Bureau of the Census (1997). The computer shipments for 1996 are from U.S. Department of Commerce (1998), and shipments for 1997 are from U.S. Department of Commerce/International Trade Association (1999). Shipments for 1998 are from U.S. Department of Commerce/International Trade Association (2000).

INTERNET

The Internet began in the 1960s with ARPAnet, or the Advanced Research Projects Agency (ARPA) network project, under the auspices of the U.S. Department of Defense. The project intended to serve the military and researchers with multiple paths of linking computers together for sharing data in a system that would remain operational even when traditional communications might become unavailable. Early users, mostly university and research lab personnel, took advantage of electronic mail and posting of information on computer bulletin boards. Usage increased dramatically in 1982 after the National Science Foundation supported high-speed linkage of multiple locations around the United States. After the collapse of the Soviet Union in the late 1980s, military users abandoned ARPAnet, but private users continued to use it, and multimedia transmissions of audio and video became available. More than 150,000 regional computer networks and 95 million computer servers hosted data for Internet users (Campbell, 2002).

Pastore (2002) cited Nielsen/NetRatings reports in reporting that 498 million people around the world had home access to the Internet during the last quarter of 2001, the rate of Internet growth reached 15 million new users in the third quarter, and that rate nearly doubled in the fourth quarter. Pastore cited eMarketer in reporting that, in 2001, 27% of the world's Internet users or 119 million people resided in the United States. Census data fail to reach such spectacular numbers, in part because they are not so recent. In keeping with the philosophy of this chapter, however, only U.S. Bureau of the Census reports contribute to the data shown in Table 2.13 and Figures 2.15 and 2.16 that report Internet usage.

Table 2.13
Internet Use by Persons 18 and Over, 1990-2001

Year	Home Access (%)	Home or Work Access (000s)	Used Internet Last 30 Days (000s)	Accessed Internet (%)	Hours/ Person/ Year	Spending for Internet Access Person/Year
1990					1	
1991					1	
1992					2	$4.39
1993					2	$5.35
1994					3	$6.20
1995					7	$11.33
1996				9.4	16	$17.13
1997					28	$25.52
1998	26.2	37,047	34,227	32.5	35	$32.18
1999		83,677	53,052		39	$37.25
2000	41.5	112,949	75,409	45.4	43	$42.92
2001	58.0				44	$48.15

Note: All data are from U.S. Bureau of the Census (1995, 1998, 1999, and 2001). Hours/person/year after 1997 and spending/person/year are estimates (U.S. Bureau of the Census, 1999).

Figure 2.15
Internet Use and Spending

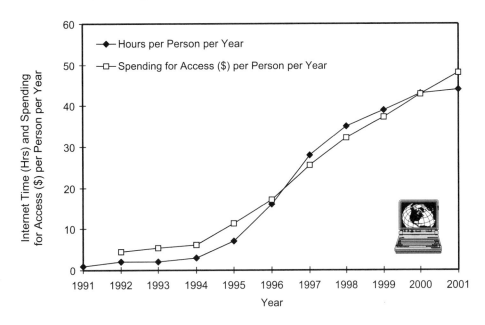

Figure 2.16
Internet Use by Persons 18 and Over

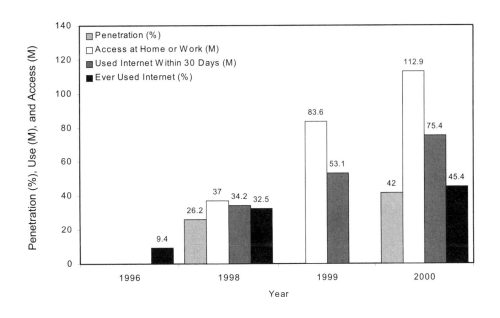

Early visions of the Internet (see Chapter 10) did not include the emphasis on entertainment and information to the general public that has emerged. The combination of this new medium with older media belongs to a phenomenon called "convergence," referring to the merging of functions of old and new media. The FCC (2002) reported that the joining of Internet services with other services constitutes the most important such convergence in providing video content. The report also noted that companies from many business areas were providing a variety of video, data, and other communications services.

As of July 2001, 58% of Americans had Internet connections at home (FCC, 2002). Among households using cable to obtain Internet access, most homes use cable modems attached to computers, although a few use set-top boxes to view the Internet over television (FCC, 2000). Nearly all major multiple system operators (MSOs) offer Internet access via cable modems.

Surfing the Internet as an adult leisure activity in 1999 reached 17.9%, comparing closely to the proportion of adults who reported going to bars or night clubs (19.9%), going to live theater (15.5%), going to museums (15.9%), or going on a picnic (16%). More than half of the adults who used the Internet reported using it two or three times weekly (U.S. Bureau of the Census, 2001).

Survey data from the National Telecommunications & Information Administration (2000) reported that 116.5 million Americans went online in August 2000. That figure represented a 37.7% increase from 84.6 million users in December 1998. Earlier statistically significant differences between men and women's Internet use disappeared, as 44.6% of men and 44.2% of women reported usage in 2000. Americans older than 50 years of age were the least likely group (29.6%) to use the Internet, but that group also demonstrated the greatest growth rate (53%) in usage since 1998. Among all users, the growth rate reached 35%.

In rural areas, 38.9% of the households in August 2000 had Internet access, a 75% increase from 22.2% in December 1998. Americans at every income level connected at far higher rates from their homes, particularly at middle-income levels. Internet access among households earning $35,000 to $49,000 rose from 29.0% in December 1998 to 46.1% in August 2000. In 2000, more than two-thirds of all households earning more than $50,000 had Internet connections (60.9% for households earning $50,000 to $74,999 and 77.7% for households earning above $75,000).

Access to the Internet expanded in 2000 across every education level, particularly for those with some high school or college education. Households headed by someone who attended college showed the greatest expansion in Internet penetration of all education levels, rising from 30.2% in December 1998 to 49.0% in August 2000 (National Telecommunications & Information Administration, 2000). Perhaps feeding some of that demand, 7,866 public libraries and 94.6% of academic libraries offered Internet access as early as 1998. By 2000, 95.7% of the nation's 16,004 public libraries offered Internet connections (U.S. Bureau of the Census, 2001).

Large gaps also remained regarding Internet penetration rates among households of different races and ethnic origins. Although growth in minority households with Internet access occurred, the rate of minority household gains was lower than the rate for all American households. With respect to individuals, while about one-third of Americans used the Internet at home in 2000, only 16.1% of Hispanics and 18.9% of blacks did so (National Telecommunications & Information Administration, 2000).

As of the end of 2000, 84.6% of households connected to the Internet accomplished the link with dial-up telephone modems (FCC, 2002). The FCC predicted that this means would remain the most frequently used connection until 2004, when an estimated 55.7% of Internet households will connect through broadband equipment.

Cable modems offer the most commonly used means of broadband access, as shown in Table 2.14 and Figures 2.17, but that share of broadband access is declining, primarily from competition of digital subscriber line (DSL) telephone service. As illustrated in Table 2.14, the actual penetration of cable modem and DSL among available households is about the same. The advantage in subscriber numbers enjoyed by cable modems (illustrated in Figure 2.17) may be a function of greater availability. Access through satellite and wireless connections accounted for only 8% of broadband subscribers by the end of 2001 (FCC, 2002). The FCC did not expect the latter market share to increase for several years.

Some companies offer very high-speed digital subscriber line (VDSL) service that includes telephone service and Internet access over previously installed copper telephone lines. Some of these services include multichannel video. The FCC (2002) reported that small companies delivering these services had 100,000 subscribers in 2001. Another 5,000 to 10,000 households in 2001 received ADSL (asymmetrical digital subscriber line), featuring video on demand, as opposed to multiple channels delivered simultaneously.

Broadband service providers and non-cable video services face difficult market hurdles to their expansion. Existing franchise agreements and contracts among cable operators, government units, utility companies, and owners of multiple dwelling unit sites often prevent new providers from offering services. Many of these contracts began before alternative communication services existed. Without legal authorization to provide services in respective communities or access to wiring in multiple dwelling units, new services often find themselves blocked from delivering communication capabilities and content.

Table 2.14
**Households with Potential Broadband
Internet Access and Subscribers, 2000-2001**

Date	Cable Modem Availability*	Cable Modem Subscribers*	% Cable Modem Subscribing	DSL Availability*	DSL Subscribers*	% DSL Subscribing
December 2000	58.5	3.9	6.7%	37.6	1.9	5.1%
June 2001	67.3	5.6	8.3%	45.0	3.0	6.7%
December 2001	81.0	7.2	8.9%	51.0	4.3	8.4%

* Millions

Note: Percentages indicate ratio of subscribers to households with the respective service available. All data are from FCC (2002). December 2001 data are estimates.

Figure 2.17
Households with DSL or Cable Modem

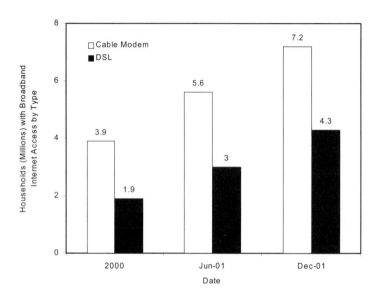

SYNTHESIS

Just as media began converging nearly a century ago when radios and record players merged in the same appliance, media in recent years have been converging at a much more rapid pace. Wireless telephony enjoyed explosive growth, and cable and telephone companies toyed with offering competing personal communications opportunities. More than 98% of American households have television, and more than 80% have either cable or satellite television, suggesting that Americans have accepted the idea of paying for television entertainment (Lee, 2002). However, just over half of American households have personal computers, and Americans seem to be more reluctant to pay for entertainment, except for pornography and gambling, delivered via the Internet. The trend toward media convergence, therefore, seems to encounter resistance in merging the television with computer (Lee, 2002).

Popularity of print media forms generally declined throughout the 1990s, perhaps in part after newspapers, magazines, and books began appearing in electronic and audio forms. Recorded music sales, including CDs, declined, although the controversies over digital music indicated no loss of public popularity of listening to music. Radio recently expanded to the Internet, although copyright questions threaten future growth in that form. Motion picture box office receipts continued their explosive growth trend, although the number of people going to theaters seems to be giving way somewhat to home viewing. New ways of recording and playing back video material encouraged home viewing of motion pictures, and wealthier patrons began reproducing the theater viewing experience at home with digital systems equipped with large screens and surround-sound. These trends illustrate that, although some ebb-and-flow occurs across various media forms, American usage of media for entertainment, information, and communication continues to expand. Thoroughly analyzing American life requires studying how Americans use media.

BIBLIOGRAPHY

Aronson, S. H. (1977). Bell's electrical toy: What's the use? The sociology of early telephone usage. In I. Pool (Ed.). *The social impact of the telephone*. Cambridge, MA: The MIT Press.

Associated Press. (2001, March 20). Audio satellite launched into orbit. *New York Times on the Web*. Retrieved March 20, 2001 from http://www.nytimes.com/aponline/national/AP-Satellite-Radio.html?ex=986113045&ei=1&en=7af33c7805ed8853.

Baldwin, T. F., & McVoy, D. S. (1983). *Cable communication*. Englewood Cliffs, NJ: Prentice-Hall.

Brown, D. (2000). Trends in selected American communications media. In A. E. Grant & J. H. Meadows (Eds.). *Communication technology update*, 7th edition. Boston: Focal Press.

Brown, D. (1998). Trends in selected American communications media. In A. E. Grant (Ed.). *Communication technology update*, 6th edition. Boston: Focal Press.

Brown, D. (1996). A statistical update of selected American communications media. In A. E. Grant (Ed.). *Communication technology update*, 5th edition. Boston: Focal Press.

Brown, D., & Bryant, J. (1989). An annotated statistical abstract of communications media in the United States. In J. Salvaggio & J. Bryant (Eds.). *Media use in the information age: Emerging patterns of adoption and consumer use*. Hillsdale, NJ: Lawrence Erlbaum Associates.

Campbell, R. (2002). *Media & culture*. Boston, MA: Bedford/St. Martins.

Federal Communications Commission. (1995, December 11). *Annual assessment of the status of competition in the market for the delivery of video programming*. CS Docket No. 95-61. Washington: FCC.

Federal Communications Commission. (1998, January 13). In the matter of assessment of the status of competition in markets for the delivery of video programming. *Fourth annual report*, CS Docket No. 97-141. Washington, DC. Available: http://www.fcc.gov/Bureaus/Cable/Reports/fcc97423.txt.

Federal Communications Commission. (1999, July 31). Broadcast station totals as of July 31, 1999. *FCC news release*. Available: http://www.fcc.gov/Bureaus/Mass_Media/News_Releases/1999/nrmm9022.wp.

Federal Communications Commission. (1999b, April 2). *1998 biennial regulatory review—Testing new technology*. CC Docket No. 98-94, Washington: FCC.

Federal Communications Commission. (2000, January 14). *Annual assessment of the status of competition in markets for the delivery of video programming*. CS Docket No. 99-230, Washington: FCC.

Federal Communications Commission. (2001, August). *Trends in telephone service*. Washington, DC: Industry Analysis Division, Common Carrier Bureau. Retrieved February 27, 2002 from http://www.fcc.gov/Bureaus/Common_Carrier/Reports/index.html.

Federal Communications Commission. (2002, January 14). In the matter of annual assessment of the status of competition in the market for the delivery of video programming. *Eighth annual report*, CS Docket No. 01-129. Retrieved February 25, 2002 from http://www.fcc.gov/csb/.

Hafner, K. (2002a, January 24). Drawn to the hearth's electronic glow. *New York Times on the Web*. Retrieved January 24, 2002 from http://www.nytimes.com/2002/01/24/technology/circuits/24SCRE.html?ex=1012907605&ei=1&en=810a5a983b94a185.

Hafner, K. (2002b, February 14). The future of cell phones is here. Sort of. *New York Times on the Web*. Retrieved February 14, 2002 from http://college3.nytimes.com/guests/articles/2002/02/14/900560.xml.

Hofstra University. (2000). *Chronology of computing history*. Retrieved March 13, 2002 from http://www.hofstra.edu/pdf/CompHist_9812tla1.PDF.

Huntzicker, W. E. (1999). *The popular press, 1833-1865*. Westport, CN: Greenwood Press.

Ink, G., & Grabois, A. (2000). *Book title output and average prices: 1998 final and 1999 preliminary figures*. D. Bogart (Ed.). (45th edition). New Providence, NJ: R. R. Bowker.

International Telecommunications Union. (1999). *World telecommunications report 1999*. Geneva, Switzerland: ITU.

Lee, A. (1973). *The daily newspaper in America*. New York: Octagon Books.

Lee, J. M. (1917). *History of American journalism*. Boston: Houghton Mifflin.

Lee, J. (2002). Interactive TV arrives. Sort of. *New York Times on the Web*. Retrieved April 4, 2002 from http://college3.nytimes.com/guests/articles/2002/04/04/910121.xml.

Murray, J. B. (2001). *Wireless nation: The frenzied launch of the cellular revolution in America*. Cambridge, MA: Perseus Publishing.

National Cable and Telecommunications Association. (2001). *Cable and telecommunications industry overview 2001*. Retrieved February 25, 2002 from http://www.ncta.com/industry_overview/ aboutIND.cfm?indOverviewID=1.

National Telecommunications & Information Administration. (2000, October 16). *Falling through the net, toward digital inclusion*. Washington: U.S. Department of Commerce. Retrieved February 22, 2002 from http://www.ntia.doc.gov/ntiahome/digitaldivide/.

Newburger, E. C. (1999, September). *Computer use in the United States*. Washington: U.S. Government Printing Office. Available: http://www.census.gov/prod/99pubs/p20-522.pdf.

Nichols, P. M. (2002, January 30). 4 studios to issue films in a high-end video format. *New York Times on the Web*. Retrieved January 30, 2002 from http://college3.nytimes.com/guests/articles/2002/01/30/ 897796.xml.

Pastore, M. (2002, March 6). At home Internet users approaching half-billion. *The Big Picture Geographics*. Retrieved March 12, 2002 from http://cyberatlas.internet.com/big_picture/geographics/article/ 0,,5911_986431,00.html.

Promotion of Competitive Networks in Local Telecommunications Markets. 15 FCC Rcd 17521 (2000).

Publishing 2002: Where the buck stops; Eight top executives talk about where the business is now and where it is going. (2002, January 7). *Publishers Weekly, 2* (49), 24-36.

Raphael, J. (2000, September 4). Radio station leaves earth and enters cyberspace. Trading the FM dial for a digital stream. *New York Times on the Web*. Retrieved September 4, 2000 from http://www.nytimes.com/library/tech/00/09/biztech/articles/04radio.html.

Reuters. (2002, January 17). Report shows VCR sales fall and DVDs surge in 2001. *New York Times on the Web*. Retrieved January 16, 2002 from http://www.nytimes.com/reuters/technology/tech-tech-television-digital.html?ex=1012367097&ei=1&en=20766a7ef2c2259d.

Selingo, J. (2002, February 14), Talking more but enjoying it less. *New York Times on the Web*. Retrieved February 14, 2002 from http://www.nytimes.com/2002/02/14/technology/circuits/ 14CELL.html?ex=1014692874&ei=1&en=0b7def632252279c.

Stellin, S. (2002, February 14). Cell phone saturation. *New York Times on the Web*. Retrieved February 14, 2002 from http://www.nytimes.com/pages/business/media/index.html.

Television & Cable Factbook, 69 (2). (2001). Washington: Warren Communications News.

U.S. Bureau of the Census. (1972). *Statistical abstract of the United States: 1972*, 93rd edition. Washington: U.S. Government Printing Office.

U.S. Bureau of the Census. (1975). *Statistical abstract of the United States: 1975*, 96th edition. Washington: U.S. Government Printing Office.

U.S. Bureau of the Census. (1976). *Statistical history of the United States: From colonial times to the present*. New York: Basic Books.

U.S. Bureau of the Census. (1978). *Statistical abstract of the United States: 1978*, 99th edition. Washington: U.S. Government Printing Office.

U.S. Bureau of the Census. (1981). *Statistical abstract of the United States: 1981*, 102nd edition. Washington: U.S. Government Printing Office.

U.S. Bureau of the Census. (1982). *Statistical abstract of the United States: 1982-1983,* 103rd edition. Washington: U.S. Government Printing Office.

U.S. Bureau of the Census. (1983). *Statistical abstract of the United States: 1984*, 104th edition. Washington: U.S. Government Printing Office.

U.S. Bureau of the Census. (1984). *Statistical abstract of the United States: 1985*, 105th edition. Washington: U.S. Government Printing Office.

U.S. Bureau of the Census. (1985). *Statistical abstract of the United States: 1986*, 106th edition. Washington: U.S. Government Printing Office.

U.S. Bureau of the Census. (1986). *Statistical abstract of the United States: 1987*, 107th edition. Washington: U.S. Government Printing Office.

U. S. Bureau of the Census. (1988). *Statistical abstract of the United States: 1989*, 109th edition. Washington: U.S. Government Printing Office.

U.S. Bureau of the Census. (1990). *Statistical abstract of the United States: 1991*, 111th edition. Washington: U.S. Government Printing Office.

U.S. Bureau of the Census. (1992). *Statistical abstract of the United States: 1993*, 113th edition. Washington: U.S. Government Printing Office.

U.S. Bureau of the Census. (1993). *Statistical abstract of the United States: 1994*, 114th edition. Washington: U.S. Government Printing Office.

U.S. Bureau of the Census. (1995a). *Consumer electronics annual 1992.* Available: http://www.census.gov/ftp/ pub/industry/ma36m92.txt.

U.S. Bureau of the Census. (1995b). *Consumer electronics annual 1994.* Available: http://www.census.gov/ftp/ pub/industry/ma36m94.txt.

U.S. Bureau of the Census. (1995c). *Statistical abstract of the United States: 1996*, 116th edition. Washington: U.S. Government Printing Office.

U.S. Bureau of the Census. (1996). *Statistical abstract of the United States: 1997*, 117th edition. Washington: U.S. Government Printing Office.

U.S. Bureau of the Census. (1997). *Statistical abstract of the United States: 1999*, 118th edition. Washington: U.S. Government Printing Office.

U.S. Bureau of the Census. (1998). *Statistical abstract of the United States: 1999*, 119th edition. Washington: U.S. Government Printing Office.

U.S. Bureau of the Census. (1999). *Statistical abstract of the United States: 1999*, 119th edition. Washington: U.S. Government Printing Office.

U.S. Bureau of the Census. (2001). *Statistical abstract of the United States: 2001*, 121st edition. Washington: U.S. Government Printing Office. Retrieved February 7, 2002 from http://www.census.gov/ prod/2002pubs/01statab/.

U.S. Copyright Office, Library of Congress, 37 C.F.R. § 201 (2002).

U.S. Department of Commerce. (1986). *U.S. industrial outlook 1986.* Washington: U.S. Department of Commerce, U.S. Bureau of Economic Analysis, and U.S. Bureau of Labor Statistics.

U.S. Department of Commerce. (1987). *U.S. industrial outlook 1987.* Washington: U.S. Department of Commerce, U.S. Bureau of Economic Analysis and U.S. Bureau of Labor Statistics.

U.S. Department of Commerce. (1988). *U.S. industrial outlook 1988.* Washington: U.S. Department of Commerce, U.S. Bureau of Economic Analysis and U.S. Bureau of Labor Statistics.

U.S. Department of Commerce. (1989). *U.S. industrial outlook 1989.* Washington: U.S. Department of Commerce, U.S. Bureau of Economic Analysis and U.S. Bureau of Labor Statistics.

U.S. Department of Commerce. (1991). *U.S. industrial outlook 1991.* Washington: U.S. Department of Commerce, U.S. Bureau of Economic Analysis and U.S. Bureau of Labor Statistics.

U.S. Department of Commerce. (1992). *U.S. industrial outlook 1992.* Washington: U.S. Department of Commerce, U.S. Bureau of Economic Analysis and U.S. Bureau of Labor Statistics.

U.S. Department of Commerce. (1994). *U.S. industrial outlook 1994.* Washington: U.S. Department of Commerce, U.S. Bureau of Economic Analysis, and U.S. Bureau of Labor Statistics.

U.S. Department of Commerce. (1998). *U.S. industry and trade outlook 1998.* New York: McGraw-Hill.

U.S. Department of Commerce. (1999). *U.S. industrial outlook 1999.* Washington: U.S. Department of Commerce, U.S. Bureau of Economic Analysis, and U.S. Bureau of Labor Statistics.

U.S. Department of Commerce/International Trade Association. (1999). *U.S. industry and trade outlook '99.* New York: McGraw-Hill.

U.S. Department of Commerce/International Trade Association. (2000). *U.S. industry and trade outlook 2000.* New York: McGraw-Hill.

White, L. (1971). *The American radio.* New York: Arno Press.

II

ELECTRONIC MASS MEDIA

Digital technologies are revolutionizing virtually all aspects of mass media. Digital video compression, interactivity, and new business opportunities are fueling an explosion in the number of mass media and the programming they provide.

The changes are most evident in multichannel video distribution services. As the following chapter indicates, cable television continues to reinvent itself, incorporating digital technology to increase channel capacity and provide new services. Chapter 6 then explains how direct broadcast satellite (DBS) services have emerged as the most aggressive competitors to cable television.

The factor shared by all of multichannel distribution services is programming. Most of these services depend, in large part, upon revenues from the pay television services explored in Chapter 4, including premium cable channels and various types of pay-per-view television, including numerous forms of video on demand. Not all technologies have been as successful as pay-TV. As discussed in Chapter 5, interactive television efforts that have been discussed as the "future of television" for more than 20 years continue to disappoint inventors and investors.

Chapter 7 explores how digital technology is forcing the biggest change ever in broadcast television as the television industry copes with the difficult reality of converting from analog to digital transmission. This chapter explores how a variety of factors, from regulation to economics, are affecting the diffusion of the eye-catching innovation that spawned digital television—high-definition television. Chapter 8 then explores broadcasting's "low-definition" challenger: streaming media.

Finally, Chapter 9 explains how radio is preparing for its own digital revolution. That revolution may take longer than the television revolution, but digital technology promises the same degree of change in radio as it has offered to all areas of television broadcasting.

In reading these chapters, you should consider two basic communication technology theories. Diffusion theory helps us understand that the introduction of innovations is a process that occurs over time among members of a social system (Rogers, 1983). Different types of people adopt a technology at different times, and for different reasons. The smallest group of adopters is the innovators: They are first to adopt, but it is usually for reasons that are quite different from later adopters. Hence, it is dangerous to predict the ultimate success, failure, diffusion pattern, gratifications, etc. of a new technology by studying the first adopters.

Diffusion theory also suggests five attributes of an innovation that are important to its success: compatibility, complexity, trialability, observability, and relative advantage (Rogers, 1983). In studying or predicting diffusion of a technology, use of these factors suggests analysis of competing technologies is as important as attributes of the new technology.

A second theory to consider is the "principle of relative constancy" (McCombs, 1972; McCombs & Nolan, 1992). This theoretical perspective suggests that, over time, the aggregate disposable income devoted to the mass media, as a proportion of gross national product, is constant. In simple terms, people spend a limited amount of their income on the media discussed in this section, and that amount rarely increases when new media are introduced. In applying this theory to the electronic mass media discussed in the following chapters, consider which media will win a share of audience income, and what will happen to the losers.

BIBLIOGRAPHY

McCombs, M. (1972). Mass media in the marketplace. *Journalism monographs*, 24.

McCombs, M., & Nolan, J. (1992). The relative constancy approach to consumer spending for media. *Journal of media economics*, 5 (2), 43-52.

Rogers, E. M. (1983). *Diffusion of innovations,* 3rd edition. New York: Free Press.

3

CABLE TELEVISION

Larry Collette, Ph.D.[*]

With cable television now in its seventh decade of existence, cable telecommunications companies find themselves facing challenges much greater than simply delivering multichannel video to subscriber households. As integral parts of an increasingly complex and evolving communications landscape, cable telecommunications companies must continue to upgrade their physical plant, offer new services, keep pace with heightened consumer expectations, and deal with competition appearing in its various forms. All of this is occurring amid a backdrop in which public policy concerns and industrial organizational changes are having profound influences on the marketplace.

This chapter examines the development of cable television, details its current status, and explores its potential future directions.

BACKGROUND

THE TECHNOLOGY

The five primary components of a traditional cable system are:

(1)   The *headend*, where television signals are brought in from a variety of satellite and over-the-air sources for distribution across the cable plant.

_____

[*] Assistant Professor, Department of Mass Communications and Journalism Studies, University of Denver (Denver, Colorado).

(2)   The *trunk cable* brings these signals to the neighborhood, with broadband amplifiers arrayed about every 2,000 feet or so to boost signal strength.

(3)   The *distribution or feeder cable* extends the signal from the trunk into the neighborhood, going past subscriber homes and providing service to feeder sections.

(4)   The *subscriber drop* taps a signal off the feeder cable and routes it to the individual subscriber residence.

(5)   The *terminal equipment* located in the subscriber's home. This can be a cable modem, a television set, or other equipment (Cicora, et al., 1999).

Figure 3.1
Traditional Cable TV Network Tree and Branch Architecture

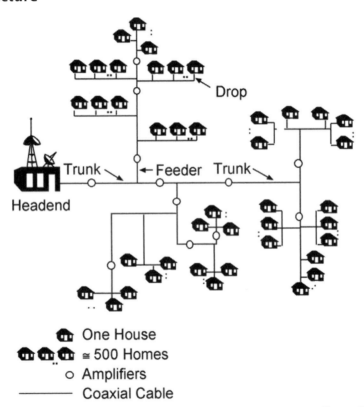

Source: Technology Futures, Inc.

Many cable systems have or are now upgrading their systems to accommodate currently expanded and future services. Fiber optic cable, with its large, seemingly limitless bandwidth capacity, is playing an important role in these upgrades. At this point, deploying a full fiber network is not regarded as economically sound because of the high capital costs and low expected return from the deployment. In the interim, most cable systems are upgrading to hybrid fiber/coax (HFC) networks.

HFC is based on a three-level hierarchy and has some limited features in common with the traditional cable architecture. In the case of HFC, a headend, a distribution hub (typically 20,000 or so homes passed), and a fiber node (500 or so homes) define the network. Rather than using the tree and branch architecture of the traditional cable system, fiber nodes may be connected to a distribution hub in a star network configuration (Adams, 2000). A fiber node converts the optical signal carried over the network into a radio frequency closer to the subscriber home. HFC makes it possible to provide a variety of advanced services, such as high-speed (broadband) Internet access, to cable subscribers.

Figure 3.2
Hybrid Fiber/Coax Cable TV System

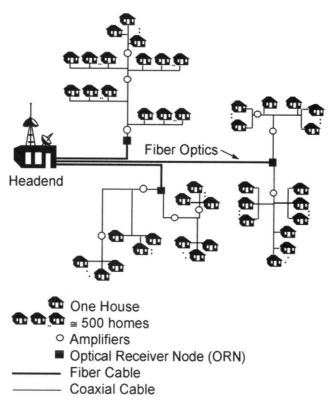

Source: Technology Futures, Inc.

HISTORICAL OVERVIEW

The initial patent for sending pictures over a wire was granted in 1937 to Louis H. Cook, a professor of aeronautical engineering at Catholic University in Washington, D.C. (Taylor, 2000). Ironically, this patent would be granted a full three years before the Federal Communications Commission (FCC) had even adopted technical standards for over-the-air transmission of television signals. The patent described a relatively primitive method for delivering "installed" television transmissions to houses through use of "an enclosed metal tube" to be received on "a conventional television receiving instrument" (Taylor, 2000).

The idea for cable television as a business enterprise and its subsequent deployment can be credited to entrepreneurs wishing to bring television services to remote, rural areas of the United States. These regions had been deprived of over-the-air television due to a shortage of licensed TV stations and because these towns were simply too far from existing stations, making television signals out of reach. Beginning in 1949, Robert Tarlton, along with other appliances dealers, first delivered retransmissions of Philadelphia television stations to the town of Landsford, Pennsylvania (Baldwin & McVoy, 1988). This early attempt was aimed at generating local demand for television sets that had been sitting in appliance dealers' showrooms. CATV (community antenna television), as it was first known, thus came into existence. Subscribers paid a few dollars a month to access a coaxial cable delivering signals from a large antenna perched on the summit of the nearby Allegheny Mountains.

At around the same time, a similar story was being played out in the Pacific Northwest. Ed Parsons created a similar television service there to import a Seattle television signal, from 125 miles away, into Astoria, Oregon (Cicora, et al. 1999). These early events, and others like them, demonstrated that people would willingly pay directly for television services they wished to receive. This important premise remains essential to the multichannel video business today.

Early systems were often a scattered patchwork of coaxial cables strung across trees, utility poles, and roofs throughout the local community. In contrast to current cable telecommunications technology, where large corporations manufacture a vast array of equipment for very specific applications, most of the early equipment consisted of ad hoc adaptations of existing equipment designed primarily for other purposes. The content on those early cable systems was limited to retransmissions of distant television stations. Agreements between the city and cable service providers were often informal "handshake deals" rather than the formalized agreements that would evolve much later.

According to the FCC, 70 communities had cable systems in 1950, serving 14,000 homes (FCC, 1996). The period from 1950 to 1955 is characterized as the "mom-and-pop" era of the cable industry, and these small enterprises faced a variety of challenges (Parsons & Frieden, 1998). Many of the early systems faced problems in finding necessary capital, locating reliable equipment suppliers, securing public rights-of-way, maintaining systems, and attracting customers. By the end of 1955, there were 400 cable systems in operation in the United States.

Most of the growth in cable during the 1960s occurred in smaller communities that were denied full access to broadcast television programming. Cable continued to have difficulties penetrating the larger metropolitan markets where up to seven VHF stations and a sprinkling of UHF stations provided "all the television anyone would ever need." There was a clear, inverse relationship between the number of channels already in a market and the demand for cable service. Improved reception, though, could motivate some to subscribe even where an adequate number of over-the-air channel options existed. The 1960s also brought the first wave of consolidation in cable systems, and with that, the appearance of the first MSOs (multiple system operators) such as Westinghouse Broadcasting, Teleprompter, and Tele-Communications, Inc. (TCI).

The development of a unique programming identity in the 1970s was, in many ways, responsible for cable first meeting some of the "blue sky" projections associated with the medium. In contrast to broadcasting, a medium where scarcity of spectrum would forever be a problem, cable and its non-radiating, sealed spectrum method of local signal delivery promised a medium of abundance.

The first major programming innovation that took advantage of the surplus channel space available over cable systems occurred in the 1970s, when Home Box Office (HBO) began using a communications satellite to make its novel service available to cable companies across the United States. (For more on HBO's history, see Chapter 4.)

Ted Turner was one of the first to realize the potential for delivery of television signals via satellite for distribution on cable television systems. He uplinked the signal of his small, independent Atlanta television station (then known as WTCG, later WTBS) to the same satellite carrying the HBO signal, allowing any cable system that picked up and distributed HBO to distribute his station's signal as well. This prescient move gave birth to the first "superstation," which inspired the creation of countless new cable channels and a number of competing superstations.

These services (and others that followed) would be instrumental in creating a unique programming identity for cable, differentiating it from broadcast competitors and driving increased consumer demand. For example, from 1976 to 1987, the number of satellite-delivered programming services grew from 4 to 70. At the close of the 1970s, 19% of American households subscribed to cable television.

The so-called cable revolution in the 1980s resulted from the creation of a unique programming identity, improved technology and capacity, and a somewhat more favorable regulatory environment. Major municipalities such as Cleveland, Detroit, Los Angeles, and others were wired for cable television. Even greater numbers of programming services came on the scene. Networks such as the USA Network (a general interest, broadly-targeted program service), CNN (a 24-hour news channel), and Cinemax (a specialized premium movie service) were created to take advantage of the new opportunity. The emergence of MTV, a partnership of Warner Bros. and American Express, gave a new definition to "niche" service, as this programming service targeted a young demographic with certain lifestyle characteristics in its programming approach.

The 1980s also saw a big change in the business dynamic within the programming business. Programmers who had once paid cable operators to carry their service now charged the operators fees for those same programming services. The Cable Communications Policy Act of 1984, a set of amendments to the Communications Act of 1934, dealt with a broad range of cable-related issues. To the cable industry's advantage, franchise fees that cable companies paid to local governments were lowered, cable rates were largely deregulated, and general franchise and re-franchising procedures were fixed in statute. By 1987, for the first time, the majority of television households in the United States would be receiving their television through a cable.

The 1990s saw cable companies move aggressively into new geographic and content markets. Importantly, this decade saw two substantial pieces of legislation that had profound impacts on the cable telecommunications industry. Congress passed The Cable Communications Consumer Protection and Competition Act of 1992, which re-regulated cable rates for a period of time in response to accusations of price gouging by cable companies. The Act also mandated signal carriage requirements, permitting local broadcasters to opt for "must-carry" or "retransmission consent." Under must-carry, the station was assured a position on the local cable system lineup. If the broadcaster chose retransmission consent, they could negotiate a fee for carriage of their signals, but the cable system was not obligated to carry the station. A "range war" of sorts flared, with many cable and broadcast interests refusing to meet each other's demands.

One cable system in Corpus Christi, Texas failed to come to terms with the Big Three network affiliates in the market in negotiating for retransmission consent and, as a result, its cable subscribers could not view the local television stations affiliated with CBS, ABC, and NBC. For the most part, however, cable operators and broadcasters recognized that mutual interests were served for the broadcaster to be carried on local cable systems. Though little money changed hands, cable operators in some cases provided "compensation" in the form of local advertisements for the broadcast stations and other forms of promotional tradeoffs. The large MSOs most often refused to pay money for retransmission consent. However, deals were struck to open channel capacity for some services on some MSO cable systems. For example, TCI provided Fox with space for its new cable channel, FX, in exchange for consent to carry the Fox broadcast network.

The Telecommunications Act of 1996 was aimed largely at increasing the prospects of competition in the telecommunications marketplace (Shaw, 2002). For the first time, telephone companies were permitted four entry paths to provide multichannel video service within their service regions. Telcos could set up a separate multichannel video subsidiary, operate an open video system (over which they could provide a set percentage of programming themselves), operate as a common carrier simply providing video carriage to others, or operate a wireless or radio frequency multichannel service. Despite a great deal of initial enthusiasm for the idea, most telephone companies have been reluctant to enter the video services market, and competition there has been slow to emerge (FCC, 2002a).

Over time, several large mergers and acquisitions have changed the face of the cable telecommunications industry. This industry consolidation has been driven, in large part, by the desire to expand the size of companies, thus giving individual companies more clout in various aspects of the business and allowing synergies that result from vertical and horizontal integration. In 1995, Time Warner acquired Turner Broadcasting System in an $8 billion stock swap. This acquisition gave Time Warner a vast array of cable networks, production units, extensive film libraries, and other assets. Several years later, that company proved to be an attractive target for America Online (AOL) in a merger of new media and old media that could eventually prove to be a model for the new century. The combined AOL Time Warner has strategic assets in production, distribution, and exhibition, and operates in a variety of markets. In another important merger, AT&T acquired TCI and Media One in the late 1990s. In so doing, the company paved the way for its reentry into the local wireline telecommunications market that it had exited in the early 1980s.

A series of cooperative "franchise swaps," in which larger MSOs attempted to cluster cable systems within a region in hopes of expanding services such as telephony, data services, and Internet access, also occurred in the late 1990s. Programmers continued international expansion, with MTV and HBO, among others, offering regional variations of their services. At the close of the 20th century, 68% of American households took cable service from one of 10,700 cable systems (NCTA, 2000). There were also 214 national cable video networks providing service to cable operators.

THE BUSINESS

The cable industry operates within a dual revenue stream world, with revenues coming from both subscribers and advertisers. In 2000, subscriber revenues within the U.S. cable industry were estimated to be $40.9 billion, with advertising yielding another $13.8 billion in revenue (NCTA, 2001). Although steadily increasing through the years in its importance, advertising still remains less lucrative than direct subscriber payments. Sixty percent of cable's subscriber revenues are derived from basic cable subscription, with 12% from pay or premium services. For cable operators, "churn" (the number of people who disconnect from cable service) is a constant concern.

Cable service is usually divided into bundled programming tiers for marketing purposes. These configurations vary by cable system, but usually include basic, extended basic, and premium or pay service tiers. The basic tier generally includes over-the-air signals, public access channels, and whatever additional channels the operator may choose to include. Expanded basic includes the basic tier, plus cable networks such as ESPN, USA, TNN, and the like. Pay or premium services may be sold per-channel or bundled in multi-pay packages, and include services such as Showtime, HBO, and their kind. Pay-per-view is offered separately from pay services and is a one-time, one-program purchase. (For more on premium cable and pay-per-view, see Chapter 4.) Some cable operators taking advantage of digital technology are also offering expanded digital programming tiers at an additional fee for subscribers. Included here may be a variety of program offerings ranging from Discovery Science to BBC America.

Cable operators pay the cable programmers for their program services. The price is calculated on a cost-per-subscriber basis, with larger cable systems (e.g., MSOs) enjoying the best price breaks. Cable operators spent $8.9 billion in the acquisition and production of programming in 2000 (FCC, 2002a). Discounts are usually given to cable operators who take multiple programming services from the same programmer, for example, ESPN, ESPN 2, ESPN News, and ESPN Classic.

RECENT DEVELOPMENTS

The rollout of digital cable is the single most important technological development for the cable industry. Digital cable allows for the expansion of the number of video program channels and greater flexibility in information service offerings. The basic digital tier usually includes around 40 or so channels of audio and video. At the close of 2000, the number of digital cable subscribers was estimated to be around 9.7 million (FCC, 2002a). That number increased to around 14.7 million in the span of one year (Special report, 2002). Large MSOs such as AT&T Broadband, Adelphia, Comcast, and Charter have increased their digital subscriber bases substantially over the past several years. At the end of 2001, AT&T had over 3.5 million digital cable subscribers, followed by Charter's 2.1 million, and Comcast with 1.9 million subscribers (Special report, 2002). Growth in this arena was particularly dramatic from 2000 to 2001, with Charter's digital customer base growing by 101%, Time Warner's 83%, and Adelphia's 105%. Morgan Stanley Research projects that, by 2005, there will be over 29 million digital cable subscribers in the United States (Special report, 2002).

In late 2001, Comcast, with a $72 billion bid, sought to acquire AT&T Broadband. This merger is undergoing regulatory scrutiny and, pending approval, the company would have 22 million subscribers in 41 states. This would make it far and away the largest MSO, with 25% of all multichannel homes in the United States (Higgins, 2001). The newly combined company would have over 5 million digital subscribers, 2.2 million high-speed broadband subscribers, and around 1 million cable telephony customers from AT&T. The company would also have serious negotiating clout in its dealings with programmers. In other MSO-related activity, Microsoft, through its subsidiaries, purchased a $500 million stake in Cox Communications in October 2001, making the computer giant a bigger player in the cable business (Higgins, 2002).

The costs associated with rapid growth claimed a prominent victim in 2001. Excite@Home, which had become the largest high-speed Internet service provider (ISP) in the U.S. market, declared bankruptcy and was forced to shut down. The company's problems were attributed to a number of poorly performing business acquisitions and a series of financial setbacks that caught up with the company in 2001. Because Excite@home's owners included a number of cable companies such as AT&T that distributed the service, its failure sent shockwaves through the industry, forcing these companies to find or create a new ISP on short notice.

Enthusiasm for interactive television via cable started to waver in mid-2001, as many MSOs scaled back deployment plans. Industry caution in this arena was prompted by a downturn in the economy and the billions of dollars lost in Internet-related businesses, suggesting that careful reviews of business models were in order (Grotticelli & Kerschbaumer, 2001). Thinking had changed, and expenditures where revenues could be clearly identified were now the desirable pattern. Still, five of the eight largest MSOs made aggressive deployment of video on demand (VOD) a priority for 2002.

In matters related to public policy, two important issues are yet to be fully resolved. Open access continues to be a contentious issue. The important policy question being "should a cable operator be required to open its network to unaffiliated Internet service providers?" This issue also brings into question a cable franchising authority to regulate non-cable or information services provided by cable operators. A series of Congressional inquiries and a number of court rulings have highlighted this issue. On March 14, 2002, the FCC announced a Notice of Public Rulemaking aimed at examining and resolving regulatory uncertainty in the area of cable modems. In doing so, the FCC designated cable modem service as an interstate information service not subject to FCC jurisdiction. Cable modems are also not a cable service as defined by the Communications Act and are not a separate telecommunications service that would be subject to common carrier provisions either. The Commission will further examine the role of state and local franchising authorities in possible regulation of cable modem service. The issue of whether to require multiple ISP access will also be investigated (FCC, 2002b). As of mid-2002, debate continues on digital must-carry following the FCC decision determining, among other things, that cable operators are obligated to carry only the primary video stream of local broadcasters (FCC, 2001). As of this writing, appeals of this FCC decision were winding their way through the courts (McConnell, 2001).

CURRENT STATUS

According to the National Cable & Telecommunications Association, there are 10,243 cable systems in operation in the United States and its territories (NCTA, 2001). Among states, Texas has the most cable systems, with 770 systems serving communities there. The largest single cable system serves New York City with 1,032,872 subscribers and is operated by Time Warner Cable. Most cable systems (56%) have channel capacity of 30 to 53 channels (NCTA, 2001). Only 13 cable systems in the United States have channel capacity of 125 or more channels. Cable service passed nearly 104 million homes by the end of June 2001, of which around 69 million were cable subscribers (FCC, 2002a). In total, there are 1,307,138 miles of cable plant in the United States and its territories, and 68% of U.S. television households subscribe. The average monthly rate paid by cable subscribers is $30.08 for basic cable, with pay channels averaging $7.96 (NCTA, 2001). The largest MSOs, excluding the proposed merger of AT&T and Comcast, are listed in Table 3.1 (NCTA, 2001). The top cable networks as of 2001 in terms of numbers of subscribers are listed in Table 3.2 (NCTA, 2001).

Table 3.1
Largest U.S. MSOs

MSO	Basic Subscribers
AT&T Broadband	16,090,000
Time Warner Cable	12,751,000
Comcast Cable Communications	7,606,800
Charter Communications	6,350,900
Cox Communications	6,193,300
Adelphia Communications	5,292,000
Cablevision Systems	2,830,800
Insight Communications	919,300
Mediacom LLC	779,000
Cable One	734,900

Source: NCTA

Table 3.2
Largest Cable Networks

Network	Subscribers
TBS Superstation	82,000,000
Discovery Channel	81,700,000
TNT	81,600,000
ESPN	81,000,000
USA Network	81,000,000
ABC Family Channel	80,500,000
A&E Network	80,358,000
TNN	80,103,000
Lifetime	79,900,000
Nickleodeon/Nick at Night	79,800,000
CNN	79,700,000
C-SPAN	79,400,000
The Weather Channel	78,900,000
The Learning Channel	78,000,000
MTV	77,300,000

Source: NCTA

In 2001, QVC, one of the two major home shopping channels, had revenues of $3.6 billion (Alleyne, 2001). This cable network generated more revenue than did the CBS and ABC broadcast networks! Other revenue totals for programmers included ESPN ($2.09 billion), HBO ($1.86 billion), TNT ($1.06 billion), and Nickelodeon ($1 billion). Lifetime enjoyed the greatest revenue gains in the year 2001, increasing 30% to $715 million (Alleyne, 2001).

High-speed Internet access via cable modem, which competes with digital subscriber line (DSL) and dial-up for business, was available to 67.3 million homes as of June 2001 (FCC, 2002a). There were approximately 5.6 million cable modem subscribers at that time. Cable's share of the broadband access market has actually shrunk over the past two years, as DSL has increased market share. (For more on cable modem service and DSL, see Chapter 19.)

FACTORS TO WATCH

> *The capacity explosion*—Digital cable will expand the potential number of cable channels available to subscribers. This will also create a rising demand for programming sources to fill those channels.

> *Original programming growth*—Programmers are very likely to produce more and more original material on their own to help control costs and assert greater control of programs for use in other windows. For example, in 2002, Disney's ESPN produced and debuted *A Season on the Brink*, its first made-for-TV original movie. HBO has enjoyed considerable success with original series productions, the latest being *Six Feet Under*, and basic cable services may follow that lead.

> *Finer "nichefication"*—The expansion of program channels will allow even more finely tuned "niche" approaches to programming. This can occur even within a channel. For example, in April 2002, Viacom's Noggin, aimed at children, devoted its evening schedule to a new service, "The N," which is aimed at tweens, children ages 9 to 12. Alternatively, more nichefication can occur through even more specialized, targeted channels, e.g., "The Choir Music Channel." Some have called this audience fragmentation strategy "boutique television" (Cairncross, 2001).

> *Leveraging the broadband plant*—High-speed Internet access and telephony services will expand as cable telecommunications companies attempt to recoup the costs of system upgrades as soon as possible. Around 4 million cable subscribers now get their telephone service via cable (Applebaum, 2002). The cable modem will continue to play an important role in the movement toward greater broadband access to the Internet. Projections for cable modem growth estimate that, by the end of 2005, there will be 17.6 million cable modem subscribers in the United States.

> *Home networking* will gain momentum as the connectivity possible with home networking technology provides better access and greater flexibility in both video and information services.

> *Made-for-cable personal video recorders (PVRs)* that allow more time shifting and flexibility in tailoring programming schedules to individual tastes will appear. Coupled with cable's expanded channel capacity, a PVR in the set-top box will enhance subscribers' potential viewing experience. This "e-ppliance," with associated software, will allow the viewer to manipulate content in a variety of ways with VCR-like functionality (Van Tassel, 2001). Details of the cable-ready PVR deployment plans were not disclosed as of

mid-2002, but there is some suggestion that deployment by several MSOs could begin as early as fall 2002.

➢ *More industry consolidation* will occur as large media/communications companies attempt to shore up their holdings. For example, the combined Comcast/AT&T company would seem likely to be interested in acquiring program networks of its own.

➢ *Program access requirements*, born of the 1992 Cable Act, require programmers to provide access to their networks to other multichannel video providers. The rules were created to prevent cable networks from striking exclusive deals with their affiliates and to prevent discriminatory pricing. These rules are set to expire in 2002. DBS (direct broadcast satellite) has grown to a point where it has become a serious threat to cable companies. Both DirecTV and EchoStar (their possible merger was in the works as this book was going to press) rely on a great many networks originally created for and by cable interests. Cable interests favor the expiration of these requirements, while DBS favors their retention and extension.

Cable is, for now, the dominant player in the multichannel video provider marketplace. The core businesses of the cable industry are the delivery of information and entertainment (cable systems) and controlling programming (programmer). Today, cable firms often exist as part of a large, integrated media/communications company. Cable's future direction as an expanded-service telecom provider, much more than a provider of television shows, will be determined, in large part, by the technology and changing patterns in audience/consumer use. Digital technology, with all its potential, is being embraced by the industry and will have profound effects. The companies within the cable sector are growing both their business holdings and their business offerings. Still, competition from other industries and services is very real and will make it imperative for cable to keep pace with change.

BIBLIOGRAPHY

Adams, M. (2000). *Open cable architecture.* Indianapolis, IN: Cisco Press.

Alleyne, P. L. (2001, November 26). Ka-ching: QVC shopping net rings up no. 2 ranking. *Broadcasting & Cable*, 46.

Applebaum, S. (2002). Technology '02 preview more good stuff. *Cablevision.* Retrieved from http:www.tvinsite.com/cablevisiont_page&doc_+articleID=CA185270.

Baldwin, T., & McVoy, D. S. (1988). *Cable communication.* Second edition. Englewood Cliffs, NJ: Prentice-Hall.

Cairncross, F. (2001). *The death of distance.* Boston: Harvard Business School Press.

Cicora, W., Framer, J., & Large, D. (1999). *Modern cable television technology: Video, voice, and data communication.* San Fransisco: Morgan Kaufmann.

Federal Communications Commission. (2001). *FCC adopts rules for cable carriage of TV signals.* Retrieved from www.fcc.gov/Bureaus/Cable/News_Releases/2001/ncrb0103.htm.

Federal Communications Commission. (2002a). *Eighth annual report on competition in video markets.* Washington: FCC.

Federal Communication Commission. (2002b). *FCC classifies modem service as information service.* Retrieved from http://www.fcc.gov.Bureaus/Cable/News_Releases/2002/nrcb0201.html.

Federal Communications Commission. (1996). *Cable television information bulletin.* Washington, DC: FCC.

Grotticelli, M., & Kerschbaumer, K. (2001, July 9). Slow and steady. *Broadcasting & Cable*, 32.

Higgins, J. M. (2001, December 31). No holiday for consolidation. *Broadcasting & Cable*, 6.

Higgins, J. M. (2002, January 28). Gates takes stake in Cox. *Broadcasting & Cable*, 13.

McConnell, B. (2001, November 5). Paxson appeals FCC carriage ruling. *Broadcasting & Cable*, 12.

National Cable and Telecommunications Association. (2000). *Cable television developments 2000*. Washington: NCTA.

National Cable and Telecommunications Association. (2001). *Cable television developments 2001*. Washington: NCTA.

Parsons, P., & Frieden, R. (1998). *The cable and satellite television industries*. Boston: Allyn & Bacon.

Shaw, J. (2002). *Telecommunication deregulation and the information economy*, Second edition. Boston: Artech House.

Special report: Digital cable. (2002, March 4). *Broadcasting & Cable,* 18.

Taylor, A. (2000). *History between their ears*. Denver: The Cable Center.

Van Tassel, J. (2001). *Digital TV over broadband: Harvesting bandwidth*. Boston: Focal Press.

4

PAY TELEVISION SERVICES

Jennifer H. Meadows, Ph.D.[*]

T he range of programming choices available to today's television viewer is increasingly enhanced by options available through pay television. Although the concept of paying for television programming seemed outrageous in the early days, pay television services have proliferated to become one of the most successful segments of the television industry. From premium channels to video on demand and direct broadcast satellites (DBS), the variety and number of pay television services and the means to receive them has increased dramatically over the past few years. This chapter will discuss traditional pay television services, such as premium channels and pay-per-view (PPV), as well as newer services such as video on demand (VOD), subscription video on demand (SVOD), and near video on demand (NVOD).

Premium channels are the most visible and familiar of the pay television services. These services, such as HBO and Showtime, offer a mix of popular movies, original programming, and sports without commercial interruption. Subscribers pay a monthly fee for each channel, usually around $10.00 per month, above the basic cable fee. In addition to these premium channels, there is a specialized group of premium channels called mini-pays. These channels carry a much lower monthly charge, usually from $1.00 to $3.00 per month, and generally focus on specific kinds of programming such as sports (The Golf Channel), science fiction (the Sci-Fi Channel), and old movies (Turner Classic Movies). In some markets, these channels are bundled in specialized cable tiers, while in other markets, they are offered on an à la carte basis.

Pay-per-view services have been offered to cable and DBS subscribers since the early 1980s. With PPV, a subscriber can order a specific program for a set price. Programming for PPV ranges

---

[*] Associate Professor, Department of Communication Design, California State University, Chico (Chico, California).

from popular movies and adult programming to sporting events and other specialized events such as rock concerts and professional wrestling. Depending on the cable or DBS system, the consumer can call and order the movie or event, or use his or her remote control to place an order. Movies are usually offered on several channels at staggered start times, while special events are usually offered on a one-time-only basis.

With PPV, the viewer must wait until a scheduled time when the cable or DBS system airs the event or movie to watch it. However, with near video on demand or enhanced pay-per-view, the same movie is scheduled on many different channels, with a different start time on each channel. This practice gives the consumer a choice of several movies starting at closely staggered times. In the past, critics of PPV have complained that the restricted start times of movies on PPV have limited its success. NVOD is more convenient, allowing the viewer more choices of start times for movies and more movie choices, but a service provider must devote many more channels to provide NVOD than ordinary PPV.

Video on demand goes one step further by allowing viewers to order from a wide variety of entertainment choices whenever they want to by just pressing buttons on their remote. The viewer is then capable of fast forwarding, rewinding, and pausing the program. VOD puts control of the programming in the hands of the viewer instead of the video service provider, making the experience much more similar to renting a home video than ordering PPV.

There are two ways to bill for VOD. The first is billing on a per-viewing basis, where a subscriber pays a fee that allows them to access a single program during a specified time period. (Most PPV VOD provides a 12-hour or 24-hour window for a single fee, allowing extra time for pausing, rewinding, etc.) The second is subscription VOD or SVOD, a service where subscribers pay a monthly fee for access to a slate of VOD programming.

Services such as VOD, SVOD, and NVOD are made possible through advances in technology such as digital video compression, fiber optics, digital video recorders, and new advanced set-top boxes. In addition, alternative video service providers have presented new opportunities for pay television services. Direct broadcast satellites, for example, presently offer NVOD services. Digital cable systems are also offering NVOD, SVOD, and VOD services.

In order for all of these pay television services to be made available, the television household must be "addressable." Addressability means that the video service provider can communicate directly with each set-top box in every household, allowing the service provider to deliver pay television programming only to consumers who request and pay for it. The set-top box decodes the blocking signal from the video service provider and unscrambles or presents the desired programming.

The incentive for local cable systems and DBS services to carry all of these forms of premium programming is economic. As a general rule, revenues from virtually all forms of premium television are split, with about half being kept by the cable or DBS company and half being paid to the programming service.

Several key factors have emerged in the past few years that will shape the future of pay television services. This chapter will discuss the issues and technology pertaining to pay television services and will highlight factors to watch in the fast-moving and quickly changing future of pay television.

BACKGROUND

Pay television has been around almost as long as television. Zenith actually began to study the possibilities of pay television while the television was still in the research and development stage (Veith, 1976). In the 1940s, Zenith introduced Phonevision, a service that supplied movies via telephone lines. Customers could choose from three movies each day for $1.00 each (Gross, 1986). The Phonevision system never progressed beyond its trial period in 1951. Also in the 1950s, Paramount tested the Telemeter system where customers inserted coins into a set-top decoding box in order to receive programming.

Frightened by the possibility of subscription television, broadcasters lobbied Congress and the Federal Communications Commission (FCC) against any form of pay television service. Those efforts delayed the authorization of pay TV until March 1959, and it was then allowed only on a trial basis. Over-the-air subscription television (STV) service was not authorized until 1968. Broadcast pay television systems were tested in subsequent years, but none of them ever took off. Part of this failure was due to FCC regulations intended to protect frightened broadcasters and theater owners. Regulations included limitations on outbidding broadcasters for programming, having to lease decoder boxes to customers instead of selling them, and being limited to one pay television service per community and then only in markets with at least four commercial stations. These regulations were later rescinded, but they served to handicap the establishment of the pay television industry.

The 1970s and 1980s saw the proliferation of over-the-air subscription television services that used scrambled UHF signals and set-top decoder boxes. These services included SelecTV and ON-TV. Signal stealing was a problem with these services because the decoder boxes used to unscramble the signals were easy to make. Further deregulation in the early 1980s opened up the subscription television market and made way for 24-hour pay services. However, competition to these over-the-air pay services arrived in the form of greater cable television penetration and the availability of premium cable channels, pay-per-view services, and home video. In addition, UHF station owners began to see that they could make more money broadcasting as independent stations rather than as subscription television services (Gross, 1986).

The first major step toward success for pay television services came in 1972 with the introduction of Home Box Office. As a cable service initially delivered to cable systems via microwave technology, HBO was originally provided to cable customers in Wilkes-Barre, Pennsylvania and quickly became a success despite a large subscriber turnover problem (known as churn).

A commitment to new technology played a major role in HBO's future success. In 1975, HBO was beamed to a communications satellite, RCA's Satcom I, to become the first "national entertainment communications network" (Mair, 1988, p. 23). The world heavyweight boxing championship fight between Mohammed Ali and Joe Frazier, the *Thrilla in Manila*, was HBO's first big national sporting event, and it launched the pay service as the leader in premium pay television services, a spot it continues to hold. After HBO made its mark, other premium channels arrived on the scene including Showtime, Cinemax, and The Movie Channel. HBO actually developed Cinemax in response to Showtime (Focus on, 2001).

At the same time as HBO was established, cable companies were experimenting with pay cable television. Warner Cable deployed the Gridtronics pay cable service in several of its cable systems in

1972 (Veith, 1976). Customers could subscribe for $5.00 a month and receive a selection of movies. Another early pay cable service was Theatrevision, which started in 1973 in Sarasota, Florida (Veith, 1976). Users of the service would buy paper tickets that they would insert into the set-top decoder, and movies were offered at scheduled times. Cox Cable Communications deployed an optical systems service called Channel 100 in 1973. The service started with the use of plastic cards and a decoder box and was upgraded in 1975 to include a touch-tone set-top decoder. Customers would call the Channel 100 office to obtain a numerical code that would unscramble the programming (Veith, 1976). These first pay cable services eventually led to PPV service.

In 1977, a federal court overturned the FCC pay cable regulation that limited the service to movies less than three years old and sports broadcasts more than five years old (Baldwin & McVoy, 1983). Many industry analysts saw PPV as the real future of cable television because hit movies could be seen earlier than on network television, and VCR penetration was still quite low. In addition, cable operators recognized the market of viewers that liked movies but did not like going to movie theaters. The movie studios were thus very interested in PPV as a distribution arm for entertainment products, and they experimented with different release windows (the time it takes for a movie to go from theatrical release to pay cable, cable, home video, and broadcast television).

Over time, the release window evolved so that movie studios earned the maximum return by releasing their movies to a pre-specified sequence of markets. Because home video proved to be much more profitable than PPV, most movies are released to home video in an exclusive 45- to 90-day window before they are released to PPV. The PPV services have argued that the window needs to be shorter because, by the time they get the movies, most viewers have already seen it on home video. Thus, buy rates (the percentage of purchase out of the total number of addressable subscribers) for PPV movies have been disappointing.

While the performance of hit movies on PPV has been disappointing, PPV has had much more success with adult services and, in the past, event programming. Boxing and professional wrestling have consistently pulled in respectable buy rates, although, in recent years, there has been declining interest in boxing following a number of disappointing major bouts.

Home video and DVD (digital video or versatile disk) have had a direct impact on PPV. Until recently, the number of movies available on PPV was constrained by the number of available channels, the limited number of start times, complicated billing and ordering procedures, lack of control of the program (you are not able to pause, fast forward, or rewind), and lack of programming choices. PPV operators have attempted to market PPV as a convenient alternative to home video, yet, to this day, home video and DVD remain a multibillion dollar business, while PPV revenues are only about one-tenth as big.

In the past several years, new video distribution services, as well as advances in communication technologies, have made new pay television services possible. In many ways, these new opportunities overcome the limitations of services such as PPV and premium pay cable. Advances in compression and bandwidth capacity and the upgrade to digital transmission systems have made it possible for cable television operators to supply more channels on their cable systems.

In response to these advances, more premium channels are multiplexing—expanding their services from one channel to several channels. For example, HBO and Cinemax began to multiplex, without an increase in subscription rates, in the early 1990s. One configuration available to cable systems and DBS providers is the Ultimate Entertainment Package with 14 channels including HBO Signature (women), HBO Family, HBO Latino, Wmax (women), OuterMax ("daring"), and ThrillerMax (Order HBO, 2002). Similarly, Showtime offers the Showtime Unlimited Package, which includes such multiplexed channels as ShoBeyond (sci-fi), ShoFamily, ShoNext (hi-tech), and ShoExtreme (action) (Showtime Unlimited, 2002). While multiplexing essentially began with premium channels offering different feeds for different time zones (delaying the west coast schedule by three hours to allow west coast viewers to watch a program at the same scheduled time as east coast viewers), now the premium multiplexes often offer different programming on each channel. One example is the Starz! Superpak that offers 13 channels including genre specific ones such as western, action, and true stories, as well as Starz Theatre (Starz! Superpak, 2002).

Additional channels also allow cable companies and alternative distribution services such as DBS to devote more channels to pay-per-view services and thus offer NVOD service where hit movies are offered every 10 to 15 minutes instead of every two hours. In addition, advanced set-top boxes make ordering easy. Where customers once had to deal with limited schedules and a complicated ordering process, viewers can now use their remote control to order, review and confirm the order, and limit ordering by children.

While NVOD is currently offered by most digital cable systems and DBS providers, video on demand is viewed by the cable and satellite industry as the service with the greatest revenue potential. With these systems, subscribers can access programming immediately and have VCR functionality, meaning they can pause, fast forward, and rewind the programming.

Time Warner's now defunct Full Service Network offered VOD. This service was deployed in Orlando, Florida from 1994 to 1997 and offered switched digital interactive multimedia services using a hybrid fiber/coax network. Customers could order movies on demand, with full VCR functionality, from a library of 100 titles for about the same price as a video rental. Ultimately, the Full Service Network was a failure as customers did not use the interactive services and thus the service could not generate enough revenue to pay for the cost of the infrastructure and the related set-top boxes.

RECENT DEVELOPMENTS

There have been some exciting developments in pay television services since 2000. The premium channels have never been stronger. HBO and Showtime continue to use multiplexing, as well as a slate of original programming, to attract subscribers. In fact, it is the original programming that is increasing subscribership for premium channels. HBO series such as *The Sopranos*, *Six Feet Under*, and *Sex and the City* have elevated the programming to the point that HBO garnered 94 Emmy nominations in 2001, up from six in 1988—unheard of for a cable network much less a premium channel (Kiska, 2001; 2001 Primetime, n.d.).

What seems to be working for premium channels is attention to a subscriber-based business plan. Unlike broadcasting, the programming is not there to sell advertising. Instead, the programming is there to attract, keep, and serve the needs of subscribers. The industry has found that viewers do not have to watch premium channels every day to feel like they are getting their money's worth. The programming just needs to be something that viewers cannot get elsewhere on broadcast or traditional cable channels. The premium television services have also been making extensive use of cross-platform marketing with Web site interactive services, and e-mail alerts. HBO chairman and Chief Executive Offer Jeff Bewkes states, "HBO is taking advantage of the interactive medium, not only to promote shows and build online community, but also to allow potential subscribers to sign up for HBO directly via HBO Express and, soon, access HBO programming exactly when they want it through a new service called HBO on Demand" (Focus on, 2001).

HBO on Demand is a subscription video on demand service (SVOD). Subscribers pay a monthly fee ($3.95) for access to a library of HBO programming and movies. Subscribers then have full VCR functionality to control the programming (HBO on Demand, n.d.). Time Warner Cable began testing this service in Columbia and Sumter, South Carolina in 2001, then began rolling out the service on other systems in 2002. The service was so popular in the first week of the South Carolina trial (when it was offered free to all HBO subscribers) that Time Warner had to limit the free access because demand was overloading the system (Stump, 2001). The system uses a Scientific-Atlanta Digital Network Control System. The most popular programs in the initial week of July 2001 were *The Sopranos*, *Sex and the City*, *Gladiator*, and *Chicken Run* (Stump, 2001).

Other programming suppliers are looking to SVOD including the Discovery Network, which is planning to offer a subscription Discovery on Demand service as well as a free Discover Choice 10 package (Stump, 2002). Not be left behind, Starz!, Showtime, and the Independent Film Channel all offer on-demand services as well. Few cable systems offer these services as of mid-2002, but those numbers should increase as systems upgrade their facilities (FAQ, n.d.). Cablevision in Western Long Island, New York is offering its subscribers a service called iO, or Interactive Optimum. This is an interactive television service that includes a long list of on-demand choices including HBO's, Showtime's, and Starz' on-demand services as well as more specialized SVOD services including kids, sports, onstage (music and theater), and TV Encore (FAQ, n.d.).

While SVOD and VOD are the most talked about new pay television services, don't discount pay-per-view. As digital cable subscribership grew to 15.2 million homes in 2001, PPV became available to more homes than ever before (Umstead, 2002b). Movies continue to be the biggest revenue generators for PPV. In Demand L.L.C., the leading PPV network, continues to grow with over 35 PPV channels. The network has expanded its lineup to include channels targeting audience niches including Urban In Demand and Cinema In Demand (Umstead, 2002a). In addition, In Demand landed the children's film *Max Keeble's Big Move* as a PPV title before it was released to home video (PPV gets, 2002). Sports may have higher revenues soon as the National Football League announced that the "Sunday Ticket" out-of-market PPV package will be available to cable by 2003. It has been limited to DBS in the past (Dressier: Cable, 2002).

One of the biggest challenges to pay-per-view is VOD and SVOD. Studios are eyeing VOD as a more profitable revenue outlet than PPV and are negotiating deals with VOD providers such as In Demand (also a leader in PPV) to obtain a larger cut of each purchase in exchange for better release windows (Umstead, 2001). Studios are also looking at distributing value-added features such as extra scenes with VOD releases similar to what is done on DVD.

A study by Jupiter Media Metrix suggests that VOD users will be primarily former PPV users (Boulton, 2001). This prediction fits with the principle of relative constancy that states that consumers will spend a constant fraction of their income on mass media, but the type of mass media will vary. If users adopt VOD, then they will likely drop something else—PPV or home video rental. The study also found that, while 45% of respondents rented videos, only 6% ordered PPV at least once a month (Boulton, 2001). PPV still has many hurdles to overcome, but the similarities of PPV to VOD also speak to the future success or failure of VOD.

Several technology factors are driving the adoption of VOD. First, cable systems are deploying advanced two-way digital networks that include fiber optics. Second, a new generation of advanced addressable set-top boxes are being used with both cable and DBS. Finally, Cable Television Laboratories, Inc. released the first VOD Content Specification V1.0 in March 2002. The specification deals with the metadata, used for title information, licensing, and display. Up to this point, each VOD provider used proprietary metadata schemes that increased the cost to VOD content owners, server system providers, and set-top box makers (Brown, 2002).

Cable VOD systems work much like NVOD except the set-top box "holds" the program for the consumer. The consumer orders the program using a remote control. A server at the cable company streams the program using MPEG-2 digital video compression to the set-top box. The consumer then has a limited time, usually a day, to watch the program. Because the program is on the set-top box, the viewer then has control of the program to fast forward, rewind, pause, etc. (Fowler, 2002).

Cable multiple system operators (MSOs), such as Comcast and AT&T Broadband, already are providing VOD services in some markets. Using VOD hardware and software provided by companies such as Diva, cable companies are aggressively working to test and implement this potential revenue stream (N2 Broadband, 2002).

Direct broadcast satellite is offering an "almost" video on demand, but in a slightly different format. DBS providers are teaming up with personal video recorder (PVR) or digital video recorder companies such as TiVo to offer subscribers "almost" video on demand services. Subscribers just order the programming using the NVOD system the DBS service provides, and it is sent to a set-top box that contains a PVR. The consumer then has control over the program. This system is more VOD than NVOD as the subscriber has control but must wait for the scheduled program time to record (Fowler, 2002). (For more information on PVRs, see Chapter 16.)

As of mid-2002, video on demand was also being offered on the Internet in a limited number of test markets. At this point, there are two major services: Intertainer and Movielink. Intertainer is partly owned by Microsoft and has distribution deals with Disney and 20th Century Fox. Subscribers pay a monthly fee and a per-movie charge of about $4.00, and then the film is streamed to the user's computer using Windows Media technology. The subscriber then has control over the film (Kary, 2001). Movielink is a joint venture of Sony, Viacom, MGM, and AOL Time Warner. The service is similar to Intertainer (Hu, 2002). The biggest inhibiting factors for Internet video on demand are that it requires a broadband connection, and the image quality is poor compared with the MPEG-2 DVD-like quality of cable and satellite VOD.

CURRENT STATUS

The future is still bright for pay television services. First, the number of digital cable households continues to grow. That number is up from 9.7 million households in 2000 to 15.2 million households in 2002 (NCTA, 2002). The number of pay cable units continues to remain stable at 51.6 million with a 71.5% ratio of pay cable units to basic cable units (NCTA, 2002).

Figure 4.1
Digital Cable Subscribers (in Millions)

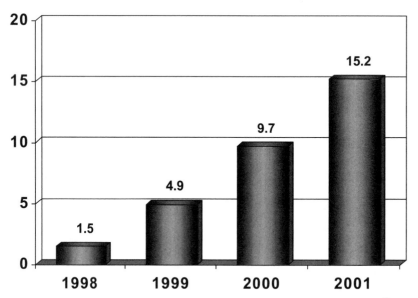

Source: NCTA (2002)

Premium channel subscribership continues to grow. As of year-end 2000, HBO was the number one premium channel, available either alone or along with Cinemax (number two) in 37 million homes in the United States (Focus on, 2001). Showtime has about half that number of subscribers, but has made great inroads with underserved populations such as Latinos, African Americans, and gay men. Showtime's original programming reflects this strategy with series such as *Soul Food*, *Queer as Folk*, and *Resurrection Boulevard*, and 20% of Showtime's subscribers are African American (Levin, 2001). Showtime's strategy is not to beat HBO but to provide a complementary or alternative premium service. Starz! Encore Group continues to be the largest provider of premium movie channels with 15 domestic channels (Starz! Encore, 2002).

Pay-per-view had its best year ever in 2001 with over $2 billion in revenue. Movies generated the greatest amount of revenue ($1.3 billion), and adult content was second with $545 million. While the movie and adult categories increased in revenue, the event category continued to slide with only $297 million in revenues. Both boxing and wrestling showed lower buy rates (Umstead, 2002b). DirecTV and EchoStar, DBS providers, generated $1.3 billion in PPV (Higgens, 2002). (By comparison, home video rentals grossed more than $8 billion in revenue in 2001 [Higgens, 2002].)

The estimates for video on demand vary widely. The Gartner Group predicts that VOD will not make its mark until 2005 and then hold only a small percentage of movie distribution revenue (about 2%) (Kelsey, 2002). The more optimistic Jupiter Media Metrix reports that video on demand was used by 300,000 people in 2001 and generated $16 million. The report predicts that, by 2006, there will be 5.3 million people using VOD, generating $641.9 million in revenue (Fowler, 2002). Finally, Cahners InStat reports that North American VOD generated $86 million in 2001 and that number will grow to $1.75 billion by 2005. As evidenced above, the future of video on demand remains up in the air. In fact, media analysts admit to difficulties obtaining good information from cable companies, and the cable companies tend to keep VOD plans under wraps (Higgens, 2002).

FACTORS TO WATCH

> *Continued growth of PPV*—Expect PPV revenues to continue to rise as more PPV channels are added to cable and DBS systems, creating and expanding NVOD service. In addition, digital cable technologies will make ordering PPV programs easier, increasing buy rates.

As more advanced digital networks and set-top boxes are ... of VOD and SVOD services on digital cable sys-... way to challenge DBS providers.

...shows—Some industry experts are predicting that ...ular that consumers will be able to watch a favorite ...all fee or the same show with commercials for free. ...ying for programming that was free in the past is the

...ional video recorders in one—Look for more video on ...ow being used on DBS with TiVo. Also look for cable ...s through their set-top boxes by subscription, so that the ...any hardware.

...in programming and content distribution—The entertain-...OD as a way to generate big revenue. As opposed to using ...to create more partnerships with VOD content providers.

...ge, pay television is finally beginning to fulfill some of its ...hip is up. Even pay-per-view movies are showing better buy ...to NVOD. The future of SVOD and VOD remains to be seen, ...ions, then pay television services will continue to flourish. The ...mers are willing to spend and how fast cable operators can up-

BIBLIOGRAPHY

2001 Primetime Emmy Nominations. (n.d.). Retrieved on March 18, 2002 from http://www.emmys.com/ awards/2001.htm.

Baldwin, T., & McVoy, D. S. (1983). *Cable communication*. Englewood Cliffs, NJ: Prentice-Hall.

Boulton, C. (2001, December 24). Movies-on-demand could harm pay per view. *ISP News*. Retrieved on March 5, 2002 from http://www.Internetnews.com/isp-news/article/0,,8_945241,00.html.

Brown, K. (2002, March 5). CableLabs issues VOD specification. *Multichannel News*. Retrieved on March, 2002 from http://www.tvinsite.com/index.asp?layout=print_page&doc_id=72451&articleID=.

Dressier: Cable will get NFL PPV. (March 8, 2002). *Multichannel News*. Retrieved on March 11, 2002 from http://www.tvinsite.com/multichannelnews/index.asp?layout=print_page&doc_id=73143&articleID=.

FAQ. (n.d.). *Interactive Optimum*. Retrieved on March 18, 2002 from http://www.io.tv/faq.html.

Focus on: HBO—Serving subscribers with unique, powerful programming. (2001, June 4). *FYI—AOL Time Warner Investor Relations Newsletter*. Retrieved on March 6, 2002 from http://www.aoltimewarner.com/investors/services/newsleter060401.html.

Fowler, G. (2002, March 5). Ask and you shall receive: Video on demand is an idea whose time has come—almost. *Wall Street Journal*, R10.

Gross, L. S. (1986). *The new television technologies*. Dubuque IA: Wm. C. Brown Publishers.

HBO on Demand. (n.d.). Retrieved May 6, 2002 from http://www.HBOonDemand.com.

Higgens, J. (2002, March 4). VOD confidential: Off the record, on the QT & very hush-hush…. *Broadcasting & Cable*. 14-18.

Hu, J. (2002, January 29). Studios step up video on demand focus. *CNET News.com*. Retrieved on March 6, 2002 from http://news.com.com/2100-1023-825477.html.

Kary, T. (2001, October 17). Microsoft in video on demand deal. *CNET News.com*. Retrieved on March 6, 2002 from http://news.com.com/2100-1023-274574.html.

Kelsey, D. (2002, January 30). Video on demand? Not so fast. *Newbytes*. Retrieved on March 6, 2002 from http://www.newsbytes.com/cgi-bin/utd/im.display.printable?client.id=newsbytes&story.id=174081.

Kiska, T. (2001, September 7). HBO hits its stride. *The Detroit News*. Retrieved on March 6, 2002 from http://detnews.com/2001/entertainment/0109/09/c01-27-87090.htm.

Levin, G. (2001, August 26). For TV diversity, it's Showtime. *USA Today*. Retrieved on March 18, 2002 from http://www.usatoday.com/life/enter/tv/2001-08-28-diversity-showtime.htm.

Mair, G. (1988). *Inside HBO*. New York: Dodd, Mead & Co.

N2 Broadband and Diva offer seamless video on demand to cable operators and content providers. (2002, January 14). *Diva Technologies*. Retrieved on March 6, 2002 from http://www.divatv.com/ ph/news_press_releases3.php3?prid=96.

National Cable and Telecommunications Association. (2002). Industry Statistics. *National Cable and Telecommunications Association*. Retrieved on March 13, 2002 at http://www.ncta.com/industry_ overview/indStat.cfm?indOverviewID=2.

Order HBO online. (2002). *HBOXpress*. Retrieved on March 4, 2002 from http://www.hbo.com/NASApp/ hboxpress/jsp/get_hbo.jsp.

PPV gets kids' flick before video. (2002, January 25). *Multichannel News*. Retrieved on March 5, 2002 from http://www.tvinsite.com/multichannelnews/index.asp?layout=print_page&doc_id=66732&articleID=.

Showtime Unlimited Package. (2002). *Showtime Unlimited*. Retrieved on March 14, 2002 at http://www.show.com/unlimited/.

Starz! Encore Chairman Sie tells cable marketers to maintain 68% of digital cable growth market share or suffer erosion of cable base. (2002). *PR Newswire*. Retrieved on March 18, 2002 from http://www.prnewswire.com/cgi-bin/storeis.pl?ACCT=105&STORY=/www/story/03-07- 2002/00016833153.

Starz! Superpak. (2002). *Starz Encore Group*. Retrieved on March 14, 2002 from http://www.starzsuperpak.com/.

Stump, M. (2002, March 8). On-demand is In Demand. *Multichannel News*. Retrieved on March 8, 2002 from http://www.tvinsite.com/multichannelnews/index=asp?layout=print_pages&doc_id=72853&articleID=.

Stump, M. (2001, July 23). Time Warner system sees SVOD demand. *Multichannel News*. Retrieved on March 5, 2002 from http://www.tvinsite.com/multichannelnews/index.asp?layout=print_page&doc_id&articleID=CA106256.

Umstead, R. T. (2002a, February 11). In Demand to roll out new channels. *Multichannel News*. Retrieved on March 5, 2002 from http://www.tvinsite.com/multichannelnews/index.asp?layout=print_page&doc_id=68951&articleID=.

Umstead, R. T. (2002b, February 5). PPV: Movie revenue up, events down. *Multichannel News*. Retrieved on March 5, 2002 from http://www.tvinsite.com/index.asp?layout=print_page&doc_id=68025&articleID=.

Umstead, R. T. (2001, August 6). MGM eyeing big VOD, PPV future. *Multichannel News*. Retrieved on March 5, 2002 from http://www.tvinsite.com/multichannelnews/index.asp?layout=print_page&docid&articleID=CA149959.

Veith, R. (1976). *Talk-back TV: Two-way cable television*. Blue Ridge Summit, PA: TAB Books.

5

INTERACTIVE TELEVISION

Paul J. Traudt, Ph.D.[*]

A profitable model for matching video with viewers actively engaging with content has long been the goal of interactive television (ITV or iTV). However, achieving the one true "killer application" continues to be elusive in the murky world of ITV. The 2001 National Association of Television Programming Executives (NATPE) convention was noteworthy for vendors touting the "interactive" features of their hardware, middleware, application platforms, and user-level applications. The same organization's subsequent 2002 convention was equally noteworthy for ITV's low profile—generally relegated to after-production game enhancements for shows such as *Real World* and *Survivor*. Historically, ITV is the wrecking ground for over one billion dollars in capital investments. Major players are still searching for viable revenue streams and ways to offset generally lukewarm consumer reception (Talwani, 2002; Human bandwidth, 2001).

Perhaps the largest impediment to successful ITV development is viewer inertia. For more than 50 years, television viewing has always been one-way analog communication. Traditional television audiences are habitually passive viewers for the simple reason that one-way technologies inhibit any real form of interactive communication. A successful ITV platform must be so compelling so as to overcome viewer inertia and generate "lean in" versus well-learned "lean back" behaviors. Another major problem with ITV is that it has come to represent so many different things. ITV was heralded in the 1970s as the new electronic forum for civic behavior, allowing participants to interact with local governments. In the mid-1990s, viewers were targeted with efforts to merge traditional television with Web-based functions. The end of the last decade is noteworthy for developments in video on demand (VOD), increasingly powerful program guides, and personal video recorders (PVRs) (Beacham, 2000a).

---

[*] Associate Professor of Media Studies, Hank Greenspun School of Communication, University of Nevada, Las Vegas (Las Vegas, Nevada).

ITV is the combination of traditional linear video and a range of interactive services (Feldman, 1997), combining entertainment programming and remote control functions with home shopping and information features of the Web (Jarvis, 2001). *Broadcasting & Cable* narrows the definition to include interactive program guides, VOD, PVRs, and Internet-type tools on television (ITV: How we, 2001). The typical configuration includes linear analog or digital video accompanied by layered, transparent or semi-transparent graphical or text elements. Some ITV providers distribute HTML (hypertext markup language) information to analog or digital television receivers, digital set-top boxes (DSTBs), or personal computers (Jarvis, 2001; wedlow, 2002). Home users can connect to ITV via the Internet, local cable television provider, direct broadcast satellite (DBS), local telephone company, or some combination of the four. Web-based interactivity takes the form of user manipulation of graphical interfaces via remote control or cordless keyboards.

The ITV industry is comprised of a multitude of players, some with overlapping functions and interests. As the boundaries map shown in Figure 5.1 illustrates, key players in the industry as of early 2002 can be broken down into two major groups (Mendelson, 2002). The first group includes interactive production services, addressable technologies, content providers, and broadband delivery systems. The second group is made up of DSTB manufacturers, middleware developers, interactive application platforms, and user-level applications for primary services. Six major service categories make up the user-level landscape (Haley, 1999; Kerschbaumer, 2000; ITV: How we, 2001; Swedlow, 2002).

Interactive gaming. This service typically shares the common characteristic of content provided by programmers or advertisers without Web browsing capability. ACTV's software in set-top boxes allows the viewer directorial control of live sporting events via remote control. For example, the user can choose from a number of ballpark cameras covering a baseball game while requesting a batter's season or game statistics.

Interactive program guides or navigators. This service is provided via satellite, cable, or other set-top box systems. The typical configuration provides programming schedules and information sent via the vertical blanking interval (VBI). Advanced DSTBs now include memory components for individual users and navigational preferences where like or similar programs will be recommended. Revenues are based on a percentage of monthly cable service fees and advertising. Gemstar-TV Guide's Web-based application and TV gateway are vying for dominance in the area seen by some as the most competitive area of ITV development (Haley, 1999).

Internet tools on TV. This technology provides Windows or browser-type interactivity in the form of graphical or text enhancements, allowing such features as e-mail, Web surfing, and chat over a broadcast video background. Originally available via cable television, Internet-on-television services are now offered through direct broadcast satellite services and even more recently via broadcast television. The typical technical configuration includes proprietary software housed in a receiver, set-top box, and tuner card. HTML data are sent via the VBI signal provided by broadcasters, but it should be noted that digital television will increasingly eliminate the need for VBI utilization. Another version of the technology utilizes software at the cable headend. Revenues are generated by means of a combination of set-top box purchases, subscription, and monthly fees. AOL TV, MSNTV, and WorldGate compete for dominance.

Figure 5.1
Interactive Television Industry Boundary Map

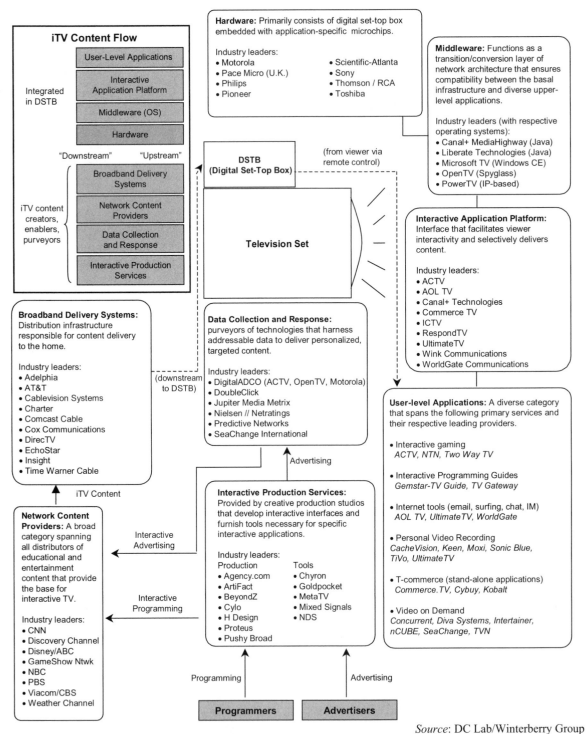

Source: DC Lab/Winterberry Group

Personal video recorders. PVRs allow for audience time shifting via set-top digital hard drive recording of video programming. Lower-priced PVRs provide 35 hours of recording capability, with some higher-priced systems providing in excess of 300 hours. They can also record, pause, playback, and resume program feeds. Revenues are generated via a combination of PVR purchases and monthly service fees. TiVo, ReplayTV, and Microsoft's Ultimate TV are major players in personal video recording services. (Chapter 16 includes an extended discussion of PVRs.)

T-commerce. With television commerce (t-commerce), users can request more information or order products or services via remote control when viewing specially produced programs or commercials. Programmers or advertisers provide content. Revenues are generated through a combination of cable subscriptions and advertising. Wink Communications' proprietary software resides in a set-top box and interprets data sent via VBI or MPEG standards. Wink enhanced commercials or programs include an icon overlay (What is Wink? 2002).

Video on demand. VOD users access programming via two-way cable, direct broadcast satellite, or DSTBs. The "bandwidth hogs of the interactive world" (Haley, 1999, p. 18), VOD requires up to four 6-MHz channels to serve the estimated 10% of cable subscribers using the service at any one time. Diva Systems and Intertainer use vacant cable television capacity and Internet connectivity, respectively, to stream broadband MPEG video to the home. VOD services can provide programming and information almost whenever the end-user chooses, either by means of real-time interactivity or previously delivered programming in residence on hard drives at the user's end. Revenues are generated through both subscriptions and advertising. Sea Change and TVN are among other competitors. (Chapter 4 contains an extended discussion of VOD.)

These six divisions are becoming increasingly fuzzy as major players converge two or more of these categories into bundled services. Some industry observers refer to all current services as *enhanced television.* The best way to understand and evaluate these current offerings may be to take a look back at the history of interactive television.

BACKGROUND

ITV development has a history of ups and downs, including simple efforts in children's programming in the 1950s, additional trials and failures in the 1960s, followed by full-scale but financially unsuccessful efforts in the 1970s and again in the 1990s. Converging communications technologies, frenzied marketplace speculation, and potentially lucrative returns on capital investment all contributed to significant expansion of ITV projects in the 1990s.

Early ITV efforts included *Winky Dink* in 1953, a CBS Television animated series designed to engage program characters and children watching the program. Scripts included audience prompts, where children were instructed to apply a transparency to the television receiver screen and use crayons to generate visual overlays and connect dots to reveal "secret" messages in order to help Winky Dink. The program aired in some markets as late as the early 1970s (Swedlow, 2002) and is now making a comeback of sorts through the sale of videos and interactive kits (Winky Dink, 2002).

Warner Amex's QUBE TV was an analog, two-way interactive cable experiment beginning in 1977 that provided interactive programming, standard programming fare, and banking services

(Rosenstein, 1994). The Columbus (Ohio) set-top box service offered 36 channels, a wired remote, and was expanded to the Dallas and Pittsburgh markets. Users could indicate preferences and vote during special programs generated for the system. Decline in investor support and technology limitations spelled QUBE TV's demise despite subscriber interest (Swedlow, 2002).

Teletext and videotex services that had been successful in Europe, such as the BBC's Ceefax, were also tested in the United States in the late 1970s and early 1980s. At about the same time, GTE's Viewdata and Knight-Ridder's Viewtron pilot programs provided electronic newspapers, weather, and agribusiness information via telephone lines. Buggy technology and limited programming contributed to user indifference. GTE later offered interactive home shopping, banking, video games, and VOD via cable television to subscribers in Cerritos (California) in the late 1980s and early 1990s. Less than 5% of system households subscribed to the expanded service (Rosenstein, 1994).

By the mid-1990s, large corporations, consortia, and alliances were testing multi-service, high-speed interactive, and enhanced television systems. Competition, growing Internet popularity, developments in MPEG compression technologies, and a general belief that ITV would result in lucrative returns fueled these efforts.

Time Warner's Full Service Network was launched in Orlando, Florida in December 1994. Time Warner offered the digital, interactive multimedia service to 4,000 test households and initially projected that over 750,000 households would be part of the system within four years. Test households were provided VOD, interactive shopping and games, a program navigator, and postal services on a pay-per-view basis (Full Service Network, 2002). In 1997, Time Warner announced the end of its $250 million dollar Full Service Network experiment, citing high cost-per-household hardware and difficulties in shifting user interests from home videocassettes. Industry observers predicted that Time Warner's was the last full-scale effort at proprietary-based ITV because of booming Internet and Web-based possibilities (Goldstein, 1997; Shiver, 1997; Marriage of, 1997).

Microsoft entered ITV in a big way in 1997 with their acquisition of WebTV, effectively leading the charge to move personal computing and Web surfing out of the home office and into home video centers (Marriage of, 1997). Bill Gates was convinced that the corporation's future depended on acquiring a larger share of the ITV market (Desmond, 1997). A significant advertising campaign helped WebTV quickly grow to claim 200,000 subscribers by November 1997. The service ultimately attracted about one million subscribers before growth leveled off in 2000.

In a separate effort that illustrates Microsoft's interest in being a part of the ITV industry, the company agreed in 1998 to provide ITV technologies to Tele-Communications, Inc. (TCI) for the production of Internet-ready set-top boxes, and included an order for five million of the boxes preloaded with the Windows CE operating system to provide e-mail, Internet access, and video on demand to subscribers. Oracle Corporation reacted with efforts to expand into enhanced TV and to speed up slow-loading Web pages (Clark, 1997). At the same time, Bill Gates also invested $1 billion in Comcast, endorsing cable system bandwidth as a digital distribution medium in an attempt to establish Microsoft's operating system as the standard for use in digital set-top boxes (Lesly, et al., 1997).

Proponents and industry analysts were predicting the deployment of digital set-tops within one to three years. Meanwhile, in establishing its new open standard, the cable television industry effectively dictated the technical requirements for ITV, including specifications for writing software in both Windows CE and Java operating system languages for servers and set-top boxes. The new standard effectively redefined the platform for ITV competitors and made proprietary software and hardware obsolete.

The Advanced Television Enhancement Forum (ATVEF), comprised of 14 of ITV's major players (including CNN, Discovery, Microsoft, PBS, Sony, and TCI), was established in 1998 and developed a platform for common technical standards (Pegg, 1999a). Those standards were diluted by industry wrangling over digital television, with every company with investments in television, the Internet, information, and entertainment hedging their bets by buying into collaborative ITV efforts. The ATVEF was recast as the Advanced Television Forum (ATF) in fall 2001 and reaffirmed its goal to find common ground for a membership of "30 content, delivery, and technology companies" directly involved in ITV (Grotticelli, 2001b, p. 28).

In 2000, Forrester Research predicted that ITV revenues in 2004 would grow to include $11 billion in advertising, $7 billion in t-commerce, $20 billion from subscriptions, and $3.1 billion from VOD (Haley, 1999; Meeting focuses, 2000). How accurate were these predictions? If the history of ITV teaches us anything, it is to put more stock into actual numbers rather than predictions.

RECENT DEVELOPMENTS

Interactive gaming. ACTV is a long-standing player in interactive gaming services and has expanded its services to include t-commerce and educational services. In May 2001, the Game Show Network announced it would be using ACTV technology to provide an interactive component to its original game show, *Mall Masters* (Game Show Network, 2001). Early in 2002, ACTV provided the technology and creative support for the National Basketball Association's launch of a live fantasy game to accompany actual game coverage on TBS and TNT cable networks (NBA Entertainment, 2002). London-based Two Way TV and its U.S. subsidiary provided enhanced ITV features for scheduled and on-demand games. Considered "service at the low end" of ITV, the service is free to customers (Pegg, 1999b, p. 14). Compared with other ITV services, interactive gaming generates low interest on the part of potential users. Results from reliable online poll data vary, but generally confirm lukewarm user reception to interactive gaming. Percentages run from a high of 42% interested in playing along with game show telecasts (Who wants, 2001) to 34% "very interested" in interaction with game shows (Bensmiller, 2002; e-poll, 2001).

Interactive program guides or navigators. Programming guides have seen the greatest growth in ITV, "partly because they are a great hook to subscribers and partly because they're cheap and easy" (Grotticelli & Kerschbaumer, 2001, p. 34). Whoever controls navigation also controls audience exposure to selected programming and information services. The dominant Gemstar-TV Guide provides a navigational tool for digital cable subscribers marketed as GUIDE Plus+ or TV Guide Interactive. The on-screen interactive guide enables viewers to access favorite programs, learn about programs up to four days in advance, use one-button scheduling, and limit access to certain programming deemed unsuitable for other household members. The system will also support VOD and PVR technologies. Gemstar boasted strong 2001 third quarter earnings in its ITV ventures, with "revenues up 339% year-over-year, and up 42% over the second quarter 2001" (Gemstar-TV Guide, 2001, p. 1). The service is free to customers. Revenues are generated via advertising.

MyDTV has developed a content-selective service designed to inform viewers when specific program material is airing, while viewing another program or channel. Some news directors at local television broadcast stations are intrigued with the service, because it affords the opportunity to entice viewers back to live broadcast television (Grotticelli, 2001a). Internet and now ITV provocateur Carl Malamud recently took NetTopBox nonprofit, in order to promote an open, non-proprietary system of programming information. Malamud argues that Internet-enabled computing provides television viewers with "core software that is in the public domain, just like the code underlying the rest of the Internet" (Bajak, 2002, p. 3D). Navigators and program guides enjoy viewer popularity and acceptance. Over 74% of respondents in a reliable online poll indicated they were "very interested" or "somewhat interested" in navigational tools, and 72% indicated they were "very interested" or "somewhat interested" in interactive program guides (e-poll, 2001, p. 8).

Internet tools on TV. In 1999, AOL invested heavily in DirecTV to create a joint service. These efforts reflected an effort to compete with cable television's ability to offer high-speed Internet access—something then unavailable via DBS. Cable television's bandwidth and high-speed return path also overshadowed DBS' reliance on telephone lines as a return path (Pegg, 2000). Fall 2000 saw the national rollout of AOL TV's set-top boxes manufactured by Philips with the goal of surpassing Microsoft's WebTV with a second-generation box designed to fully integrate Internet tools on television. The box sold for $249, with existing AOL subscribers paying an additional $15 per month and new subcribers a total of $25. Industry watchdogs praised the system's wireless keyboard and instant messaging or chat group features, but criticized the system's program directory and interactive gaming features (Mossberg, 2000). A separate unit designed to work with DirecTV has been announced, but was not yet on the market as of mid-2002.

Digital television broadcast viewers in the Cincinnati (Ohio) market that can receive WKRC-TV's DTV signal can now receive over-the-air Internet access. Called Delta V, subscribers pay a $40 monthly fee and need a digital personal computer tuner card and standard television antenna attached to the computer. The connection speed of the Delta V service is 256 Kb/s (kilobits per second), five times faster than a traditional phone line connection. Developers estimate that only 6,000 subscribers are needed to generate over $1 million in annual station revenues (Kerschbaumer, 2002).

Other ventures are attempting to marry Internet features with PVRs. The Moxi Media Center technology introduced at the 2002 Consumer Electronics Show in Las Vegas featured a DSTB with an 80-gigabyte hard drive capable of delivering four separate video or audio channels, as well as instant messaging and e-mail. Moxi Media's developer is Steve Perlman, founder of WebTV, whose business model for the DSTB is to license the technology to cable and satellite TV operators. Currently, the technology is expensive compared to other DSTBs, coming in at $425 for one household television receiver and an additional $250 for the second receiver (Wong, 2002). AOL Time Warner, EchoStar Communications, and Cisco Systems have all invested in Moxi Media. Deployment by cable system operators of DSTBs containing Moxi software is projected for mid-2002.

Microsoft's much-heralded WebTV was merged with MSN to become MSN TV in spring 2001. One year later, the same corporation's Ultimate TV unit was also folded into MSN TV, after attracting only 100,000 subscribers. Microsoft argued that the mergers were designed to eliminate duplication of efforts and create better efficiency. Currently, the MSN TV system is a satellite receiver DSTB with 35 hours of PVR capability and Internet on television features. The system is only offered via DirecTV (Johnston, 2002). Internet on television remains a hard sell, particularly for potential off-line customers. Only 21% of 944 off-line respondents indicated a desire for e-mail on ITV, and only 16% of 400 online respondents expressed interest in using ITV for chatting with other viewers (Who wants, 2001).

Personal video recorders. PVR technology grew slowly through 2001 and 2002, but industry analysts expect the growth rate to increase once word-of-mouth regarding the technology begins to spread. (See Chapter 16 for a full discussion of recent developments regarding PVRs.)

T-commerce. Wink Communications had penetrated about 3.5 million U.S. households with its set-top box software by August 2001. Wink competes with RespondTV and WorldGate Communications for clients intent on developing interactive forms of television advertising. Ford Motor Corporation, Glaxo-Smith Kline, and Lands' End have all produced and aired interactive commercials. The typical format includes a pop-up icon when interactive commercials are aired. Home viewers then answer a set of questions about their interest in the product or service and their likelihood to purchase in the near future. Pertinent contact information is already contained in a system database set up at service initiation. Most typically, the interactive event is followed by personalized direct marketing mail to the viewer's residence (Jarvis, 2001). Less than 1% of all commercials aired on television are enhanced for interactive commerce. A major impediment to growth is incompatibility between competing proprietary platforms. DBS holds some promise for t-commerce, but the return path remains a major weak link (Wink tops, 2000). ITV-based commerce remains at the very low end of consumer interest. Only 9% of respondents from off-line households expressed any interest in ITV shopping compared with 17% of online recipients (Who wants, 2001).

Video on demand. In 2000, many cable operators claimed that VOD was still unproven, citing cumbersome servers and modulators, and complicated integration with cable system engineering (Higgins, 2000). Another major problem is the overlap between PVRs and VOD. SeaChange initiated a significant undertaking in VOD development in 2001 with Comcast, Cablevision, Time Warner Cable, and Rogers of Canada, fueled by a $10 million dollar investment by Comcast. Diva, in collaboration with MSOs (multiple system operators) Charter, Insight, and AT&T Broadband, claimed 4.5 million subscribers in 2001 (Brown, 2001). (See Chapter 4 for more about recent developments in VOD.)

CURRENT STATUS

Despite robust predictions in the late 1990s, ITV remains a world of lofty promises and expectations, with little to show for considerable technological innovation and capital expenditures. ITV penetration remains very low, with estimates ranging around eight million ITV-enabled DSTBs, most providing a limited range of functions (Jarvis, 2001; Mendelson, 2002). A major stumbling block in ITV development is the cable television industry. Cable operators remain reluctant to upgrade DSTBs with advanced capabilities because of considerable costs incurred in replacing older, more conventional set-top boxes (Gruszka, 2001). Mendelson (2002) notes that the problem is circular, because "end users won't pay unless there is compelling content, the programmers won't provide enhanced content unless the advertisers pay extra, and … advertisers won't pay unless there are enough iTV-enabled boxes" (p. 14). Television audience interest in ITV remains varied. Figure 5.2 shows the results from a study conducted in late 2001 of online viewers (e-poll, 2001). The results, reported at a 95% confidence level, indicated that PVR, VOD, and viewing guides or navigator functions were of greatest interest to respondents. Ironically, the ability to skip commercials was the most popular of all potential ITV features—a compelling obstacle for t-commerce advocates.

FACTORS TO WATCH

In August 2001, Jupiter Media Metrix optimistically projected that "iTV service will reach a critical mass [in December 2002], when some 17% of U.S. households are expected to have subscribed to iTV service" (Jarvis, 2001, p. 20). Recall that Forrester Research optimistically predicted revenues in the billions of dollars for major ITV applications by 2004. Such estimates appear whimsical in the face of economic and penetration realities. The shakeout in ITV is clearly underway. Late 1990s efforts to converge the one "killer application" have run headlong into incompatible end-user behaviors—passive viewing of television utilizing enhanced ITV features versus active Internet and Web-related behaviors via Internet-on-TV services.

Interactive gaming will probably survive and continue to be offered as a sidecar to game show, sporting, and primetime genres where a small percentage of the at-home audience might feel compelled to engage in real-time program interaction. Interactive programming guides and smart navigators, a very popular item with television viewers, will remain a central feature of any DSTB system and will continue to be a feature of more and more powerful PVRs. The future of t-commerce remains shaky. Major advertisers will continue to participate in t-commerce experiments, enticed by platform developers, but the application's lucrative role in generating future sales and revenues has yet to overcome home-user preferences for other forms of shopping and commerce.

Figure 5.2
Level of Interest in Various ITV Features

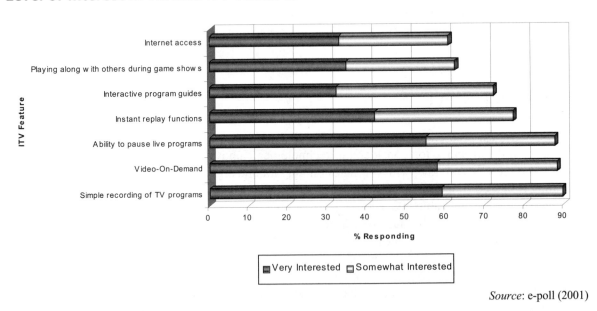

Source: e-poll (2001)

BIBLIOGRAPHY

Bajak, F. (2002, January 28). Gadfly seeks interactive media guide. *Las Vegas Review Journal,* 1D, 3D.

Beacham, F. (2000a, February 9). TV getting personal. *TV Technology, 18,* 24.

Beacham, F. (2000b, October 18). The (re)selling of interactive TV. *TV Technology, 18,* 24-25.

Bensmiller, K. (2002). What consumers want from iTV. *NAPTE Media Trends, 4* (1), 17-18.

Brown, P. (2001, March 26). VOD: Coming slowly as existing model is questioned. *TV Technology, 19,* 18, 26.

Clark, D. (1997, August 13). Oracle plans to integrate TV programs with data from the World Wide Web. *Wall Street Journal,* B7.

Desmond, E. (1997, November 10). Set-top boxing. *Fortune, 136,* 91-93.

e-Poll. (2001, December). *Interactive television: Consumer usage and attitudes.* Encino, CA: A. Marks.

Feldman, T. (1997). *An introduction to digital media.* London: Routledge.

Full Service Network (FSN) in Orlando, Florida. (n.d.). *Hong Kong University of Science & Technology.* Retrieved April 26, 2002 from http://www.ust.hk/~webiway/content/USA/Trial/fsn.html.

Game show network synchs up with spotmagic and ACTV to take *Mall Masters* interactive. (2001, May 15). *ACTV, Inc.* Retrieved from http://www.actv.com/actvfinal/mallmay15.html.

Gemstar-TV Guide International, Inc. reports financial results for the quarter ended September 30, 2001. (2001, November 11). *Gemstar-TV Guide.* Retrieved from http://gemstartvguide.com/pressroom/display_pr.asp?prId=95.

Goldstein, S. (1997, May 17). Time Warner proves it's RIP for VOD; Direct mail, DVD tentative bedfellows. *Billboard, 109,* 58.

Grotticelli, M. (2001a, August 20). Expanding on EPG. *Broadcasting & Cable, 131* (35), 24.

Grotticelli, M. (2001b, September 13). To advance ITV. *Broadcasting & Cable, 131* (34), 28.

Grotticelli, M., & Kerschbaumer, K. (2001, June 9). Slow and steady. *Broadcasting & Cable, 131* (34), 32, 34, 36, 38, 40-42.

Gruszka, M. (2001, September 5). IBM banks on interactive television. *TV Technology, 19,* 22, 26.

Haley, K. (1999, September 6). New direction: Forget the superhighway; Many roads lead to interactive TV. *Broadcasting & Cable, 129,* 18-19, 22, 24, 26, 28, 30, 32, 34, 36.

Higgins, J. (2000, March 6). VOD: Is it a business yet? *Broadcasting & Cable, 130,* 23, 26, 28.

Human bandwidth still a limiting factor. (2001, January 23). *NAPTE Daily,* 29.

ITV: How we define it. (2001, June 9). *Broadcasting & Cable, 131* (29), 34.

Jarvis, S. (2001, August 27). iTV finally comes home. *Marketing News,* 1, 19, 20.

Johnston, C. (2002, February 20). Microsoft keeping ITV. *TV Technology, 20* (4), 1, 17.

Kerschbaumer, K. (2000, April 12). Ballmer: Place your bets. *Broadcasting & Cable, 130,* 12-14.

Kerschbaumer, K. (2002, January 7). A clear DTV Internet strategy. *Broadcasting & Cable, 132* (1), 25-26.

Lesly, E., Cortese, A., Reinhardt, A., & Hamm, S. (1997, November 24). Let the set-top wars begin. *Business Week, 3554,* 74-75.

Marriage of convenience. (1997, November). *Time Digital,* 60-64.

Meeting focuses on interactive TV. (2000, May 13). *Las Vegas Review Journal,* D6.

Mendelson, B. (2002). Interactive TV: Hope or hype? *NAPTE Media Trends, 4* (1), 14-15.

Mossberg, W. (2000, July 16). AOL TV now has too many rough edges to fulfill promise. *Las Vegas Review-Journal,* 3L.

NBA Entertainment teams with TBS Superstation, TNT and Sprite to launch first-ever live fantasy game on NBA.com. (2002, January 10). *ACTV, Inc.* Retrieved from http://www.actv.com/actvfinal/january10.html.

Pegg, J. (1999a, April 21). Interactivity needs standards. *TV Technology, 17,* 16.

Pegg, J. (1999b, April 21). Major interactive players emerge. *TV Technology, 17,* 14, 16.

Pegg, J. (2000, March 8). What's next for DBS. *TV Technology, 18,* 1, 14.

Rosenstein, A. (1994). Interactive television. In A. E. Grant (Ed.). *Communication technology update,* 3rd edition. Newton, MA: Focal Press.

Shiver, J. (1997, June 2). Time Warner's interactive TV project blinks. *Los Angeles Times*, D1.

Swedlow, T. (2002, April 2). 2000 interactive enhanced television: A historical and critical perspective. *Interactive TV Today*. Retrieved March 15, 2002, from http://www.itvt.com/etvwhitepaper.html.

Talwani, S. (2002, February 20). NAPTE: Executives debate advanced TV. *TV Technology, 20* (4), 1, 18.

What is Wink? (n.d.). Retrieved April 26, 2002, from http://wink.com/contents/tech_diagram.shtml.

Who wants their ITV? (2001, January 15). *The Industry Standard*, 138-139.

Wink tops the ITV pecking order. (2000, June 28). *TV Technology, 18* (3), 10, 12.

"Winky Dink" series making a comeback. (2002, March 12). *Las Vegas Review Journal*, E1.

Wong, M. (2002, January 7). Set-top box ready to shine. *Las Vegas Review Journal,* D1.

6

DIRECT BROADCAST SATELLITES

Ted Carlin, Ph.D.[*]

W̶hat is the best way to get multichannel video programming today? The answer is becoming more complicated. As discussed in Chapter 3, cable television is changing, upgrading to digital service that offers more channels, better picture quality, and extras such as Internet service. Recent cable price hikes, primarily due to increased programming costs, have given consumers more reasons to consider another option for their television service. Already frustrated by the typical cable problems of signal quality, channel capacity, and customer service, higher cable prices from 1999 to 2002 have pushed many consumers to search for an alternative.

Waiting for these consumers are the communication technologies of direct broadcast satellite (DBS) systems, wireless cable systems (MMDS and LMDS), home satellite dishes (HSDs), the telephone, and the Internet. With the increasing development of digital production, transmission, and reception equipment, the multichannel video program distribution (MVPD) field is growing crowded as companies race to develop digital systems that will entice consumers away from firmly established cable television providers. The digital revolution in television technology is starting to give consumers what the Federal Communications Commission (FCC) has always promoted: a level playing field of multichannel video program distribution services from which consumers can pick and choose.

This chapter will focus on the fastest growing multichannel video program distribution service: DBS systems. Providers of DBS systems are actively competing with cable television for subscribers and currently offer the most comprehensive programming alternatives for consumers.

---

[*] Associate Professor and Chair, Department of Communication and Journalism, Shippensburg University (Shippensburg, Pennsylvania).

BACKGROUND

When it was originally conceived in 1962, satellite television was never intended to be transmitted directly to individual households. After the FCC implemented an "open skies policy" to encourage private industry to enter the satellite industry in 1972, satellite operators were content to distribute programming between television networks and stations, cable programmers and operators, and business and educational facilities (Frederick, 1993). The FCC assigned two portions of the Fixed Satellite Service (FSS) frequency band to be used for these satellite relay services: the low-power C-band (3.7 GHz to 4.2 GHz) and the medium-power Ku-band (11.7 GHz to 12.2 GHz).

In late 1975, Stanford University engineering professor Taylor Howard was able to intercept a low-power C-band transmission of the Home Box Office (HBO) cable network on a makeshift satellite system he designed (Parone, 1994). In 1978, Howard published a "low-cost satellite-TV receiving system" how-to manual. Word spread rapidly among video enthusiasts and ham radio operators, and, by 1979, there were about 5,000 of these television receive-only (TVRO) HSD systems in use.

Today, these 6- to 12-foot TVRO satellite dishes are still commonplace throughout the United States, especially in rural areas not served by cable television services, with just under 1.8 million in use today (FCC, 2000a). The large receiving dish is required to allow proper reception of the low-power C-band transmission signal. A number of factors, including the high cost of the HSD system (around $2,000), the large size of the dish, city and county zoning laws, and the scrambling of C-band transmissions by program providers, prevented HSD systems from becoming a realistic, national alternative to cable television for MVPD service.

In the 1980s, a few entrepreneurs turned to the medium-power Ku-band to distribute satellite transmissions directly to consumers. By utilizing unused transmission space on existing Ku-band relay satellites, these companies aimed to be the first to create a direct-to-home (DTH) satellite transmission service that would use a much smaller receive dish than existing HSDs (Whitehouse, 1986). The initial advantages of these DTH systems over HSD systems included the higher frequencies and power of the Ku-band, which resulted in less interference from other frequency transmissions and stronger signals to be received on smaller, 3-foot dishes.

Many factors contributed to the failure of these medium-power Ku-band DTH services in the 1980s, including:

> High consumer entry costs ($1,000 to $1,500).

> Potential signal interference from heavy rain and/or snow.

> Limited channel capacity (compared with existing cable systems).

> Restricted access to available programming (Johnson & Castleman, 1991).

DTH ventures by Comsat, United Satellite Communications, Inc., Skyband, Inc., and Crimson Satellite Associates failed to get off the ground during this period.

Between 1979 and 1989, the World Administrative Radio Conference (WARC) of the International Telecommunications Union (ITU) authorized and promoted the use of a different section of the FSS frequency band. The ITU, as the world's ultimate authority over the allocation and allotment of all radio transmission frequencies (including radio, TV, microwave, and satellite frequencies), allocated the high-power Ku-band (12.2 GHz to 12.7 GHz) for "multichannel, nationwide satellite-to-home video programming services in the Western hemisphere" (Setzer, et al., 1980, p. 1). These high-power Ku-band services were to be called direct broadcast satellite services. Specific DBS frequency assignments for each country, as well as satellite orbital positions to transmit these frequencies, were allocated at an ITU regional conference in 1983 (RARC, 1983).

A basic description of a DBS service, based on the ITU's specifications, was established by the FCC's Office of Plans and Policy in 1980:

> A direct broadcast satellite would be located in the geostationary orbit, 22,300 miles above the equator. It would receive signals from earth and retransmit them for reception by small, inexpensive receiving antennas installed at individual residences. The receiver package for a DBS system will probably consist of a parabolic dish antenna, a down-converter, and any auxiliary equipment necessary for encoding, channel selection, and the like (Setzer, et al., 1980, p. 7).

The FCC then established eight satellite orbital positions between 61.5°W and 175°W for DBS satellites. Only eight orbital positions are available for DBS transmission to the United States because a minimum of 9° of spacing between each satellite is necessary to prevent interference among signal transmissions. The FCC also assigned a total of 256 analog TV channels (transponders) for DBS to use in this high-power Ku-band, with a maximum of 32 DBS channels per orbital position. Only three of these eight orbital positions (101°W, 110°W, and 119°W) can provide DBS service to the entire continental United States. Four orbital positions (148°W, 157°W, 166°W, and 175°W) can provide DBS service only to the western half of the country, while the orbital position at 61.5°W can provide service only to the eastern half.

The FCC received 15 applications for these DBS orbital positions and channels in 1983, accepted eight applications, and issued conditional construction permits to the eight applicants. The FCC granted the construction permits "conditioned upon the permitee's due diligence in the construction of its system" (see 27 C.F.R. Sect. 100.19b). DBS applicants had to do two things to satisfy this FCC "due diligence" requirement:

(1)   Begin construction or complete contracting for the construction of a satellite within one year of the granting of the permit.

(2)   Begin operation of the satellite within six years of the construction contract.

The original eight DBS applicants were: CBS, Direct Broadcast Satellite Corporation (DBSC), Graphic Scanning Corporation, RCA, Satellite Television Corporation, United States Satellite Broadcasting Company (USSB), Video Satellite Systems, and Western Union. During the 1980s, some of these applicants failed to meet the FCC due diligence requirements and forfeited their construction permits. Other applicants pulled out, citing the failures of the medium-power Ku-band DTH systems, as well as the economic recession of the late 1980s (Johnson & Castleman, 1991).

In August 1989, citing the failures of the eight DBS applicants to launch successful services, the FCC revisited the DBS situation to establish a new group of DBS applicants (FCC, 1989). This new group of applicants included two of the original applicants, DBSC and USSB. These were joined by Advanced Communications, Continental Satellite Corporation, Direcsat Corporation, Dominion Satellite Video, EchoStar Communications Corporation, Hughes Communications, Inc., and Tempo Satellite Services.

From 1989 to 1992, none of these DBS services was able to launch successfully. In addition to raising capital, most were awaiting the availability of programming and the development of a reliable digital video compression standard. Investors were unwilling to invest money into these new DBS services unless these two obstacles were overcome (Wold, 1996).

Cable operators, fearing the loss of their own subscribers and revenue, were placing enormous pressure on cable program networks to keep their programming off the new DBS services. Cable operators threatened to drop these program networks if they chose to license their programming to any DBS service (Hogan, 1995). These program networks were seen as the essential programming needed by DBS to launch their services because the DBS companies had little money or expertise for program production of their own (Manasco, 1992).

In late 1992, DBS companies had this programming problem solved for them through the passage of the Cable Television Consumer Protection and Competition Act. The Act guaranteed DBS companies access to cable program networks, and it "[forbade] cable television programmers from discriminating against DBS by refusing to sell services at terms comparable to those received by cable operators" (Lambert, 1992, p. 55). This provision, which has since been upheld in the Telecommunications Act of 1996, finally provided DBS companies with the program sources they needed to attract investors and future subscribers.

The other obstacle—the establishment of a digital video compression standard—was solved by the engineering community in 1993 when MPEG-1 was chosen as the international video compression standard. By using MPEG-1, DBS companies could digitally compress eight program channels into the space of one analog transmission channel, thus greatly increasing the total number of program channels available on the DBS service to consumers. (For example, the FCC had assigned DirecTV 27 analog channels. Using MPEG-1, DirecTV could actually provide their subscribers with 216 channels of programming.) In 1995, DBS companies upgraded their systems to MPEG-2, an improved, broadcast-quality compression standard.

With these obstacles behind them, two DBS applicants, Hughes Communications and USSB, were the first to launch their DBS services in June 1994. Utilizing the leadership and direction of Eddie Hartenstein (DirecTV) and Stanley Hubbard (USSB), Hughes established a subsidiary, DirecTV, to operate its DBS system, and then agreed to work with USSB to finance, build, deploy, and market their DBS systems together (Hogan, 1995). Hughes launched three satellites to the 101°W orbital position from 1993 to 1995. Both companies then signed a contractual agreement with Thomson Consumer Electronics to use Thomson's proprietary Digital Satellite System (DSS) to transmit and receive DirecTV and USSB programming (Howes, 1995).

To ensure its availability in rural areas, DirecTV signed an exclusive agreement with the National Rural Telecommunications Cooperative (NRTC) that allowed NRTC affiliates the right to market and distribute DirecTV in rural NRTC markets. A number of affiliates implemented the agreement in 1996, offering sales, installation, billing, collection, and customer service. Consolidation among affiliates became commonplace, and two affiliates emerged as the major players: Pegasus Satellite Television and Golden Sky Systems, Inc. In May 2000, Pegasus purchased Golden Sky for $1 billion in stock to create the third largest provider of DBS and the 10th largest MVPD in the United States. Pegasus is now the only MVPD focused exclusively on rural and underserved areas of the country and serves 1.5 million households in 41 states (Pegasus Communications, 2002; Golden Sky, 2000).

After the successful launch of DirecTV and USSB in 1994, the FCC tried to force the other DBS applicants into bringing their DBS services to the marketplace. In late 1995 and early 1996, the FCC once again re-evaluated the DBS applicants for adherence to its due diligence requirements. After several hearings, the FCC revoked the application of Advanced Communication Corporation and stripped Dominion Satellite Video of some of its assigned channels for failing to meet these requirements.

The FCC denied appeals by both companies and auctioned the channels in January 1996. MCI and News Corporation, working together in a joint venture, obtained the Advanced DBS channels, while EchoStar obtained the Dominion channels to add to its previously assigned channels. EchoStar also acquired the channels from two other applicants, DirecSat and DBSC, through FCC-approved mergers in 1995 and 1996 (FCC, 1996).

In another merger, R/L DBS, a subsidiary of Loral Aerospace Holdings and Rainbow Media Holdings, acquired the DBS channels of Continental Satellite Corporation. Continental was forced to turn over the channels to R/L DBS after failing to meet previous contractual obligations with Loral Aerospace for the launch of Continental's proposed DBS satellite (FCC, 1995). In December 2000, the FCC granted an extension to the R/L DBS construction permit. The extension requires the launch of a satellite and commencement of DBS service by not later than December 29, 2003. R/L DBS has entered into an agreement with Lockheed-Martin for the construction of a satellite scheduled to be delivered in May 2003. Rainbow Media continues to evaluate the scope of its direct broadcast satellite business, including exploring opportunities for strategic partnerships for R/L DBS (Cablevision contracts, 2001).

On March 4, 1996, EchoStar launched its high-power DBS service, the Digital Sky Highway (DISH) Network, using the EchoStar-1 satellite at 119°W to become the third U.S. DBS applicant to successfully begin operations. Similar to DirecTV, EchoStar launched a second satellite in September 1996, at the 119°W orbital position, to increase its number of available program channels to 170 on the DISH Network.

The DISH Network does not use the same DSS transmission format used by DirecTV. Instead, it uses the international satellite video transmission standard, Digital Video Broadcasting (DVB), which was created after the DSS standard. Like DSS equipment, the DISH Network's DVB equipment uses MPEG-2 for digital video compression. What this means is that DISH Network subscribers can receive only DISH Network transmissions, and DirecTV subscribers can receive only DirecTV transmissions. Despite differences in transmission standards, both the DSS and DVB systems employ an 18-inch receiving dish, a VCR-sized integrated receiver/decoder (IRD), and a multi-function remote control.

Similar to DirecTV, the DISH Network requires subscribers to purchase the DVB system, and then pay a separate amount for monthly or yearly programming packages. Also, like DirecTV, the DISH Network offers a choice of professional installation or a do-it-yourself installation kit.

The cable television industry did not ignore the implementation and growth of these DBS companies. In 1994, Continental Cablevision, intent on establishing a cable "headend in the sky" for consumers living in non-cabled areas of the United States, was able to enlist the support of five other cable operators (Comcast, Cox, Newhouse, Tele-Communications, Inc. [TCI], and Time Warner) and one satellite manufacturer (GE Americom) to launch a successful medium-power Ku-band DTH service (Wold, 1996). The service, named Primestar, transmitted 12 basic cable channels from GE Americom's medium-power K-1 satellite to larger 3-foot receive dishes. Primestar offered far fewer channels than any of the cable operators' own local cable systems, so they believed that their cable subscribers would not be interested in Primestar as a replacement for cable service. (Primestar was not a true DBS service because it did not use FCC-assigned high-power DBS channels, although most consumers were unaware of this discrepancy.)

As DirecTV and USSB began to prove that DBS was a viable service in late 1994, Primestar decided to change its focus and expand and enhance its offerings to compete directly with DBS. Primestar converted its 12-channel analog system to a proprietary DigiCipher-1 digitally compressed service capable of delivering about 70 channels. In 1997, Primestar moved its service to GE Americom's medium-power GE-2 satellite at 85°W, and increased its channel capacity to 160. To differentiate itself from the DBS companies, Primestar decided to market its service just like a local cable TV service does by leasing the equipment and the programming packages together in one monthly fee. Subscribers were not required to purchase the Primestar dish, IRD, and remote, although equipment purchase was an option.

RECENT DEVELOPMENTS

As of 1998, there were four companies operating DTH services in the continental United States. There were three high-power DBS services (DirecTV, USSB, and the DISH Network) and one medium-power DBS service (Primestar). In April 1999, DirecTV strengthened its position in the market with the purchase of both Primestar and USSB, leaving only two players in the U.S. DBS market. These mergers not only strengthened DirecTV's subscriber base, but also provided satellites and orbital slots that enabled DirecTV to expand its services.

The completion of these transactions gave DirecTV:

➢ More than seven million U.S. subscribers.

➢ More than 370 entertainment channels delivered through five high-power DBS space-craft.

> ➢ The broadest distribution network in the DBS industry, combining more than 26,000 points of retail sale with Primestar's rural and small urban-based distribution network.

> ➢ High-power DBS frequencies at each of the three orbital slots that provide full coverage of the continental United States: 101°W, 110°W, and 119°W.

> ➢ The opportunity for DirecTV to begin "local-into-local" broadcast signal carriage.

> ➢ For the first time, DBS service to Hawaii, which had not been previously served by any DBS or DTH provider.

As DirecTV was increasing its market dominance, the DISH network was using aggressive pricing strategies for programming and DVB equipment. These efforts enabled DISH in December 1997 to reach one million subscribers, reaching that milestone faster than any other DBS/DTH service (Hogan, 1998a). Marketing itself as the best value in satellite television, the DISH Network offered its equipment for only $199, plus installation. DirecTV responded by lowering its DSS prices, resulting in a price war that continues today.

In mid-1998, the DISH Network announced plans to transmit local broadcast stations back into their local markets. By early 1999, the DISH Network finalized an agreement with MCI to acquire its 28 DBS channels at 110°W. This provided the DISH Network with enough channel capacity to provide local-into-local service in major U.S. television markets. On May 19, 1999, the FCC granted the application of MCI and EchoStar for transfer of MCI's license to construct, launch, and operate a DBS system at the 110°W location. On June 16, 1999, EchoStar was also granted authority to temporarily relocate one of its satellites to a new orbital slot in order to improve DBS service to Alaska and to initiate service to Hawaii.

On May 17, 1999, the FCC granted Dominion Video Satellite, Inc. authority to commence operation of a DBS service using the EchoStar III satellite currently in orbit at 61.5°W. This authorization waived Dominion's due diligence requirement to build and launch a satellite of its own because the lease arrangement with EchoStar was viewed as an efficient method of commencing service, as long as Dominion maintained control over the programming (FCC, 1999). Dominion then launched its SkyAngel religious programming service in fall 1999 with 16 channels of video and 10 channels of audio for $9 per month. Subscribers must use DISH Network equipment to access SkyAngel's programming.

Table 6.1 summarizes the status of the DBS licensees as of mid-2002.

Table 6.1
U.S. DBS Licensees

Orbital Position	61.5°W	101°W	110°W	119°W	148°W	157°W	166°W	175°W	Total
Satellite Deployment	ES-3	DTV-1R DTV-2 DTV-3 DTV-4S	DTV-1 ES-5 ES-6 ES-8	ES-2 ES-4 ES-7 DTV-5 DTV-6	ES-1				
DirecTV Channels		32	3	11					46
DISH Channels	11		29	21	24			22	107
SkyAngel Channels	8								8
R/L DBS Channels	11								22
Unassigned Channels	2				8	32	32	10	73
Total	32	32	32	32	32	32	32	32	256

Source: *T. Carlin*

Notes: (1) ES = EchoStar satellites; DTV = DirecTV satellites
(2) 101°W, 110°W, 119°W are the only full-CONUS slots
(3) R/L DBS surrendered its 11 channels at 166°W back to the FCC on 9/15/98.
(4) SkyAngel has applied to the FCC for 8 channels at 166°W.
(5) Launch of DTV-5 and ES-9 (to 121°W) is expected during late-2002.

CURRENT STATUS

UNITED STATES DBS/DTH

The United States is the world's number one user of DBS/DTH services. As of January 2002, there were 17.61 million DBS/DTH subscribers in the United States. Table 6.2 summarizes the subscription figures for the industry since its inception in July 1994.

Most of these 17.61 million subscribers are located in rural areas that are not served by a local cable system. Total multichannel video household penetration in the continental United States, including 69 million cable television subscribers and 842,500 TVRO users, is about 88 million households. This means "the U.S. is running out of unserved homes to pitch, particularly in the boonies" (DBS knockin', 1998, p. 2). Cable television is still the dominant technology for the delivery of video programming to consumers in the MVPD marketplace, although its market share continues to decline. As of June 2001, 78% of all MVPD subscribers received their video programming from a local franchised cable operator, compared with 80% in 2000, 82% in 1999, and 85% in 1998 (FCC, 2002; FCC, 2000a).

Table 6.2
U.S. DBS/DTH Subscribers

Date	Total DTH	DirecTV	Primestar	DISH
7/1/94	70,000		70,000	
12/1/94	390,000	200,000	190,000	
7/1/95	1.15 mil.	650,000	500,000	
12/1/95	1.98 mil.	1.1 mil.	880,000	
7/1/96	2.95 mil.	1.6 mil.	1.27 mil.	75,000
12/1/96	4.04 mil.	2.13 mil.	1.60 mil.	285,000
7/1/97	5.04 mil.	2.64 mil.	1.76 mil.	590,000
12/1/97	5.95 mil.	3.12 mil.	1.90 mil.	965,500
7/1/98	6.60 mil.	3.45 mil.	2.01 mil.	1.14 mil.
12/1/98	8.48 mil.	4.36 mil.	2.27 mil.	1.85 mil.
7/1/99	9.93 mil.	5.42 mil.	1.94 mil.	2.57 mil.
12/1/99	10.97 mil.	7.84 mil.		3.13 mil.
7/1/00	11.74 mil.	8.24 mil.		3.50 mil.
12/1/00	14.80 mil.	9.54 mil.		5.26 mil.
7/1/01	16.30 mil.	10.13 mil.		6.17 mil.
12/1/01	17.61 mil.	10.75 mil.		6.86 mil.

Source: *SkyReports*

It also means that DirecTV and the DISH Network are aggressively pursuing current cable customers for their services. Various new marketing campaigns by DirecTV and the DISH Network are constantly implemented to attack the cable industry's most observable weaknesses: rate hikes (due largely to increased programming costs that continue to rise faster than inflation), lack of channel variety (due to limited analog systems just now being replaced by digital upgrades), and customer service problems (due to past monopolistic practices still plaguing the cable industry) (FCC, 2002).

While attacking these cable industry problems, DBS companies are also trying to solve two main issues, as well as several related issues. The two main issues being addressed are (1) multiple television setups in the home and (2) subscriber access to local broadcast television stations and broadcast networks. Both issues have been considered major impediments to the development and growth of DBS systems as true competitors for cable television customers (Hogan, 1998b).

MULTIPLE TV SETUPS

When DBS companies began operations in the mid-1990s, the goal was to get a basic one-TV DBS system into as many rural subscriber homes as possible (Boyer, 1996). Due to declining costs, increased technology, and a new effort to attract cable customers, the focus has shifted to providing more user-friendly DBS service. According to Bill Casamo, executive vice president for DirecTV, the cost of a second receiver has always been a barrier to entry for some first-time subscribers, "As we go more into cabled markets, that becomes more of a factor," because cable customers are accustomed to seeing cable in multiple rooms within a home (Hogan, 1998e, p. 18).

As a result, DirecTV and the DISH Network have been marketing multiple TV setups for new subscribers. There are now three types of dishes available to consumers. They include:

(1)   A single-LNB (low noise block) dish, which receives programming from one satellite orbital location.

(2)   A dual-LNB dish, which also receives programming from one satellite orbital location.

(3)   A multi-location dish, which receives programming from multiple satellite orbital locations.

The single-LNB model is the most basic and allows only one DBS receiver to be connected. It receives signals only from the DBS provider's primary orbital location. That means it can receive most of the mainstream programming but not some of the less common programming such as foreign language programming or high-definition television (HDTV) programming. Also, depending on a subscriber's location, a single-LNB model may not allow the reception of local network affiliates, since not all local affiliates are broadcast from the primary orbital location. Use of a single-LNB dish requires a single coaxial cable from the dish into the home to connect the single receiver, so it is the easiest to install.

The dual-LNB model receives programming from a single orbital location, but viewers can connect one, two, or more receivers to it. This may be a better choice for those who want only the most common mainstream programming, but who want more than one receiver connected either now or sometime in the future. Like the single-LNB models, dual-LNB also may not allow the reception of local network affiliates. If subscribers want more than one receiver in the household, two coaxial cables must be run from the dish into the home.

Multi-location dishes are required by those who want to receive signals from satellites in different orbital slots. These dishes have only recently become available, but they are becoming more popular, especially in those areas where the local affiliates are not carried on the primary orbital location. Like the dual-LNB model, they actually have two antennas, but they are focused at different positions in the sky. Multi-location dishes require up to four coaxial cables to be run from the dish into the home.

LOCAL-INTO-LOCAL

The second issue, access to local television stations, has been much more difficult to overcome for DBS in urban and most suburban communities. A provision in the Satellite Broadcasting Act of 1988 prohibited DBS subscribers who live *within* the Grade B coverage area of local broadcast television stations from receiving any broadcast television stations or networks via their DBS system (SHVA, 1988). Subscribers in these areas were forced to connect an over-the-air television antenna to their DBS system or subscribe to a local cable TV system to receive any broadcast television stations. (DBS providers are allowed to provide broadcast stations that are available via satellite to those subscribers living *outside* of local station Grade B coverage areas [i.e., rural, non-cabled areas], and each offers various à la carte packages of independent stations and network affiliates.)

This issue was finally resolved when the Satellite Home Viewer Improvement Act of 1999 (SHVIA) was signed into law on November 19, 1999. SHVIA significantly modifies SHVA from 1988, the Communications Act, and the U.S. Copyright Act (SHVIA, 1999). SHVIA was designed to promote competition among MVPDs such as DBS companies and cable television operators while, at the same time, increasing the programming choices available to consumers.

Most significantly, for the first time, SHVIA permits DBS companies to provide local broadcast TV signals to *all* subscribers who reside in the local TV station's market (also referred to as a designated market area [DMA]), as defined by Nielsen Media Research. This ability to provide local broadcast channels is commonly referred to as local-into-local service.

The DBS company has the option of providing local-into-local service, but is not required to do so. In addition, a DBS company that has chosen to provide local-into-local service is required to carry, upon request, all TV stations in markets where the DBS company carries at least one local TV station. This is now referred to as the "carry one, carry all" rule. However, a DBS company is not required to carry a local broadcast TV station that substantially duplicates the programming of a local broadcast TV station already being carried. A DBS company also is not required to carry more than one local broadcast TV station that is affiliated with a particular TV network unless the TV stations are licensed to communities in different states (FCC, 2001).

SHVIA allows satellite companies to provide distant network broadcast stations to eligible satellite subscribers in unserved areas. (A distant signal is one that originates outside of a satellite subscriber's DMA.) The FCC created a computer model for DBS companies and television stations to use to predict whether a given household is served or unserved. Congress incorporated this model into SHVIA, but also required the FCC to improve the accuracy of the model by modifying it to include vegetation and buildings among the factors to be considered. The DBS company, distributor, or retailer from which subscribers obtained their satellite system and programming are to be able to tell subscribers whether the model predicts that they are served or unserved. (The FCC does not provide these predictions.) If unserved, the subscriber would be eligible to receive no more than two distant network affiliated signals per day for each TV network. For example, the household could receive no more than two *ABC* stations, no more than two *NBC* stations, etc.

SHVIA also permitted DBS companies to distribute a national PBS signal to all subscribers—served and unserved—until January 1, 2002. DBS companies may now choose to provide the local PBS affiliate or another noncommercial station within a local market or may provide the national PBS signal to subscribers that are eligible to receive distant signals.

There are seven other DBS issues that have attracted attention since the beginning of 2002:

➢ Retransmission consent agreements.

> ➢ Antenna restrictions.

> ➢ Public interest obligations.

> ➢ Transmission of HDTV signals.

> ➢ Internet access.

> ➢ Available programming.

> ➢ Strategic alliances.

Retransmission consent agreements. In order to deliver local-into-local service, DBS companies were mandated by the SHVIA to seek retransmission consent agreements with the owners of local television stations. SHVIA required the FCC to revise the existing cable television rules surrounding retransmission consent agreements to encompass all MVPDs including DBS providers. SHVIA prohibits a TV station that provides retransmission consent from engaging in exclusive contracts for carriage or failing to negotiate in "good faith" until January 1, 2006, allowing DBS companies time to fully implement their local-into-local services.

In March 2000, the FCC, by First Report and Order (FCC, 2000b), established a two-part test for good faith negotiations. The first part consists of a brief, objective list of procedural standards applicable to television broadcast stations negotiating an agreement. The second part allows an MVPD to present facts to the FCC that constitute a TV station's failure to negotiate in good faith. The order directs the FCC staff to expedite resolution of good faith and exclusivity complaints, and notes that the burden of proof is on the MVPD complainant.

Antenna restrictions. As directed by Congress in Section 207 of the Telecommunications Act of 1996, the FCC adopted the Over-the-Air Reception Devices Rule concerning government and non-government restrictions on subscribers' ability to receive video programming signals from DBS systems, wireless cable providers, and television broadcast stations. The rule (47 C.F.R. Section 1.4000) has been in effect since October 14, 1996. It prohibits restrictions (including deed restrictions and local ordinances) that impair the installation, maintenance, or use of antennas to receive video programming. The rule applies to video antennas including DBS dishes that are less than one meter (39.37 inches) in diameter (or of any size in Alaska), TV antennas, and wireless cable antennas. The rule prohibits most restrictions that:

(1)   Unreasonably delay or prevent installation, maintenance, or use.

(2)   Unreasonably increase the cost of installation, maintenance, or use

(3)   Preclude reception of an acceptable quality signal.

The rule applies to viewers who place video antennas on property that they own and that is within their exclusive use or control, including condominium and cooperative owners who have an area where they have exclusive use, such as a balcony or patio, in which to install the antenna. The rule applies to townhomes and manufactured homes, as well as to single-family homes.

On November 20, 1998, the FCC amended the rule so that it applies to rental property where the renter has exclusive use, such as a balcony or patio (OTARD, 1999). Exclusive use means an area of the property that only the renter may enter and use to the exclusion of other residents. For example, a condominium or apartment may include a balcony, terrace, deck, or patio that only the renter can use, and the rule applies to these areas. The rule does not apply to common areas, such as the roof, the hallways, the walkways, or the exterior walls of a condominium or apartment building. In essence, this amendment greatly increases the potential customers available to receive DBS service.

Public interest obligations. Seeking to further level the competitive environment between DBS and cable, the FCC also adopted rules (FCC 98-307) implementing Section 25 of the Cable Television Consumer Protection and Competition Act of 1992, which imposed certain public interest obligations on DBS providers. The statute requires DBS companies to set aside 4% of their channel capacity exclusively for noncommercial programming of an educational or informational nature. DBS companies cannot edit program content, but must simply choose among qualified program suppliers for the reserved capacity. In 2000, DirecTV challenged these public interest obligation rules by asserting that they cause noncommercial station carriage to occupy a much larger percentage of DBS provider channel capacity relative to any cable system operator. The FCC solicited public comments on the impact of the rules and, in 2001, decided to deny DirecTV's complaint. The FCC retained the rules in an Order on Reconsideration (FCC, 2001) and stated that the rules provide the same degree of carriage by satellite carriers as is provided by cable systems. Basically, the public interest obligation rules promote parity between DBS and cable by assuring that consumers receive via satellite essentially the same local channels they would receive if they subscribed to cable.

As of January 2002, DirecTV offered nine channels of public interest programming to fulfill this obligation: Clara+Vision, C-SPAN, EWTN, Inspirational Life, NASA TV, PBS YOU, StarNet, TBN, and WorldLink TV. The DISH Network offered 14 channels: BYUTV, C-SPAN, Educating Everyone, EWTN, FSTV, HITN, Mayerson Academy, NASA TV, PBS YOU, Research TV, TBN, UCTV, Universityhouse Channel, and WorldLink TV.

DBS companies must also comply with the political broadcasting rules of Section 312 of the Communications Act, granting candidates for federal office reasonable access to broadcasting stations. They must also comply with Section 315's rules granting equal opportunities to federal candidates at the lowest unit charge.

Transmission of HDTV signals. DirecTV and the DISH Network are now using similar approaches to their HDTV equipment plans. DirecTV HDTV reception is provided by multi-location dishes and receivers that are able to receive both the standard compressed NTSC and HDTV signals from the satellite. Both the NTSC and HDTV signals are sent to an HDTV set with a built-in DirecTV receiver or a DirecTV-enabled HDTV receiver and a slightly larger than normal triple-LNB, 18 inches × 24 inches DirecTV multi-satellite dish antenna. HDTV programming on DirecTV includes:

> ➤ HDNet, the first national all-HDTV television network, broadcasting 16 hours per day of live sports, movies, concerts, and more.

> ➤ The HBO HDTV channel, which airs continuously and delivers blockbuster movies and original programming from Home Box Office, all in HDTV format.

➢ Pay-per-view movies in HD format, available from 2 A.M. to 10 A.M.

The DISH Network is using the integrated NTSC/HDTV Model 6000 receiver, which can deliver HDTV programming onto a 16:9-ratio HDTV screen and supports both 720p and 1080i HD formats. It is intended to provide seamless switching between HDTV and standard TV with accompanying Dolby Digital surround-sound. HDTV programming on the DISH Network consists of HBOHD and ShowtimeHD. (For details on HDTV formats, see Chapter 7.)

Internet access. Similar to cable systems and phone companies, DirecTV and the DISH Network have been actively upgrading their systems to provide Internet services to subscribers. Because the DBS signals are sent as digital information, the systems can send video, audio, and computer data in any combination to the receivers. Each DBS channel has a large amount of bandwidth, some of which the DBS companies are using for data services such as Internet or interactive TV services.

DirecTV's first Internet offering, DirecPC, provided DirecTV subscribers Internet service through a separate, second dish and receiver. In 2001, DirecTV upgraded the capabilities of DirecPC and renamed the service DirecWay. The DirecWay service comes in two modes: a one-way system that allows users to receive large files with fast satellite speed while uploading information back to the Internet through a phone line, and a two-way (satellite return) system that offers users the power of the satellite for both uploads and downloads, eliminating the need to monopolize a phone line to surf the Internet.

A second option is the DirecDuo system, which allows subscribers to receive up to 400 Kb/s (kilobits per second) Internet access and over 200 DirecTV channels on a single elliptical dish. A third option is DirecTV/AOL TV, a strategic alliance with America Online begun in 2000 to provide a combination DirecTV and AOL service. A fourth option is a partnership with Microsoft's Ultimate TV, allowing users to receive DirecTV programming, Internet access, and digital video recording in a specially configured Ultimate TV receiver. DirecTV's newest Internet access option is DirecTV DSL, an expensive, DSL (digital subscriber line) Internet connection that uses digital phone lines. (For more on DSL, see Chapter 19.)

DISH Network's one Internet option is called StarBand. Comparable to DirecTV DSL, StarBand is a two-way satellite-only service that offers high-speed access that is always-on, does not tie up a phone line, but is fairly expensive ($60 per month in addition to DISH TV programming costs).

Available programming. In terms of programming, DirecTV and the DISH Network have been able to acquire all of the top cable program networks, sports channels and events, and pay-per-view (PPV) events as envisioned by the 1992 Cable Act. What differentiates one service from the other is how the program services are priced, packaged, and promoted. Each service has on-screen program guides, parental control features, preset PPV spending limits, instant PPV ordering using the remote control and a phone line hookup, favorite channel lists, equipment warranties, and 800 phone numbers for customer service.

DIRECTV. Programming on DirecTV consists of packages costing between $31.99 and $85.99 per month that consist of different combinations of basic cable channels and packages of premium movie channels such as HBO, Cinemax, Showtime, and Starz! It also offers individual PPV movies, concerts, and sporting events as available through DirecTicket (i.e., movies for $2.99; boxing for

$14.95). Using the remote control, subscribers can search the interactive program guide to access desired channels or to request PPV events. DirecTV also offers unique packages of college and professional sports (MLB Extra Innings, MLS Shootout, NBA League Pass, NFL Sunday Ticket, NHL Center Ice, and ESPN College Basketball & Football). It also offers 31 CD-quality, commercial-free digital audio channels as part of its Total Choice packages.

DirecTV offers three packages of Spanish-language service, DirecTV Para Todos, to subscribers. DirecTV Para Todos offers more than 22 Spanish language national and international channels including Univision, Discovery en Español, Fox Sports World Español, Galavisión, MTV-S, TVN Chile, and Canal Sur, among others. All packages include seven Music Choice channels of commercial-free Spanish-language music.

To receive any DirecTV programming, subscribers must purchase a DSS equipment package from a variety of retailers and have the DSS system installed. DirecTV has authorized 17 different companies to manufacture DSS equipment (including GE, HNS, Panasonic, RCA, Sanyo, and Sony), hoping to entice consumers with familiar, reliable brands. Prices vary according to individual retailers, the brand name chosen, and the complexity of the DSS system selected. Equipment prices can range from $99 to over $499, plus $49 to $199 for installation (see http://www.directv.com for the most current information).

THE DISH NETWORK. The DISH Network is to satellite TV what Saturn is to automobiles: The service is highly practical, the packages are the most inexpensive yet comprehensive, and prices range from $22.99 to $72.99 per month. With its large channel capacity and deployed satellites, the DISH Network is marketing itself as the only satellite service to deliver more than 500 video and audio channels to subscribers, using many of these channels to deliver a range of Latino programming as well as pay-per-view movies, sporting events, etc. This capacity has also allowed the DISH Network to be the first satellite provider to supply local channels to selected cities via a second dish antenna ($5.99 per month per city). Most of these locals can now be obtained with a dual-LNB dish. The optional second dish can also provide subscribers with international programming in 10 languages, specialty religious or science programming, and data services.

Despite the fact that the receiver specifications are developed exclusively by EchoStar, the units offer a wide range of options and are competitively priced with similar DSS units. Some of the higher-grade receivers offer features such as RF remotes, timed remote control of VCRs, seamless integration with off-air signals and local listings in the channel guide, on-screen caller-ID, and even an integrated D-VHS recording deck. All receivers are feature-upgradeable via satellite. Equipment is available directly from EchoStar (via phone or the Web) or through local distributors. Like DirecTV, programming sign-up and/or changes are implemented immediately via a 24-hour 800 number. Technical support for installation or hardware issues is also available (for the most current information, see http://www.dishnetwork.com).

Strategic alliances. What once was an industry in search of reliable distribution technology and attractive programming is now an industry focused on brand awareness, marketing strategies, and strategic alliances. DirecTV and the DISH Network are consistently using promotions such as freeviews of various channels and sponsorships of major events (i.e., World Wrestling Federation [WWF], college basketball tournaments) and concerts (i.e., Shania Twain, Tom Petty).

DirecTV has established strategic alliances with SMATV and MMDS services, such as CS Wireless, Wireless One, and Heartland Communications, to provide DirecTV to multiple dwelling units (MDUs) such as apartments and townhomes. (SMATV or satellite master antenna television is a closed-circuit cable TV system that distributes television programming from a satellite to a single building or small campus. MMDS is multichannel multpoint distribution system.) In addition, DirecTV has also formed a distribution alliance with GTE, SBC Communications, and Bell Atlantic to allow these telecom companies to offer DirecTV program packages through their phone lines via digital set-top converter boxes to MDUs and single-family homes.

Other recent DirecTV alliances include several with various business establishments (i.e., bars, restaurants, hotels, hospitals, private offices, malls, and fitness clubs) to provide customized packages of DirecTV services. Clients include Applebee's, Ruby Tuesday's, American Airlines, and Marriott Hotels. Business travelers can even receive DirecTV on selected commercial airlines (Alaska Airlines and JetBlue Airways) and private corporate jets via DirecTV Airborne. Launched in 1999, DirecTV Airborne offers passengers 24 channels of real-time DirecTV sports, news, children's, and general programming. Two of the 24 channels are reserved for specialty programming including concerts and special events. The service is viewed on in-flight entertainment equipment supplied by LiveTV. The low-profile LiveTV antenna, located in the top center of the aircraft's fuselage, maintains constant communication with the DirecTV satellites located at the 101°W orbital slot (DirecTV, 2002).

Becoming more involved in interactive enhancements, DirecTV and Wink Communications announced an alliance to provide Wink-enhanced DirecTV receivers in July 1999. Wink's technology allows advertisers and television networks to create interactive enhancements to accompany traditional television advertisements and programs. By clicking the remote control during an enhanced ad or program, viewers receive program-related information such as local weather, sports updates, and product samples and coupons, and viewers can make purchases instantly for no charge (Wink Communications, 2002).

In January 2000, DirecTV announced a partnership with TiVo personal television to get even more involved in interactive television. The new DirecTV/TiVo combination digital receiver will store up to 30 hours of recorded programming. Additional features include high-quality, all-digital video, high-end Dolby Digital (AC-3) audio, live TV pause, instant replay, slow motion, frame forward/back, variable speed rewind, and fast forward (TiVo, 2000). Then, in June 2000, DirecTV, Microsoft, and Thomson Multimedia announced a partnership to create a new advanced RCA DirecTV System with the Microsoft Ultimate TV service. Ultimate TV is the first and only DBS television platform that integrates the following features in one receiver: DirecTV programming, digital video recording, live TV controls, and interactive television, including Internet access from the television set (Microsoft, 2000).

The DISH Network has also jumped into the interactive services arena quite aggressively. Working with interactive services company, OpenTV, the DISH Network is developing "DISH Home" for rollout in 2002 (DISH Network, 2002). DISH Home will provide interactive TV services such as video replay, interactive TV advertising, and entertainment services including movie information and music news. The digital VCR-type product will also have the ability to record OpenTV-enabled interactive programs and services, allowing the viewer to interact even with recorded TV programs. OpenTV set-top box software is used by more than 40 million viewers in over 50 countries, including subscribers to British Sky Broadcasting (BSkyB) in the United Kingdom and TPS in France (OpenTV, 2002).

Like DirecTV, the DISH Network has also started to provide its equipment and programming to MDUs. In early 2000, it signed an agreement with Castle Cable Services, Inc., a private cable TV service provider, to install the first of EchoStar's quadrature amplitude modulation (QAM) systems for MDUs in San Francisco. Residents of Avalon Towers will be able to select various packages of the DISH Network's digital television services—up to 500 channels—as well as Castle Cable's 60-channel analog service (Castle Cable, 2000).

Used for the last five years in Europe, QAM offers an affordable way to provide MDUs with hundreds of channels of programming. Residents can receive their personalized DISH Network programming using shared dishes instead of individual dish antennas. QAM technology can be applied using a building's existing cable, which saves thousands of dollars in installation expense, and converts the QPSK signal used by the direct broadcast satellite industry to QAM, a digital modulation scheme used by the cable television industry. This conversion, which takes place at the headend, allows the delivery of DISH Network programming using only 300 MHz of bandwidth, less than half the bandwidth of a typical cable system.

INTERNATIONAL DBS/DTH

Although other countries have used satellites to transmit television signals to stations and cable television systems, Japan was the first country to launch a DBS service in 1984 (Otsuka, 1995). In October 1996, Japan's largest satellite operator, JSAT, launched the country's first digital DBS system, PerfecTV. In March 1998, PerfecTV merged with a competitor, News Corporation's JSkyB. The combined digital DBS service, SkyPerfecTV, delivered about 200 channels to its subscribers.

DirecTV Japan, a digital DBS competitor launched by Hughes Electronics Corporation in December 1997, ended service in 1999 by merging its operations into SkyPerfecTV. This added about 400,000 subscribers to SkyPerfecTV's base of 1.7 million. As part of the transaction, Hughes and other shareholders of DirecTV Japan received an equity stake in SkyPerfecTV (JSkyB, 2000). As of mid-2002, SkyPerfecTV had 2.62 million subscribers accessing 300 channels of television and audio (Sky Perfect Communications, 2002).

STAR (formerly STAR-TV) was launched in 1991 in Hong Kong, and is still the driving force for satellite television in Asia. Within six months of STAR-TV's launch, eight million viewers had tuned in. Today, STAR covers 53 countries, spanning an area from Egypt to Japan and the Commonwealth of Independent States to Indonesia, reaching an estimated audience of 300 million viewers (STAR, 2002). In 2001, the company launched the first 24-hour commercial FM radio network in India. In 2002, the company rebranded from STAR-TV to STAR, reflecting the company's evolution from a television brand to a multi-service, multi-platform brand. STAR offers both subscription and free-to-air television services, using AsiaSat 1 as its primary satellite platform with additional services available on the AsiaSat 2 and Palapa C2 satellites.

In Europe, satellite consortia SES Astra and Eutelsat continue to dominate the European MVPD market, grabbing a 92% share of total MVPD households, including cable TV. According to SES Astra, one out of every two households in Europe receives satellite television. SES Astra delivers over 1,000 digital and analog radio and television channels to over 89 million households (SES-Astra, 2002). Eutelsat's Hot Bird Satellite TV service delivers 900 channels to over 100 million households (Eutelsat, 2002).

Primary competition to SES Astra and Eutelsat has been from a number of recent national/ regional DTH systems including News Corporation's England-based BSkyB, France's CanalSatellite and TPS, Germany's DF-1, Italy's Telepiu, Norway's Canal Digital, and Spain's CSD and Via Digital. Digital DBS/DTH operators have done extremely well in France, Japan, and Malaysia, but have been playing catch up in Germany and Italy (Worldwide DTH, 2002).

Closer to the United States in Latin America and Canada, DBS systems are also expanding. In Latin America, as deregulation and privatization of the telecommunications markets continue to spread through the region, the result has been fierce competition in satellite services. The leader in DBS in Latin America is DirecTV Latin America, which provides DBS service to over 1.2 million subscribers in 27 markets, including Brazil, Costa Rica, Mexico, and Panama (Hughes, 2001). DirecTV Latin America, which commenced service in mid-1996, is a multinational company owned by Hughes Electronics Corporation, Venezuela's Cisneros Group of Companies, Brasil's Televisao Abril, and Mexico's MVS Multivision.

In Canada, Star Choice Television is one of two firms that have satellite systems in operation. Star Choice delivers Canada's largest channel selection with more than 300 channels to its more than 600,000 subscribers. Star Choice also launched Canada's first elliptical dish, which facilitates multiple satellite reception. Canadians can purchase Star Choice equipment at more than 4,000 locations across the country, including Radio Shack, Future Shop, Canadian Tire, Sears, Leon's, and The Brick. The other company, Bell ExpressVu, uses EchoStar's DISH Network equipment to operate a 275-channel DBS service. Subscribers must purchase the DVB equipment and a basic tier of programming, and then can add a wide range of specialty programming tiers including the Sports Bar, Kids Size, the Network Platter, and Film Feast. Now owned by the largest telecom company in Canada (BCE, Inc.), Bell ExpressVu serves over one million subscribers (Bell ExpressVu, 2002). Both Bell ExpressVu and Star Choice offer programming in English and French.

FACTORS TO WATCH

The primary factor that members of the DBS and cable industries are focused on as of mid-2002 is the proposed merger of EchoStar Communications and Hughes Communications, which was announced on December 3, 2001. EchoStar is the parent company of the DISH Network, and Hughes is the parent company of DirecTV. The combined company would use the EchoStar name and adopt the DirecTV brand for its services and related products. The merger would create the nation's second-largest MVPD platform with more than 16.7 million subscribers, of which 1.8 million subscribers are NRTC affiliates, and 14.9 million subscribers are owned-and-operated by the combined company. EchoStar and Hughes have pledged that the merger, for which they hope to secure approval during 2002, will not cause disruption of service or additional expense to existing customers of either DirecTV or DISH. Cable TV companies presently control more than 80% of the U.S. MVPD market, while a combined EchoStar-Hughes would provide service to about 17% of the market.

The $26 billion merger is expected to require approximately $5.5 billion of total financing. Completion of this financing is being supported by a commitment of approximately $2.75 billion from Deutsche Bank, and a commitment of approximately $2.75 billion from General Motors. The GM commitment is secured by a pledge of $2.75 billion of EchoStar stock held by a trust controlled by EchoStar Chairman and Chief Executive Officer Charles Ergen.

The transaction is subject to a number of conditions, including approval by a majority of each class of GM shareholders. Approval of the majority of EchoStar's voting shares has already been given by written consent. The merger is subject to regulatory clearance under the Hart-Scott-Rodino Act and approval by the Federal Communications Commission. The transaction is also contingent upon the receipt of a favorable ruling from the Internal Revenue Service that the separation of Hughes from GM will qualify as a tax-free spin-off for U.S. federal income tax purposes.

FCC Chairman Michael Powell appointed W. Kenneth Ferree, chief of the Cable Services Bureau, to head an intra-agency team of FCC officials reviewing whether the merger is in the public interest. Up-to-date FCC-related information on the merger can be obtained at a special merger Web site operated by the Cable Services Bureau at http://www.fcc.gov/csb/echoditv/.

The combination of EchoStar and Hughes is expected to generate savings from the elimination of costly duplicate satellite bandwidth and infrastructure, and strong management offering more effective fundamental business practices. The new company would also have enhanced scale to compete more effectively against the dominant U.S. cable and broadband providers—a critical factor given the increasing consolidation in the cable industry. Consumers across the country will receive programming from the merged direct broadcast satellite service via one small satellite dish and will pay the same nationwide price for services. New set-top boxes and satellite dishes capable of receiving satellite signals from multiple orbital slots will be made available free of charge to all existing DirecTV and EchoStar customers who will require new equipment in order to receive local channels from the combined company (EchoStar, 2002).

The merger will enable the combined company to deliver local broadcast TV channels in all 210 DMAs in the United States, including Alaska and Hawaii. This "Local Channels, All Americans" plan will allow the merged company to offer every consumer in the continental United States, Alaska, and Hawaii access to satellite-delivered local television signals. This will be made possible by the elimination of duplicate channels currently carried by both DirecTV and DISH and by launching three new spot-beam satellites. Currently, DirecTV and DISH deliver local broadcast channels via satellite to consumers in only 42 metropolitan markets. Implementation of the "Local Channels, All Americans" plan is slated to begin immediately upon regulatory approval of the merger, and the roll-out is to be completed as soon as 24 months thereafter (EchoStar, 2002). Many interested parties have already started to comment on the proposed merger:

> ➢ The Senate Judiciary Subcommittee on Antitrust, Competition and Business, and Consumer Rights held hearings on the merger in early 2002 as part of a larger examination of consolidation in the pay television marketplace.

> ➢ R/L DBS and Cablevision told the FCC in February 2002 that the government should provide DBS spectrum to competing satellite TV entities if it approves the pending merger between EchoStar and DirecTV.

> ➤ The National Rural Telecommunications Cooperative and Pegasus Communications, the two top DirecTV resellers in rural America, have argued loudly against the merger citing the fear of monopolistic practices in rural areas by the new company.

> ➤ Several state governors, including governors from rural states such as Colorado and Louisiana, have made statements supporting the merger as a way to deliver much-needed competition to cable.

A secondary factor of interest is a proposal by Northpoint Technologies to share spectrum space currently being used by DBS providers. The frequency band between 12.2 GHz and 12.7 GHz is allocated to the Fixed and Broadcasting-Satellite radio services on a co-primary basis. The ITU permits the operation of services that provide "terrestrial radiocommunication services" in the same band, subject to the restriction that they "shall not cause harmful interference to the space services (DBS in the United States) operating in the band" (FCC, 2000c, p. 4). In 1999, Broadwave USA, a subsidiary of Northpoint Technologies, filed a petition with the FCC seeking an authorization to operate terrestrial stations delivering Multichannel Video Distribution and Data Service (MVDDS) in the 12.2–12.7 GHz band. Subsequently, two other companies, PDC Broadband Corporation and Satellite Receivers, Ltd. filed similar applications with the FCC (FCC, 2000c).

The FCC issued a Notice of Proposed Rulemaking on November 24, 1998, and a First Report and Order and a Further Notice of Proposed Rulemaking as ET Docket 98-206 on December 8, 2000. These documents address the issues associated with permitting MVDDS in the band, and conclude that sharing the band between MVDDS and DBS systems is possible, subject to certain precautions that must be taken to prevent interference to DBS systems. In January 2001, the FCC selected an independent engineering firm, the MITRE Corporation, to perform an analysis to determine whether these two services can share the band without harmful interference to DBS systems.

Following extensive testing and analysis, MITRE provided the FCC with several recommendations and questions for further study. Chief among them were the following:

> ➤ MVDDS/DBS band sharing appears feasible if—and only if—suitable mitigation measures are applied. Different combinations of measures are likely to prove "best" for different locales and situations.

> ➤ Potential MVDDS design changes may reduce the interference impact on DBS. These include real-time power control, multiple narrow transmitting-antenna beams, the use of circular polarization, and increasing the size of MVDDS receiving antennas.

> ➤ Corrective measures can be applied at DBS receiver installations. These include relocation and retrofitting of existing DBS antennas, the use of alternative antenna designs, and the replacement of older DBS set-top boxes (MITRE, 2001).

The FCC has been analyzing the report, as well as comments supplied by companies on both sides of this issue. A decision from the FCC is expected in mid-2002. Unlike the 2000-2002 period in the DBS industry, which was focused on corporate innovation, marketing, and deal making, the 2002-2004 period is likely to be dominated by legislators, regulators, and lawyers in Washington, D.C.

Direct broadcast satellite service has proven itself a viable alternative to cable television. Its future success will be determined, in large part, by the impacts of the factors explored above, especially consolidation, technological innovation, and new business models.

BIBLIOGRAPHY

Bell ExpressVu. (2002, Feb. 8). Bell ExpressVu customers upgrading to new digital channels. *Bell Canada Enterprises*. Retrieved March 14, 2002 from http://www.bce.ca/en/news/releases/bev/2002/02/08/6785.html.

Boyer, W. (1996, April). Across the Americas, 1996 is the year when DBS consumers benefit from more choices. *Satellite Communications*, 22-30.

Cablevision contracts with Lockheed Martin for DBS satellite. (2001, May 30). *Andrews Space & Technology*. Retrieved March 14, 2002 from http://www.spaceandtech.com/digest/flash2001/flash2001-041.shtml.

Castle Cable, Inc. (2000). *New developments*. Retrieved March 14, 2002 from http://www.castlecable.com.

DBS knockin' on cable's doors. (1998, March). *Cableworld*. Retrieved March 5, 1998 from http://www.mediacentral.com/magazines/cableworld.

DirecTV. (2002, March). *DirecTV for business*. Retrieved March 14, 2002 from http://www.directv.com/DTVAPP/buy/Business_PvtJets.jsp.

DISH Network. (2002, March). *When will I get DISH Home?* Retrieved March 14, 2002 from http://www.dishnetwork.com/content/technology/itv/dish_info/index.asp.

EchoStar. (2002, February 26). *Merged EchoStar and Hughes will deliver local broadcast channels to all 210 U.S. television markets.* Retrieved March 14, 2002 from http://www.echostarmerger.com/5030/wrapper.jsp?PID=5030-10&CID=5030-030402A.

Eutelsat. (2002, March). *The fleet*. Retrieved March 14, 2002 from http://www.eutelsat.org/satellites/4_1_1.html.

Federal Communications Commission. (1989). *Memorandum opinion and order*. MM Docket No. 86-847.

Federal Communications Commission. (1995). *Memorandum opinion and order*. MM Docket No. 95-1733.

Federal Communications Commission. (1996, February 14). *Status report*. Report No. SPB-37.

Federal Communications Commission. (1999, May 14). Dominion Video Satellite, Inc. application for minor modification of authority to construct and launch order and authorization. *Report and order*. CS Docket No. 98-102.

Federal Communications Commission. (2000a, January 14). Annual assessment of the Status of competition in the market for the delivery of video programming. *Sixth annual report*. CS Docket No. 99-230.

Federal Communications Commission. (2000b, March 14). Implementation of the Satellite Home Viewer Improvement Act of 1999/retransmission consent issues, good faith negotiation, and exclusivity. *First report and order*. FCC 00-99.

Federal Communications Commission. (2000c, December 8). In the matter of amendment of parts 2 and 25 of the Commission's rules to permit operation of NGSO FSS systems co-frequency with GSO and terrestrial systems in the Ku-band frequency range. *First report and order/further notice of proposed rulemaking*. ET Docket 98-206.

Federal Communications Commission. (2001, September 4). Implementation of the Satellite Home Viewer Improvement Act of 1999 (CS Docket 00-96). *Order on reconsideration*. FCC 01-249.

Federal Communications Commission. (2002, January 14). Annual assessment of the status of competition in the market for the delivery of video programming. *Eighth annual report*. CS Docket No. 01-129.

Frederick, H. (1993). *Global communications & international relations*. Belmont, CA: Wadsworth.

Gilat Networks. (2000). *What is Gilat-To-Home?* Retrieved March 10, 2000 from http://www.gilat2home.com/faq/index.html.

Golden Sky Systems, Inc. (2000). *Golden Sky to merge with Pegasus Communications Corporation in $1 billion transaction*. Retrieved March 10, 2000 from http://www.gssdirectv.com/ press/60X-story.html.

Hogan, M. (1995, September). US DBS: The competition heats up. *Via Satellite*, 28-34.

Hogan, M. (1998a, January 19). Demand remained strong for DBS in 1997. *Multichannel News, 19* (3), 33.

Hogan, M. (1998b, February 2). Digital cable not immediate threat, says DBS. *Multichannel News, 19* (5), 12.

Hogan, M. (1998e, March 2). DBS discounts 2nd receivers. *Multichannel News, 19* (9), 3, 18.

Howes, K. (1995, November). US satellite TV. *Via Satellite*, 28-34.

Hughes. (2001). *Galaxy Latin America, LLC changes name to DirecTV Latin America, LLC*. Retrieved March 14, 2002 from http://www.hughes.com/ir/pr/00_11_14_directvla.xml.

Johnson, L., & Castleman, D. (1991). *Direct broadcast satellites: A competitive alternative to cable television?* Santa Monica, CA: Rand.

JSkyB, PerfecTV to combine operations on May 1. (1998, March). *SkyREPORT*. Retrieved March 5, 1998 from http://www.skyreport.com/jskyb.htm.

Lambert, P. (1992, July 27). Satellites: The next generation. *Broadcasting & Cable, 124* (31), 55-56.

Manasco, B. (1992, April). The U.S. multichannel marketplace in the year 2000. *Via Satellite*, 44-49.

Microsoft. (2000, June 12). Microsoft, DirecTV and Thomson Multimedia join forces to make television more personal and interactive. *Press release*. Retrieved March 14, 2002 from http://www.microsoft.com/PressPass/press/2000/Jun00/DirecTVpr.asp.

MITRE Corporation. (2001, April*). Analysis of potential MVDDS interference to DBS in the 12.2–12.7 GHz band*. Retrieved March 14, 2002 from http://www.fcc.gov/oet/info/mitrereport/mitrereport_4_01.pdf.

OpenTV. (2002). *The company*. Retrieved March 14, 2002 from http://www.opentv.com/company/.

Otsuka, N. (1995). Japan. In L. Gross (Ed.). *The international world of electronic media*. New York: McGraw-Hill.

Over-the-Air Reception Devices Rule (OTARD Rule). (1999, January). 47 C.F.R. Section 1.4000.

Parone, M. (1994, February). Direct-to-home: Politics in a competitive marketplace. *Satellite Communications*, 28.

Pegasus Communications. (2002, March). *Pegasus Communications Corporation completes acquisition of Golden Sky*. Retrieved March 14, 2002 from http://www.pegsattv.com/company_info/set.html.

Regional Administrative Radio Conference. (1983). *Final report and order*. Geneva: ITU.

Satellite Home Viewer Act of 1988. (1998). 17 U.S.C. § 119.

Satellite Home Viewer Improvement Act of 1999. (1999, November 19). 17 U.S.C. § 122.

SES-Astra. (2002, March). *Broadcasting*. Retrieved March 14, 2002 from http://www.ses-astra.com/corporate/company/index.shtml.

Setzer, F., Franca, B., & Cornell, N. (1980, October 2). *Policies for regulation of direct broadcast satellites*. Washington, DC: FCC Office of Plans and Policy.

Sky Perfect Communications, Inc. (2002, March). *Change in the number of subscribers*. Retrieved March 14, 2002 from http://www.skyperfectv.co.jp/skycom/e/ir/frame1e.html.

SkyPerfectTV gets DirecTV Japan. (2000, March). *SkyREPORT*. Retrieved March 10, 2000 from http://www.skyreport.com/skyreport/mar2000/030300.htm#dtv.

STAR. (2002, March). *Corporate information*. Retrieved March 14, 2002 from http://www.startv.com/eng/corporate/index.html.

TiVo, Inc. (2000). *What is TiVo?* Retrieved March 10, 2000 from http://www.tivo.com/flash.asp?page=discover_index.

Whitehouse, G. (1986). *Understanding the new technologies of the mass media*. Englewood Cliffs, NJ: Prentice-Hall.

Wink Communications. (2002, March). *What is Wink?* Retrieved March 14, 2002 from http://www.wink.com/contents/whatiswink.shtml.

Wold, R. N. (1996, September). U.S. DBS history: A long road to success. *Via Satellite*, 32-44.

Worldwide DTH platforms. (2002, March). *SkyREPORT*. Retrieved March 14, 2002 from http://www.skyreport.com/globaldth.htm.

7

DIGITAL TELEVISION

Peter B. Seel, Ph.D. & Michel Dupagne, Ph.D.[*]

The period between 2002 and 2004 is a crucial watershed stage for the diffusion of digital television (DTV) in North America, Europe, and Japan. In the United States, two key deadlines imposed during this period by the Federal Communications Commission (FCC) for broadcast stations will become a benchmark for the adoption of this technology. All full-power commercial broadcasters were expected to commence DTV operations by May 1, 2002. Non-commercial stations have an extra year to comply with this obligation. Simply mandating these deadlines to begin transmitting a digital television signal does not automatically guarantee that there will be a significant volume of DTV programming delivered to American homes, but it does signify that the transition to DTV broadcasting in the United States is moving forward. This is equally true for Japanese and European digital television, although European broadcasters have not all rolled out DTV service at the same pace and have, for now, ruled out HDTV transmission.

This very expensive conversion from analog to digital television broadcasting is the most significant change in global broadcast standards since color images were added in the 1950s and 1960s. However, the dream of one global television standard that might replace the incompatible analog NTSC, PAL, and SECAM regional standards has failed to materialize. In fact, there is competition between the proponents of the various digital standards as they seek to enlist non-aligned nations to adopt one system over others.

It is only a matter of time until digital TV broadcasting is adopted around the world. For better or worse, television is a ubiquitous global medium. The glow of the tube can be found from rustic cabins

---

[*] Peter Seel is Associate Professor, Department of Journalism and Technical Communication, Colorado State University (Fort Collins, Colorado). Michel Dupagne is Associate Professor, School of Communication, University of Miami (Coral Gables, Florida).

near the Arctic Circle to open-air shelters on remote Pacific islands. It is a primary source of news and entertainment for viewers from New York to New Delhi. The diffusion of digital television will provide a movie-friendly display with a wider image, improved sound quality, and the higher resolution needed for projecting a sharp image on a screen. However, the most significant change that it will bring about is that DTV is a computer-friendly technology. For example, the display will be capable of presenting a film from a broadcast channel while simultaneously providing a picture-in-picture sports Web site. The DTV screen of the near future will be able to seamlessly merge broadcast programs with Internet-delivered content.

One key attribute of digital technology is "scalability"—the ability to produce audio/visual quality as good (or as bad) as the viewer desires (or will tolerate). This does not refer to program content quality—that factor will still depend on the creative ability of the writers and producers. Within the constraints of available transmission bandwidth, digital television facilitates the dynamic assignment of sound and image fidelity in a given transmission channel. Within the quality universe of DTV, there are a wide variety of possible display options. The three most common options are:

➢ HDTV (high-definition television).

➢ SDTV (standard-definition television).

➢ LDTV (low-definition television).

High-definition television represents the highest pictorial and aural quality that can be transmitted through the air. It is defined by the Federal Communications Commission (FCC) as a system that provides image quality approaching that of 35mm motion picture film, has an image resolution of approximately twice that of conventional National Television System Committee (NTSC) television, and has a picture aspect ratio of 16:9 (FCC, 1990). At this aspect ratio of 1.78:1 (16 divided by 9), the television screen is wider in relation to its height than the 1.33:1 (4 divided by 3) of NTSC. It is closer in aspect ratio to the images seen in movie theaters that are 1.85:1 or even wider. Figure 7.1 compares a 16:9 HDTV set with a 4:3 NTSC display—note that the higher resolution of the HDTV set permits the viewer to sit closer to the set, resulting in a wider angle of view.

Figure 7.1
Wider Viewing Angle with HDTV

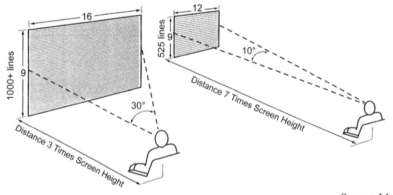

Source: M. Dupagne & P. Seel

SDTV, or standard-definition television, is another type of DTV technology that can be broadcast *instead of* HDTV. Digital SDTV transmissions will offer lower resolution than HDTV, but they will be available in both narrow- and wide-screen formats. Using digital video compression technology, it will be feasible for U.S. broadcasters to transmit four to six SDTV signals instead of one HDTV signal in the allocated 6-MHz digital channel. Thus, a television station would be able to deliver a day-time soap opera while simultaneously broadcasting a newscast, sports highlights, stock market updates, and a children's program in SDTV. Some stations may reserve true HDTV single-channel programming for evening prime-time hours. The development of multichannel SDTV broadcasting, called "multicasting," is an approach that Public Broadcasting Service (PBS) affiliates are actively investigating. It would permit PBS stations to fulfill their daytime educational mandate to children and adult learners, while providing a variety of programming choices for other audiences at the same time.

LDTV is low-definition television. Using scalable digital technology, image and sound quality can be compressed to the point that obvious defects can be seen and heard (as with streamed Internet media with a slow, dial-up modem). Some cable operators are using digital compression to expand their channel offerings without extensively rebuilding their facilities. They sometimes use compression ratios as high as 12:1, thereby fitting 12 digital channels in the space formerly required by a single analog channel. While such image quality is acceptable for a talk show, obvious visual defects may be seen in fast-paced sports programming. Video streamed over the Internet is another example of LDTV. Video streaming refers to the process of delivering video clips in real time to online users (Pavlik, 2000). The video is heavily compressed, and the frame rate is often 15 fps (frames per second) or less. A typical result is video playing in a small window with visible quality degradation. This shortcoming is primarily due to the limited bandwidth of dial-up connections. As broadband Internet access becomes more widely available in the home, streamed audio and video quality will improve. (For more on streaming technology, see Chapter 8.)

BACKGROUND

In the 1970s and 1980s, Japanese researchers at NHK developed two related analog HDTV systems: an analog "Hi-Vision" *production* standard with 1,125 scanning lines and 60 fields (30 frames) per second; and an analog "MUSE" *transmission* system with an original bandwidth of 9 MHz designed for direct broadcast satellite (DBS) distribution to the Japanese home islands. Japanese HDTV transmissions began in 1989 and steadily increased to a full schedule of 17 hours a day by October 1997 (Nippon Hoso Kyokai, 1998).

The decade between 1986 and 1996 was a significant era in the diffusion of HDTV technology in Japan, Europe, and the United States. There were a number of key events during this period that shaped advanced television technology and related industrial policies:

> ➤ In 1986, the Japanese Hi-Vision system was rejected as a world HDTV production standard by the CCIR, a subgroup of the International Telecommunications Union (ITU), at a Plenary Assembly in Dubrovnik, Yugoslavia. European delegates lobbied for a postponement of this initiative that effectively resulted in a *de facto* rejection of the Japanese technology. European governments and their high-technology industries were still recovering from Japanese dominance of their VCR markets, and resolved to create their own

distinctive HDTV standard that would be intentionally incompatible with Hi-Vision/ MUSE (Dupagne & Seel, 1998).

➤ By 1988, a European R&D consortium, EUREKA EU-95, developed a system known as HD-MAC that featured 1,250 wide-screen scanning lines and 50 fields (25 frames) displayed per second. This analog 1,250/50 system was used to transmit many European cultural and sporting events, such as the 1992 summer and winter Olympics in Barcelona, Spain and Albertville, France.

➤ In 1987, the FCC in the United States began a series of policy initiatives that led to the creation of the Advisory Committee on Advanced Television Service (ACATS). This committee was charged with investigating the policies, standards, and regulations that would facilitate the introduction of advanced television (ATV) services in the United States (FCC, 1987).

➤ U.S. testing of analog ATV systems by ACATS was about to begin in 1990 when the General Instrument Corporation announced that it had perfected a method of digitally transmitting a high-definition signal. This announcement had a bombshell impact since many broadcast engineers were convinced that digital television transmission would be a technical impossibility until well into the 21st century (Brinkley, 1997). The other participants in the ACATS competition soon developed digital systems that were submitted for testing. Ultimately, three competitors in the testing process with digital systems (AT&T/Zenith, General Instrument/MIT, and Philips/Thomson/Sarnoff) decided to merge into a common consortium known as the Grand Alliance. With the active encouragement of the Advisory Committee, they combined elements of each of their proponent systems in 1993 into a single digital Grand Alliance system for ACATS evaluation.

The FCC adopted a number of key decisions during the ATV testing process that defined a national transition process from NTSC to an advanced broadcast television system:

➤ In August 1990, the FCC outlined a *simulcast* strategy for the transition to an ATV standard (FCC, 1990). This strategy required that U.S. broadcasters transmit *both* the new ATV signal and the existing NTSC signal concurrently for a period of time, at the end of which the NTSC transmitter would be turned off. Rather than try and deal with the inherent flaws of NTSC, the commission decided to create an entirely new television system that would be incompatible with the existing one. This was a decision with multibillion dollar implications for broadcasters and consumers since it meant that all existing production, transmission, and reception hardware would have to be replaced with new equipment capable of processing the ATV signal.

➤ The FCC originally proposed a transition window of 15 years (now set at 10 years) from the adoption of a national ATV standard to the shutdown of NTSC broadcasting (FCC, 1992). This order caused consternation on the part of the television broadcast industry at what they perceived as too short a transition period. The nation was in the midst of a recession in 1992, and broadcasters were concerned about the financial cost of replacing their NTSC facilities. The transition cost for a single station is estimated to range from $2 million for simply retransmitting the network signal to $10 million for complete digital production facilities (Ashworth, 1998).

> In the summer of 1995, the Grand Alliance system was successfully tested, and a digital television standard based on that technology was recommended to the FCC by the Advisory Committee on November 28 (ACATS, 1995).

> In May 1996, the FCC proposed the adoption of the *ATSC digital television standard* based upon the work accomplished by the Advanced Television Systems Committee (ATSC) in documenting the technology developed by the Grand Alliance consortium (FCC, 1996a). The proposed ATSC DTV standard specified 18 digital transmission variations as outlined in Table 7.1 below. Stations would be able to choose whether to transmit one channel of high-resolution, wide-screen HDTV programming, or four to six channels of standard-definition programs during various dayparts.

Table 7.1
**U.S. Advanced Television Systems
Committee (ATSC) DTV Formats**

Format	Active Lines	Horizontal Pixels	Aspect Ratio	Picture Rate *	U.S. Adopter
HDTV	1,080 lines	1,920 pixels/line	16:9	60I, 30P, 24P	CBS & NBC (60I)
HDTV	720 lines	1,280 pixels/line	16:9	60P, 30P, 24P	ABC (30P)
SDTV	480 lines	704 pixels/line	16:9 or 4:3	60I, 60P, 30P, 24P	Fox (30P)
SDTV	480 lines	640 pixels/line	4:3	60I, 60P, 30P, 24P	None

* "I'" indicates interlace scan in *fields*/second and "P" means progressive scan in *frames*/second.

Source: ATSC

Note that the standard allows for both interlace and progressive scanning. Interlace scanning is a form of signal compression that first scans the odd lines of a television image onto the screen, and then fills in the even lines to create a full video frame every 1/30th of a second. The present NTSC standard uses interlace scanning to reduce the bandwidth needed for broadcast transmission. While interlace scanning is spectrum-efficient, it creates unwanted visual artifacts that can degrade image quality. Progressive scanning—where each complete frame is scanned on the screen in only one pass—is utilized in computer displays because it produces fewer image artifacts than interlace scanning. Progressive-scan DTV receivers are capable of displaying small-font text (e.g., from a World Wide Web site) that would be illegible on a conventional NTSC television set.

In December 1996, the FCC finally approved a DTV standard that deleted *any* requirement to transmit any of the 18 transmission video formats listed in Table 7.1 (FCC, 1996b). It also did not mandate any specific aspect ratios, or any requirement that broadcasters must transmit true HDTV on their digital channel. The FCC resolved a potential controversy over the image aspect ratio by leaving this basic decision up to broadcasters as well. They were free to transmit digital programming in narrow- or wide-screen ratios as they wished.

The ATSC standard did specify the adoption of the Dolby AC-3 (Dolby Digital) multichannel audio system. The AC-3 specifications call for a surround-sound, six-channel system that will approximate a motion picture theatrical configuration with speakers screen-left, screen-center, screen-right, rear-left, rear-right, and a subwoofer for bass effects. This audio system will enhance the diffusion of "home theater" television systems with multiple speakers and either a video projection screen, a direct-view CRT monitor, or flat-panel display that can be mounted on the wall like a painting.

In April 1997, the FCC defined how the United States would make the transition to DTV broadcasting and set December 31, 2006 as the target date for the phase-out of NTSC broadcasting (FCC, 1997). However, the U.S. Congress passed a bill in 1997 that would allow television stations to continue operating their analog transmitters as long as more than 15% of the households in a market cannot receive digital broadcasts through cable or DBS, or if 15% or more of local households lack a DTV set or a converter box to display digital images on their older television sets (Balanced Budget Act, 1997). This law gave broadcasters some breathing room if consumers do not adopt DTV technology as fast as set manufacturers would like.

To demonstrate their good faith in an expeditious DTV conversion, the four largest commercial television networks in the United States (ABC, CBS, NBC, and Fox) made a voluntary commitment to the FCC to have 24 of their affiliates in the top 10 markets on the air with a DTV signal by November 1, 1998 (Fedele, 1997). All of their affiliates in the top 30 markets were expected to broadcast digital signals by November 1, 1999, and, except for delays in a few markets caused by tower construction, this deadline was met. Table 7.2 outlines the rest of the rollout mandated by the FCC in 1997 (FCC, 1997).

Table 7.2
U.S. Digital Television Broadcasting Phase-in Schedule

Phase	# of Stations	Market Size	Type of Station	DTV Transmission Deadline	% of U.S. HHs*
1	24	Top 10	Voluntary	November 1, 1998	--
2	40	Top 10	Network Affiliates	May 1, 1999	30%
3	80	Top 30	Network Affiliates	November 1, 1999	53%
4	~ 1,037	All	All Commercial	May 1, 2002	~100%
5	~ 365	All	Non-Commercial	May 1, 2003	--
6	~ 1,500	All	All	December 31, 2006	Planned NTSC Reversion Date

* Television households capable of receiving at least one local DTV broadcast signal.

Sources: FCC and *TV Technology*

CURRENT STATUS

UNITED STATES

Receiver costs. In 1998, the first HDTV receivers went on sale in the United States at prices ranging from $5,000 to $10,000 or more (Brinkley, 1998), but prices have since dropped to between $1,200 and $3,500. Many of these early models were only "HDTV-capable" monitors and required a separate $1,000 decoder box to receive HDTV signals. More sets are now available as fully integrated models (decoder included), but they are still a distinct minority of available models. According to the Consumer Electronics Association (CEA), factory-to-dealer sales of DTV sets and set-top decoder boxes totaled 1,459,731 units in 2001, including 97,157 integrated DTV receivers (CEA releases, 2002). CEA predicts that sales of DTV products will number 2.1 million in 2002, 4 million in 2003, and 10.5 million in 2006.

These statistics are a bit misleading because they aggregate DTV monitors that lack decoding for HDTV signals along with DTV receivers that include built-in decoders. The CEA estimates that only about 16% of all DTV sets that have been sold in the United States since 1998 include standalone or integrated HDTV decoding functionality (CEA releases, 2002). Table 7.3 is a summary of available DTV sets in a typical retail electronics store in a large western U.S. city as of March 2002.

Table 7.3
Sample Prices for DTV Monitors and
Receivers in March 2002

Manufacturer	Type*	Screen Size**	Aspect Ratio	HDTV Decoder	Price
Hitachi	DV	32"	4:3	No	$1,200
Hitachi	DV	36"	4:3	No	1,600
Zenith	DV	32"	4:3	No	1,400
Zenith	DV	36"	4:3	No	1,900
Panasonic	DV	36"	4:3	No	2,000
RCA	DV	38"	16:9	Yes	2,500
RCA	PJ	61"	16:9	Yes	3,500
Sony	PJ	53"	4:3	No	1,900
Sony	PJ	57"	16:9	Yes	5,000
Hitachi	PJ	53"	4:3	No	2,500
Hitachi	PJ	61"	4:3	No	3,000
Panasonic	PJ	51"	4:3	No	2,000
Panasonic	PJ	65"	4:3	No	3,500

* DV = direct view sets, PJ = projection models
** Diagonal

Source: P. B. Seel

As of 2002, this snapshot confirms that only a minority of the sets include an HDTV decoder, and these are usually the most expensive models available. Some digital sets without decoders have descriptions touting they are "HDTV-ready" (the Zenith 32-inch), and offer "Virtual HD" quality (the Hitachi 53-inch), which may be somewhat misleading to potential consumers. What is also surprising is that many of the digital sets have a 4:3 aspect ratio, which indicates that they will have to letterbox true HDTV programs transmitted in a 16:9 aspect ratio. No wide-screen 16:9 HDTV sets that included a decoder were available for less than $2,000, despite the predictions of some set manufacturers that integrated DTV models would be available at this price during 2002 (Johnston, 2001). These less expensive models are available in narrow-screen 4:3 and require a tuner/decoder to receive HDTV broadcast signals. These decoders are now available from a variety of manufacturers at prices that range from $390 to $800.

It is abundantly clear that the success of digital television will rely on consumer perceptions of the technology and the cost of the receivers. A summary of the research conducted in the 1980s and 1990s reveals that consumers prefer HDTV to NTSC in overall picture quality. They also consider improved picture quality as the most desirable characteristic of HDTV, followed by wider screen size, improved sound quality, and larger receiver size (Dupagne, 2002). This finding was reconfirmed in a recent consumer acceptance study (Book, 2002), which indicated that 90% of the respondents expected digital television to provide better picture quality. After viewing HDTV programming, 73% of the same respondents reported that the HDTV picture was "a lot better" than the picture they currently received on their home television set. This is an encouraging finding for consumer electronics manufacturers, although nearly 40% of the respondents remained concerned about the cost of the new DTV sets (Book, 2002).

Programming. The major television networks are dealing with the chicken/egg DTV diffusion dilemma by providing a gradually increasing amount of DTV programming. Japanese consumer electronics companies have underwritten production and analog-to-digital conversion costs on ABC and CBS (by Panasonic and Mitsubishi, respectively) in an effort to jumpstart DTV set sales. ABC was a pioneer in transmitting feature films, *Monday Night Football*, and the 2000 Super Bowl in HDTV (Dickson, 1999). In the fall of 2001, both ABC and CBS were simulcasting the bulk of their prime-time schedule in both NTSC and HDTV with financial support from consumer electronics companies (Table 7.4). In addition to over-the-air broadcasting, viewers can watch movies in HDTV on HBO and Showtime, distributed by DirecTV, EchoStar, and selected cable operators. Sports and entertainment programming on the new HDNet service is also delivered by DirecTV (Grotticelli, 2001). In all, nearly 500 hours of HDTV sports and entertainment programming a week are now available to the American public, and this output will continue to rise in 2002 and 2003 as the broadcast networks prepare to simulcast their entire schedule in HDTV format.

Station conversion: As of March 6, 2002, 87% (1,413) of U.S. television stations in all markets have filed DTV construction permit applications with the FCC (http://www.fcc.gov). Of these stations, 272 are transmitting a DTV signal, primarily in the top 30 markets. (WNBC-DT and WABC-DT's New York digital transmission facilities were located in tower 1 of the World Trade Center and were destroyed in the September 11, 2001 attack—five broadcast engineers were among the lives lost.)

Table 7.4
HDTV Programming Schedule on U.S. Broadcast, Cable, and Satellite Networks (October 2001)

Network	Hours Per Week	Type of Programming
ABC	14.0	6 dramas, 8 sitcoms, 2 movies
CBS	24.5	11 dramas, 7 sitcoms, 1 football game, 1 soap opera (5 days)
NBC	5.0	1 late-night talk show (5 days)
PBS	0.75	2 or 3 documentaries per month
HB0	168.0	movies, 24/7
Showtime	168.0	movies, 24/7
HDNet	112.0	sports and entertainment

Source: Broadcasting & Cable

As noted in Table 7.2, all U.S. commercial stations are supposed to air a DTV signal by May 1, 2002, but, as of March 25, 2002, 862 stations—nearly two-thirds of all commercial stations—have requested an extension of this deadline (McConnell, 2002). Based on a survey of the broadcasters conducted by the National Association of Broadcasters, the FCC had estimated that about one-third of all commercial stations would miss the 2002 deadline and said that it would consider financial hardship as a mitigating circumstance, especially for small-market broadcasters (FCC, 2001c). In the same Memorandum Opinion and Order, the Commission declined to adopt a blanket extension policy for all the stations missing the 2002 DTV construction deadline and instead decided to review waiver applications on a case-by-case basis.

Cable deployment. Because nearly 70% of U.S. television households receive their signals from a cable provider, digital conversion of cable facilities is essential to pass along DTV programming from broadcasters to consumers. However, the cable industry has been ambivalent about providing HDTV/SDTV service for two fundamental reasons:

(1)   They would have to provide an extra DTV channel for every existing analog channel under FCC simulcast rules, constraining system capacity without increasing revenue.

(2)   SDTV technology would permit terrestrial broadcasters to transmit four to six of these channels simultaneously over the air, in effect transforming them into multichannel providers in direct competition with cable news and sports channels.

For these reasons, the cable industry has been decidedly unenthusiastic about the advent of DTV in the United States, preferring to use their bandwidth to add SDTV channels to their line-ups (Grotticelli, 2001). On the other hand, DTV proponents, such as broadcasters and consumer electronics manufacturers, insist that cable systems must carry DTV simulcasts of broadcast programming for the new format to succeed. During the first months of 2002, some cable operators, such as Time Warner, Comcast, Cox, and Charter, became more proactive, perhaps to thwart the competitive advantage of DBS providers, announcing plans to deliver DTV signals to their subscribers (Kramer, et al., 2002).

For instance, Comcast, which had made available HDTV programming to 1.3 million customers in parts of Pennsylvania, New Jersey, and Delaware in late 2001, will now offer HDTV signals from local broadcast stations and pay-cable networks such as HBO and Showtime in northern Virginia in mid-2002, suburban Maryland before the end of 2002, and Washington, D.C., in 2003 (Albiniak, 2002). It may be coincidental that Comcast was seeking federal approval of their $72 billion purchase of AT&T Broadband during this period.

Dual carriage: In July 1998, the FCC (1998) launched a Notice of Proposed Rulemaking to address the contentious issue of carriage of local broadcast digital signals during the DTV conversion. It proposed seven must-carry options, running the gamut from the Immediate Carriage Proposal to the No Must Carry Proposal. (For an explanation of the must-carry issue, see Chapter 3.) At issue is not so much whether the (analog) must-carry rules, upheld constitutionally in *Turner Broadcasting System v. F.C.C.* (1997), are applicable to the digital world as it is whether the Commission could mandate cable operators to carry *both* analog and digital signals until 2006 or later. Despite pressure from broadcasters, the Commission has declined to require must-carry for both the analog and digital signals of any one broadcast television station (FCC, 2001a).

JAPAN

On December 1, 2000, NHK and several commercial broadcasters began transmitting digital HDTV programs via the BSAT satellite, along with SDTV and datacasting services. While NHK delivers one HDTV channel and two SDTV channels, the five commercial broadcasters must choose between transmitting one HDTV signal and three SDTV signals within their allotted spectrum. Thus far, most of them have opted for the HDTV format (Nagaya, 2001). In January 2002, the Japan Electronics and Information Technology Association reported that shipment of digital HDTV sets and set-top boxes barely numbered about 800,000 in 2001, which does not bode well for the industry's anticipated goal of 10 million units by December 2003 (Japan's digital, 2002). In March 2002, digital broadcasting services began on a new CS satellite. Current digital broadcasting subscribers will be able to point their dish antennas to the CS satellite and receive CS-based services.

Japanese broadcasters are expected to launch digital terrestrial television, including HDTV and interactivity services, in the three largest cities (Tokyo, Osaka, and Nagoya) by 2003 and then in the rest of the country by 2006. The technical features of the Japanese ISDB-T (Integrated Services Digital Broadcasting-Terrestrial) standard, a variant of the European DVB-T standard, are summarized in Table 7.5. As of March 2002, the Japanese government has suggested that it may delay the introduction of digital terrestrial broadcasts in some regions due to the high cost of preventing interference between analog and digital signals (Digital terrestrial, 2002). The cost of the digital terrestrial transition is now expected to top 200 billion yen ($1.6 billion). Analog television broadcasting may cease by 2011 provided that 85% of Japanese households own the necessary equipment to receive digital signals.

Table 7.5
International Terrestrial DTV Standards

System	ISDB-T	DVB-T	ATSC DTV
Region	Japan	Europe	North America
Modulation	OFDM	COFDM	8-VSB
Aspect Ratio	1.33:1, 1.78:1	1.33:1, 1.78:1, 2.21:1	1.33:1, 1.78:1
Active Lines	480, 720, 1080	480, 576, 720, 1080, 1152	480, 720, 1080*
Pixels/Line	720, 1280, 1920	Varies	640, 704, 1280, 1920*
Scanning	1:1 progressive, 2:1 interlace	1:1 progressive, 2:1 interlace	1:1 progressive, 2:1 interlace*
Bandwidth	6-8 MHz	6-8 MHz	6 MHz
Frame Rate	30, 60 fps	24, 25, 30 fps	24, 30, 60 fps*
Field Rate	60 Hz	30, 50 Hz	60 Hz
Audio Encoding	MPEG-2 AAC	MUSICAM/Dolby AC-3	Dolby AC-3

* As adopted by the FCC on December 24, 1996, the ATSC DTV image parameters, scanning options, and aspect ratios were not mandated, but were left to the discretion of display manufacturers and television broadcasters (FCC, 1996b).

Source: Seel & Dupagne

EUROPE

While U.S. and Japanese broadcasters have put a premium on HDTV programming, their European counterparts have continued to resist it. To date, no European satellite operator or terrestrial broadcaster airs programs in HDTV format, and Swedish-based HD-Divine, once the hope for disseminating HDTV broadcasting hardware in Europe, is no longer active. Instead of emphasizing high picture quality, European broadcasters have chosen to supply more channels to viewers using SDTV-type (625 lines) technology. Philip Laven, Director of the European Broadcasting Union's Technical Department, explains, "Broadcasters in Europe are reluctant to embrace HDTV because they do not perceive any demand from the public for improved picture quality" (Laven, 1998, p. 2). Interestingly, he predicts a return of HDTV and the eventual rejection of SDTV in the European marketplace when affordable 50-inch flat panel displays become available to consumers in the next decade.

Meanwhile, virtually all European broadcasters have conducted experimental terrestrial field trials using the DVB-T standard of the Digital Video Broadcasting Group (see Table 7.5). As of March 2002, four European countries have formally begun digital terrestrial television (DTTV) operations: United Kingdom (1998), Sweden (1999), Spain (2000), and Finland (2001); Denmark and Norway have delayed their DTTV launches (see Table 7.6). The other countries will follow in 2002 and 2003. Many DTTV services will offer 16:9 television programming and interactive features (Wood, 2002). While it is difficult to generalize DTTV policies across the European Union, almost every member state follows these procedures:

(1)   They allocate spectrum for digital services and grant licenses to public service and/or commercial broadcasters.

(2)   They determine the initial coverage of digital services.

(3)   They set a target date for ceasing analog television operations (Bajon & Fontaine, 2001).

Yet, not everything is rosy for European DTTV broadcasting, and several broadcasters are struggling financially, casting doubt on the commercial viability of digital terrestrial television in Europe. For instance, Spain's two-year-old Quiero TV has only signed up about 130,000 subscribers and is for sale (Crawford & Harding, 2002). Even the pioneering ITV Digital (formerly ONdigital) in the United Kingdom with an estimated 1.3 million subscribers is under such severe financial pressures that it is threatening to shut down its sports channel in August 2002 unless it is able to renegotiate the £315 ($449) million television rights contract with the U.K. Football League (Harris, 2002).

Table 7.6
Existing and Planned Digital Terrestrial Television Services in Western Europe (2001)

Country	Launch Date	Initial Coverage*	Number of Multiplexes**	Analog Shut-off Date
Belgium	?	<50	<4	?
Denmark	2001	60-70	<4	?
Finland	08/27/01	<50	<4	2005-2010
France	Q4 2002	60-70	6	?
Germany	2001 in selected areas	<50	6	2005-2010
Ireland	Q3 2002	>90	>6	2010-2015
Italy	2003	<50	<4	2005-2010
Netherlands	Q1 2002	<50	5	Before 2005
Norway	09/01/01	60-70	<4	?
Spain	05/05/00	60-70	5	01/01/12
Sweden	04/01/99	<50	4	?
United Kingdom	11/15/98	60-70	6	2006-2010

* Percentage of the population.
** Multiplexes are assigned frequencies to public service and commercial broadcasters for providing digital terrestrial television services. Each multiplex/frequency can support multiple DTTV channels.

Source: European Broadcasting Union

RECENT DEVELOPMENTS

Prior to the 2002 National Association of Broadcasters convention, FCC Chairman Michael Powell created quite a stir by sending a letter to key telecommunications companies urging voluntary action to speed the DTV transition (Proposal for, 2002). In his letter, Powell encouraged the following steps, among others:

➢ The top four broadcast networks (ABC, CBS, Fox, and NBC), joined by HBO and Showtime, would simulcast HDTV or other "value-added DTV programming" during at least 50% of their primetime schedules starting in fall 2002.

➢ By January 1, 2003, affiliates of the top four networks in U.S. markets 1 through 100 would have digital transmission equipment installed to "pass through" network DTV programming to viewers "without degradation of signal quality."

➢ Cable systems with at least 750 MHz of channel capacity would carry the DTV signals of "up to five" broadcast or other digital programming services (e.g., HBO). Powell also urged that cable MSOs provide their customers (via sale or lease) digital set-top boxes capable of displaying HDTV content on the attached set.

➢ Direct broadcast satellite (DBS) companies would carry the signals of up to five networks that provided DTV content during at least 50% of primetime hours.

➢ Television set manufacturers would include DTV tuners in all new receivers with a phase-in schedule of 100% of sets 36-inches or larger by January 1, 2005, 100% of sets 25-inches or larger by January 1, 2006, and 100% of sets 13-inches or larger by December 31, 2006. Piracy-secure digital inputs from DTV cable boxes would be available on HD-capable receivers by January 1, 2004.

The letter did not carry the force of a ruling from the Commission requiring these actions, but it represents what is known as a "raised eyebrow" technique to cajole the companies to speed the DTV transition. The underlying message is that if the commission is not pleased with the voluntary actions of the involved industries, the FCC will use its rulemaking authority to enforce the steps outlined in Powell's letter. It is telling that the document was footnoted with the statement, "Nothing contained in this Proposal for Voluntary Industry Action in intended to prejudge any issue in pending or future Commission proceedings" (p. 1). The not-so-subtle subtext is that if telecommunications firms do not work together, the FCC has the authority to create transition regulations that the companies may like even less than these proposals.

FACTORS TO WATCH

The global diffusion of DTV technology will evolve over the first decade of the 21st century. Led by the United States, Japan, and nations of the European Union, digital television will gradually replace analog transmission in technologically-advanced nations. It is reasonable to expect that many of these countries will have converted their cable, satellite, and terrestrial facilities to digital technology by 2010. In the process, it will influence what people watch, especially as it will offer easy access to the Internet and other forms of entertainment that are more interactive than traditional television.

The global development of digital television broadcasting is entering a crucial stage in the coming decade as terrestrial and satellite DTV transmitters are turned on, and consumers get their first look at HDTV and SDTV programs. Among the issues that are likely to emerge in 2002 and 2003:

> *Receiver prices*—Since 1998, about 2.2 million DTV sets have been sold in the United States, and retail prices have declined by more than 40% in three years (CEA, 2002). But how much further do prices of DTV sets have to drop before stimulating consumer demand beyond affluent innovators and early adopters? Recent research shows that the least expensive HDTV set available on the U.S. market (without a decoder) still costs more than $1,200. As the Japanese experiment with Hi-Vision demonstrated, high retail prices can thwart the successful diffusion of a new technology despite software availability and consumer interest (Dupagne & Seel, 1998). After more than 10 years of diffusion, sales of analog HDTV sets in Japan still only numbered 831,000 by January 2000.

> *Station conversion*—While two-thirds of commercials stations were expected to be on the air with a digital signal by May 1, 2001, the exact opposite happened. As of March 25, 2002, nearly two-thirds of these stations requested an extension. The FCC is concerned about this outcome and may levy fines at stations, especially medium-sized stations, if the Commission determines that DTV operational delays are unwarranted (McConnell, 2002). These delays, whether they will translate into months or years, make the 2006 target date for switching off NTSC broadcasting increasingly unlikely.

> *Indoor reception*—Another serious technological issue affecting the DTV transition concerns problems with the indoor reception of DTV signals. In urban areas such as New York City, DTV signals reflect off tall buildings and can create multi-path reception problems with small indoor "rabbit-ear" antennas. A *New York Times* journalist best summarizes his experience with antenna reception of HDTV broadcasts as follows: "If you are the kind of person who would have loved owning a car in 1910, believing that the new worlds a vehicle would open outweighed the need to change a tire every 10 miles or crank the engine by hand, then HDTV is for you" (Taub, 2001, p. G7).

> In October 1999, the Sinclair Broadcast Group filed a petition with the FCC asking that the Commission reconsider its selection of the 8-VSB modulation technology for DTV transmission (McConnell, 1999). Sinclair's tests indicated that the European COFDM modulation scheme was superior to 8-VSB DTV for indoor reception in urban areas. However, the FCC rejected the Sinclair petition in February 2000 after its Office of Engineering and Technology reviewed the evidence and decided that the merits of COFDM did not justify modifying the DTV transmission standard (FCC, 2000). It reaffirmed its support for the 8-VSB modulation system in 2001 (FCC, 2001a). Nevertheless, the Commission remains acutely aware of the poor performance of DTV indoor reception and has urged the industry to improve the reception characteristics of the U.S. DTV system, as well as the reception capability of DTV sets, as quickly as possible. Of course, once it moves beyond experimentation, cable and satellite carriage of HDTV

signals will alleviate over-the-air reception problems and accelerate the diffusion of HDTV technology into American homes.

➢ *Copyright protection*—In July 2001, two major studios, Warner Brothers and Sony, approved a licensing agreement with the so-called 5C digital equipment manufacturers (Hitachi, Intel, Matsushita, Sony, and Toshiba) to protect and encrypt digital transmission of their works against unauthorized copying (Romano, 2001). As of March 2002, however, the other studios have steadfastly refused to sign the agreement, contending that it does not cover protection of broadcast digital programs. They have called upon the FCC and Congress to mandate the inclusion of the broadcast protection component in the 5C specification. This system would embed digital "watermarks" into broadcast digital programs that carry specific copying instructions (Hearn, 2002). When fully endorsed, the 5C copy-protection technology will be incorporated into digital home entertainment appliances, such as television sets and set-top boxes.

In the coming decade, the merger of the television and the computer will create interesting programming options for viewers. Television programming will offer a level of potential viewer interactivity that has been just a pipe dream for the past two decades. DTV telecomputers with high-bandwidth connections to the Internet will have two-way transmission capabilities (Pavlik, 2000). Viewers will have the choice of passively watching what is being telecast, or they may opt to actively participate in the program through a chat room in a program-related website. Pundits have been predicting the arrival of interactive television for over two decades—perhaps it is finally here.

In the United States, the volume of simulcast HDTV programming is increasing each year. This programming will entice affluent consumers to consider paying a premium for new digital television sets to access this HDTV content. Since nearly 70% of U.S. television viewers are cable subscribers, they will start to implore their cable companies to devote additional channel capacity for DTV programming. The speed of the transition to DTV broadcasting in the United States is now in the hands of television manufacturers who must find innovative ways to reduce the cost of digital sets, and cable operators who must balance the need to serve a minority of DTV subscribers with the demands of their NTSC customers seeking additional analog channels. There is a venerable Chinese saying, "may you live in interesting times." This will certainly apply during the next five years of the transition from analog to digital television broadcasting.

BIBLIOGRAPHY

Advisory Committee on Advanced Television Service. (1995). *Advisory Committee final report and recommendation.* Washington, DC: ACATS.

Albiniak, P. (2002, March 15). Major Comcast markets to get HDTV. *Broadcasting & Cable.* Retrieved March 15, 2002 from http://tvinsite.com/broadcastingcable/index.asp?layout=story&articleId=CA201622&pubdate=03/15/2002&stt=001&display=searchResults.

Ashworth, S. (1998, March 23). Finding funds for the transition. *TV Technology, 16* (10), 12.

Bajon, J., & Fontaine, G. (2001). *Development of digital television in the European Union.* Montpellier, France: IDATE.

Balanced Budget Act of 1997, Pub. L. No. 105-33, § 3003, 111 Stat. 251, 265 (1997).

Book, C. L. (2002, April). Consumer response to the PBS model for digital television. Paper presented at the annual meeting of the Broadcast Education Association, Las Vegas, NV.

Brinkley, J. (1997). *Defining vision: The battle for the future of television.* New York: Harcourt Brace & Company.

Brinkley, J. (1998, January 12). They're big. They're expensive. They're the first high-definition TV sets. *New York Times,* C3.

CEA releases final 2001 DTV sales figures. (2002, January 17). Retrieved January 29, 2002 from http://www.ce.org/newsroom/newsloader.asp?newsfile=8744.

Consumer Electronics Association. (2002). *DTV guide.* Arlington, VA: CEA.

Crawford, L., & Harding, J. (2002, February 19). Beleaguered Quiero TV put on the market. *Financial Times,* 27. Retrieved March 23, 2002 from Lexis-Nexis Academic Universe.

Dickson, G. (1999, August 30). HDTV rolling at CBS, ABC. *Broadcasting & Cable, 128,* 10.

Digital terrestrial TV delayed. (2002, March 21). *Asahi Shimbun.* Retrieved March 23, 2002 from Lexis-Nexis Academic Universe.

Dupagne, M. (2002). Adoption of high-definition television in the United States: An Edsel in the making? In C. A. Lin & D. J. Atkin (Eds.). *Communication technology and society: Audience adoption and uses.* Cresskill, NJ: Hampton Press.

Dupagne, M., & Seel, P. B. (1998). *High-definition television: A global perspective.* Ames, IA: Iowa State University Press.

Fedele, J. (1997, September 25). DTV schedule breeds apprehension. *TV Technology, 15,* 16.

Federal Communications Commission. (1987). *Formation of Advisory Committee on Advanced Television Service and announcement of first meeting.* 52 Fed. Reg. 38523.

Federal Communications Commission. (1990). Advanced television systems and their impact on the existing television broadcast service. *First report and order,* 5 FCC Rcd. 5627.

Federal Communications Commission. (1992). Advanced television systems and their impact on the existing television broadcast service. *Second report and order/further notice of proposed rulemaking,* 7 FCC Rcd. 3340.

Federal Communications Commission. (1996a). Advanced television systems and their impact upon the existing television broadcast service. *Fifth further notice of proposed rulemaking,* 11 FCC Rcd. 6235.

Federal Communications Commission. (1996b). Advanced television systems and their impact upon the existing television broadcast service. *Fourth report and order,* 11 FCC Rcd. 17771.

Federal Communications Commission. (1997). Advanced television systems and their impact upon the existing television broadcast service. *Fifth report and order,* 12 FCC Rcd. 12809.

Federal Communications Commission. (1998). Carriage of the transmissions of digital television broadcast stations. *Notice of Proposed Rulemaking,* 13 FCC Rcd. 15092.

Federal Communications Commission. (2000, March 3). *Letter denying petition for expedited rulemaking.* Retrieved March 23, 2002, from http://www.fcc.gov/Bureaus/Engineering_Technology/ News_Releases/2000/nret002a.txt.

Federal Communications Commission. (2001a). Carriage of digital television broadcast signals. *First report and order and further notice of proposed rulemaking,* 16 FCC Rcd. 2598.

Federal Communications Commission. (2001b). Review of the Commission's rules and policies affecting the conversion to digital television. *Report and order and further notice of proposed rulemaking,* 16 FCC Rcd. 5946.

Federal Communications Commission. (2001c). Review of the Commission's rules and policies affecting the conversion to digital television. *Memorandum opinion and order on reconsideration,* 16 FCC Rcd. 20594.

Grotticelli, M. (2001, October 8). Nets increase HDTV output. *Broadcasting & Cable, 130* (35), 37.

Harris, N. (2002, March 23). Football: ITV Digital "will close" rather than pay in full. *The Independent.* Retrieved March 23, 2002 from http://globalarchive.ft.com/globalarchive/article.html?id= 020323001421&query=Football%3A+ITV+Digital.

Hearn, T. (2002, February 12). Hollywood seeks copy-protection aid. *Multichannel News*. Retrieved March 23, 2002 from http://tvinsite.com/multichannelnews/index.asp?layout=story&articleId= CA196762&pubdate=02/12/2002&stt=001&display=searchResults.

Japan's digital numbers hit two million. (2002, January). *Television Asia*, 12. Retrieved March 23, 2002 from Lexis-Nexis Academic Universe.

Johnston, C. (2001, February 7). DTV sets on the floor, but is anyone buying? *TV Technology, 18*, 1, 19.

Kramer, S. D., Brady, S., Neel, K. C., Cole, R., & Figler, A. (2002, February 25). Is it prime-time for HDTV? *Cable World, 14,* 16-25.

Laven, P. (1998). HDTV or not. *EBU Technical Review*, *276*, 2-3.

McConnell, B. (1999, October 11). Sinclair hurls TV gauntlet. *Broadcasting & Cable, 128*, 19-20.

McConnell, B. (2002, March 25). DTV: Put up or pay up. *Broadcasting & Cable*. Retrieved March 25, 2002 from http://tvinsite.com/broadcastingcable/index.asp?layout=print_page&doc_id= 77889&articleID=CA202936.

Nagaya, T. (2001). Good start to digital HDTV broadcasts. *Broadcasting Culture & Research, 16*. Retrieved January 27, 2002 from http://www.nhk.or.jp/bunken/BCRI-news/h16-n1.html.

Nippon Hoso Kyokai. (1998). *NHK factsheet '98.* Tokyo: Nippon Hoso Kyokai.

Pavlik, J. V. (2000). TV on the Internet: Dawn of a new era? *Television Quarterly, 30* (3), 31-47.

Proposal for voluntary industry actions to speed the digital television transition. (2002, April 4). *Letter from Federal Communications Commission Chairman Michael Powell*. Retrieved May 2, 2002 from http://www.fcc.gov/commissioners/powell/mkp_proposal_to_speed_dtv_transition.pdf.

Romano, A. (2001, July 23). Studios' digital deal. *Broadcasting & Cable*, *130*, 45.

Taub, E. A. (2001, February 15). High-definition TV: All or nothing at all. *New York Times*, G7. Retrieved February 19, 2001 from Lexis-Nexis Academic Universe.

Turner Broadcasting System, Inc. v. F.C.C., 117 S.Ct. 1174 (1997).

Wood, D. (2002). Bits "R" Us—New economics and approaches for digital broadcasting. *EBU Technical Review*, *289*. Retrieved March 23, 2002 from http://www.ebu.ch/trev_289-wood.pdf.

8

STREAMING MEDIA

Jeffrey S. Wilkinson, Ph.D. [*]

The technology of "*streaming*" media over the Internet is one of the most exciting multimedia developments at the beginning of the 21st century. Streaming media is the transmission of audio and video from one computer to another (McEvoy, 2001, p. 118). Streaming is heavily dependent on bandwidth, and creating streaming media "always involves tradeoffs between image quality and playback versus fast download times and accessibility" (Schenk, 2001, p. 340). But just as the capabilities of consumer electronics have increased with broadband, the content quality has also improved dramatically. Right now, millions of users are regularly listening or watching content provided by countless Web sites and portals around the world every day.

WHAT STREAMING IS AND IS NOT

Streaming occurs when audio or video content is sent across a network and displayed in real time, either as a self-contained segment or displayed through animation using a program such as Flash. Typically, streaming entails little or no waiting; just the few seconds needed to buffer the file before it is played back on the receiving computer using a suitable player. This action is accomplished through a special streaming server that uses Real-Time Streaming Protocol (RTSP) and Real-Time Transport Protocol (RTP). This constitutes "true streaming" because the content is not downloaded onto a client hard-drive.

A second type—what is called "progressive streaming"—uses a regular HTTP Web server. It is not viewed as true streaming, however, because the file material is downloaded onto the user's hard

---

[*] Associate Professor, Communication Studies Department, Hong Kong Baptist University (Kowloon Tong, Hong Kong, China).

drive before playback. Since video files are so large, this is normally not the first choice for users, but it is the easiest (maybe the only) way around firewalls.

The main advantages of true streaming are speed, control, and flexibility. Streamed material is played back quickly—almost instantaneously. Control is maximized because the original content remains on the server—it is not (usually) downloaded onto a user's hard drive. Also, the provider can control access using a password, registration, or some other security feature. Streaming is also flexible, as streamed segments are placed individually on a host server, and updated or changed as needed.

STREAMING APPLICATIONS

There are two basic types of streaming: "live" and "on-demand." Hundreds of radio stations now offer live audio streaming over the Internet (see, for example, http://broadcast.yahoo.com). In addition, countless organizations are now using live streaming to deliver events and speeches to interested parties.

The most common video streaming application is video on demand. Public, private, commercial, and non-commercial firms have adopted this technology to provide information and entertainment to clients and consumers alike (Seidner, 2001). Accessing streamed material can be extremely simple— click on the link (which launches the player), and watch the program.

THE IMPORTANCE OF BANDWIDTH

Bandwidth determines how much data can pass through a network connection per second, and it is affected by network traffic, connection speed, and the size of the file being streamed. If the stream rate (the rate at which the packets are sent to your computer) exceeds that of your connection, playback experience is degraded by constant stopping or stalling. This happens, for example, if you have a 56 Kb/s (kilobits per second) modem and wish to watch something streamed at 80 Kb/s or higher. The stopping or stalling of an Internet stream can also happen because of bandwidth limitations further upstream; for example, when too many people are requesting streams from a site at the same time, or when the Internet, in general, is experiencing congestion.

BACKGROUND

Streaming technology was introduced in 1995 by Progressive Networks (now called RealNetworks). Their product, RealAudio, was a way for people to listen to high-quality audio signals over the Internet through a computer (using 14.4 Kb/s and 28.8 Kb/s modems at the time). When streaming video was offered shortly thereafter, global interest took off. Now, streaming is a common feature of a number of Web sites.

COMPONENTS NEEDED TO STREAM ON-DEMAND

To stream content, there are some key components that must be considered. First, *content must be produced.* Avoid the cynical "they'll watch anything," and try to have something worthwhile to stream. Compare the interest shown between the low-quality black-and-white film clip of the

Kennedy assassination versus the latest Hollywood big-budget flop. Try to have something interesting before you go to the trouble of recording, editing, and compressing the audio and/or video.

Figure 8.1
Streaming Media Diagram

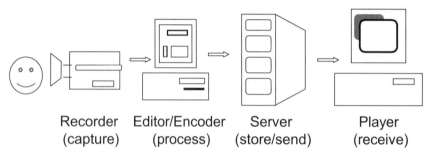

| Recorder | Editor/Encoder | Server | Player |
| (capture) | (process) | (store/send) | (receive) |

Source: J. S. Wilkinson

Second, *the content must be encoded and stored on a server*. All content must be converted to digital form, compressed, and encoded in one of the streaming formats (depending on what players you want people to use). Use the highest-quality source materials for the best streamed images. Musical performances may be encoded to favor the audio; a "talking head" might need maximum video quality (a talking head is bad enough—but a *jerky/blurry* talking head?). Know the purpose of the message you're streaming as well as who your audience is and what type of connection they have.

The third component to consider is the *Internet connection itself.* High-quality streamed images demand high bandwidth. The user also needs enough bandwidth to view the streamed content. The higher the bandwidth, the better the detail and the motion that is viewed by the user. For example, streams of 300+ Kb/s can handle many action scenes, and 80 Kb/s to 150 Kb/s work fine for things such as keynote speeches. But dial-up modems over telephone lines (56 Kb/s modems) can only handle bit rates of 45 Kb/s and less (28.8 modems typically handle 20 Kb/s). At these lower rates, even a talking head can look like a sloppy slideshow.

The fourth component needed is a compatible player. All the dominant streaming formats provide free players, as well as low-cost, higher-end versions. Free players are now bundled with computer operating systems, and to upgrade or add another player, a user can simply go to the host site on the Web and follow the steps for installation.

COMPONENTS FOR LIVE STREAMING:

To do live streaming, you need to consider all of the above, but with the addition of a dedicated computer (called a "broadcaster") to create the streams from the live inputs (such as a microphone or camera) by digitizing, compressing, and sending out the content as an RTP stream.

STREAMING PLATFORMS

To play streamed material, the consumer has to install a compatible player on his or her computer. Several types of proprietary systems have come and gone, and many new ones continue to be launched. By far, the three most common platforms for providing and playing streamed content are Real (RealNetworks), Windows Media Player (Microsoft), and QuickTime (Apple). Each has some technical differences (making it proprietary), each has its merits, and all three are regularly upgraded.

For example, both Real and Windows formats offer a feature that enables the players to shift the streaming rate up or down according to the best available bandwidth. Real's SureStream and Windows Intelligent Streaming work by encoding content at multiple stream rates. This allows the user to enjoy the content without delays or stopping, even though the quality of the picture may vary throughout.

Other important formats are MP3, MPEG-1, and MPEG-2. MP3 has become extremely popular for distributing audio files over the Internet. MP3 (which is "MPEG Audio Layer 3" and not "MPEG-3") is related to MPEG, and is a highly compressed format, which means the files are extremely small—typically 128 Kb/s for two-channel stereo (Halsall, 2001, p. 965). Nevertheless, the quality is extremely high and the sound of MP3 files may be indistinguishable from uncompressed CD-audio (Terran Interactive, 1995-2000, p. 211). A wide variety of players can handle MP3 files. (For more on MP3, see Chapter 17.)

MPEG is a standards-based format specifically designed to handle digital audio and video. MPEG-1 produces high-quality audio and video streams that can be played on a number of players, including QuickTime and Windows Media Player. MPEG-2 is broadcast-quality video that, as of this writing, is still too demanding and unstable for popular streaming applications.

RealNetworks/RealOne. The early dominance of RealNetworks has been attributed to it being the first on the scene, and having the foresight to give away the players rather than sell them. Since 1995, several generations of improvements have been made in its products. The overall pricing structure for Real continues to be stable. A basic player is a free download, and new models (with tech support) are offered for a price. As of March 1, 2002, the RealOne player was listed on the Web site as the free player for users to download.

Real offers a wide variety of servers and other products, as streaming applications/needs expand. Real continues to offer a free "starter" streaming server that allows up to 25 concurrent Internet viewers. Upgrading to the 60-stream server (Internet and intranet) with technical support lists for $1,995.

Two categories of servers were offered in March 2002, depending on whether your need is Internet or intranet. Intranet streaming servers (200 to 500 concurrent viewers) started at $3,995, and Internet streaming servers (100 to 2,000) started at $5,995. Although prices are always "subject to change," a 2,000 stream server was priced at $79,995 just two years earlier in March 2000.

Microsoft/Windows Media Player. Shortly after RealNetworks began streaming audio, Microsoft became interested in the technology and, for a time in the late 1990s, the two were partners. After the partnership dissolved, Windows Media Player was offered as a free plug-in to users. Since then, improved versions of Windows Media Player have been regularly released. Windows Media Player 7 was very highly regarded, and helped the platform gain an upper hand by virtue of being a good player that was self-compatible (operating system, browser, media player, various tools).

Because streaming is just one of the many things Microsoft does, it is easy to get lost in the vast array of products and services. The cost of the server and the streaming capabilities can be buried (offered for free), because it is just one part of the larger whole. As you buy into the full range of services Microsoft provides, costs begin to accumulate. But once you've purchased a package, you can efficiently control content and Web transactions, including pay-per-view (e.g., pay-per-download or pay-per-stream), registration, subscription, and digital rights management.

Apple/QuickTime. In 1999, Apple began offering streaming with the QuickTime 4 platform. QuickTime is often preferred by many engineers because it plays on both PC and Macintosh computers and can be delivered from any kind of file or Web server such as UNIX, Solaris, Linux, Mac OS, or Windows NT. The QuickTime file format is an open standard and was adopted as part of the MPEG-4 specification (Apple, 2002, p. 6).

Apple has an open-source code approach where they allow vendors to configure the QuickTime format for their use, then send the changes back to Apple for incorporation into new versions of the server software. QuickTime has worked out many bugs in earlier versions and has proven itself to be very stable with high-quality playback.

RECENT DEVELOPMENTS

TECHNICAL ADVANCEMENTS

A number of groups have endorsed MPEG-4 as the technical standard for streaming media, and QuickTime, Real, and Windows Media all now offer at least some degree of MPEG-4 compatibility.

In addition, streaming is moving into the domain of wireless access (Alvear, 2002). Real announced a mobile wireless streaming system, and PacketVideo has begun testing a system in France, China, Korea, and Switzerland. In addition, Nextel announced a deal with XSVoice to allow live wireless audio. Billed as "better than streaming," Nextel's service brings content immediately to consumers through their wireless phones.

Perhaps the most widely reported technological advancement comes from companies such as Akamai that use *edge-caching* to enable companies to deliver Web content at a substantially reduced cost. The Akamai platform works by placing the applications close to the end-user, thus avoiding congestion points on the Internet. This, in turn, offloads significant amounts of traffic from origin Web sites, allowing the clients to reduce their costs in terms of Web and application servers, reverse proxy caches, switches, and load balancers. For example, Akamai Public Relations Manager Caryn Brownell says edge-caching enabled one client to reduce their number of servers by 60. In addition, Akamai was able to save Tower Records over $2 million, and Victoria's Secret $1 million in infrastructure costs in 2001 (personal communication, April 19, 2002).

APPLICATIONS—ENTERTAINMENT, CORPORATE COMMUNICATION, EDUCATION

The Internet is changing, reflecting changes in the way people use it. According to a survey by the Pew Center (2002), the novelty is wearing off and Internet use is becoming more stable. Users report slight increases in using the Net to listen to music and watch video clips and, at the same time, report slightly lower viewing of regular television. It is no surprise, then, that this is precisely the type of content that's been pushed en masse to consumers. In December 2001, both Real and Listen.com unveiled fee-based music services. RealOne announced a charge of $9.95 a month for its service, and Listen.com offered three packages for either $5.95 or $7.95 a month. A month later, Liquid Audio offered its service and the option to use either a free player or for $19.95, Player Plus 6.1, that allowed consumers to "stream, download, purchase, playback, and record digital music as well as burn CDs" (Pruitt, 2002b).

All told, some industry forecasts predict that these subscriber-music services will pull in between $1 billion and $1.6 billion dollars by 2005 (Pruitt, 2001; Pruitt, 2002a). However, it remains to be seen if consumers will be willing to pay for a service that they have received for free through peer-to-peer networks. The combination of broadband and video has also resulted in companies launching VOD services. For example, Intertainer has blended cable and broadband in the top 35 U.S. markets to offer content from the major studios for a basic $7.99 monthly fee (with additional charges for pay-per-view). (For more on Intertainer, see Chapter 19.)

Corporations have also adopted streaming as a more efficient way to train staff and announce new services or product launches. One survey reported that 80% of the Fortune 200 companies were now streaming content—primarily for training, corporate communication, advertising/marketing, and entertainment (Seidner, 2001). Overall, the median investment was $400,000 annually, and "stream hour per person" costs were judged to be efficient at roughly $20. Another report cited by an industry executive suggested that corporate spending on streaming in 2000 was $140 million and will leap to $2.8 billion by 2005 (Virata, 2002). In addition it was predicted that "product launches and marketing applications will take up the biggest slice of the streaming budget pie by 2005, accounting for $567 million" (Virata, 2002).

New ways of delivering educational materials are also being adapted to streaming. While a number of professors have experimented with streaming lecture materials for individual classes (see for example http://www.hkbu.edu.hk/~wilkinso), more formal structures have also been launched to make streaming an integral part of modern higher education (see for example, University of Cincinnati's Streaming Media project at http://stremedia.uc.edu). Many online course development software packages, such as WebCT, include streaming capabilities.

CURRENT STATUS

As of April 2002, the latest versions of the big three are RealOne, Windows Media Player 8, and Apple QuickTime 6. Since free versions of all three platforms are available, it is probably best just to make sure all three are installed. You can select one to be preferred, but that seems primarily an individual choice. Paid players include the fully supported RealOne and Liquid Audio's 6.1.

In terms of platform popularity, both Real and Microsoft report 240 million to 250 million users, and Apple QuickTime has over 100 million (RealNetworks.com, 2002; Apple, 2002). Real also announced it had 500,000 active paying subscribers for RealOne and GoldPass media subscription service (RealNetworks.com, 2002).

Figure 8.2
Streaming Media Screen Shot

This screen shot includes the three most popular players: QuickTime 4.0, RealOne, and Windows Media Player (left to right). In addition to the standard look of each player, a user can modify the appearance of most players by applying a "skin" that changes the layout of the player, graphics, colors, etc. Technology does not yet, however, allow us to control the appearance of the people in the videos.

Source: J. S. Wilkinson

All three player platforms are solid and well-used. Many popular sites for streaming content allow users either Real or Windows Media Player. QuickTime is also extremely popular and countless sites stream it as well.

Measures of the most popular streaming content are more difficult to come by. As of March 2002, there was no single, authoritative measure of the most popular audio or video streams comparable to the Nielsen ratings for television. Both Arbitron and Measurecast provide estimates of the audience for Internet radio services, but no comparable measure exists for video streams. The existing Internet audience ratings systems have another limitation: these measures are limited to companies requesting measurement, so they do not include closed-circuit Webcasts and similar content.

Arbitron has, however, provided measures of average use among U.S. consumers. As of January 2002, they estimate that 40 million Americans (17% of the population) listened or watched online streams in the previous month, with 30% of those having broadband Internet access at home and more than half being 34 or younger (Arbitron, 2002).

FACTORS TO WATCH

DIGITAL RIGHTS MANAGEMENT (DRM).

The year 2002 will be a year of upheaval for radio stations that stream over the Internet. In February, a panel appointed by the U.S. Copyright Office announced proposed licensing rates for Webcasters. The Copyright Arbitration Royalty Panel (CARP) stated that basic royalty rates should be 0.07 cents per performance for streaming radio stations (0.02 for non-commercial broadcasters) and 0.14 cents for other Internet transmissions. Here is another way to view these rates:

One listener of one song/performance	=	.07 cents
100 listeners of one song/performance	=	7 cents
1,000 listeners of one song/performance	=	70 cents
10,000 listeners of one song/performance	=	7 dollars

As Web listenership increases and becomes more like broadcast radio listening, the royalty rates can become a sizable line item expense. For example, a broadcaster with only 1,000 regular Web listeners (five days per week, one hour per day, 12 songs per hour) faces the following additional yearly expense:

(Listeners)	×	(days/year)	×	(avg # songs/hour)	×	(royalty rate)	
1,000	×	260	×	12	×	.07	= $2,184.00

Internet radio stations without a broadcast license issued by the Federal Communications Commission (FCC) would pay twice this amount. In addition, licensing fees would be either $500 or 9% or 10% of gross revenues (depending on the business). Many Webcasters were unhappy, and charged that this would force many to quit and leave music streaming to the music label companies (Alvear, 2002). The Recording Industry Association of America (RIAA) viewed the rates as appropriate and a way for artists to get paid. A final decision by the Librarian of Congress (to accept or reject the report) must be made no later than May 21, 2002 (see the CTU Web site at www.tfi.com/ctu for an update).

PRIVACY

In order to thrive (or at least survive), streamed services—even from broadcasters—may accelerate a shift toward charging for content. As this occurs, the struggle between personal privacy and copyrights will escalate. When it was released, Windows Media Player 8 immediately came under fire by privacy watchers because it could keep detailed logs of DVDs played on the user's PC, then share that information with Microsoft (Evers, 2002). Microsoft says the technology is for maintaining security and ensuring that copyright regulations are being adhered to. This reason will undoubtedly be the rallying cry for all businesses streaming content. But as more people pay for services such as streamed entertainment, the more willingly we accept that "Windows Media Administrator can also create *detailed logs* that include data such as individual client connection information" (emphasis added, Birney, 2000).

FREE OR FEE?

The Web may have reached a point where a users' ability to roam the Net freely for content is coming to an end. For years, Microsoft (and others) have advocated a pay-per-click model for the Internet, and streaming seems to be the application to make it happen. Sporting events have a long history of pay-per-view, and pay-per-view on Internet is just waiting to happen (Finke, 2001). As American radio stations strive to cover performer royalties, the U.S. media will join other countries such as Norway in shifting to a pay-for-content model (CNN.com, 2002). There remains widespread consumer resistance (Pew Internet, 2001), but the battle for streamed content seems already to have been decided, and the biggest issue will now become "at what price."

BIBLIOGRAPHY

Alvear, J. (2002, February 21). Copyright panel announces rates for Internet radio. *Streaming Media Newsletter*. Retrieved February 26, 2002 from http://www.streamingmedia.com/newsletter/072601e.html.

Apple. (2002). *QuickTime for the Web: For Windows and Macintosh*. San Francisco: Morgan Kaufmann.

Arbitron. (2002). *The current state of streaming: January 2002*. Retrieved May 8, 2002 from http://www.arbitron.com/webcast_ratings/streaming_media_audiences.htm.

Birney, B. (2000, October). *Streaming from a Web server*. Retrieved March 1, 2002 from http://msdn.microsoft.com/library/default.asp?url=/library/en-us/dnwmt/html/webserver.asp.

CNN.com. (2002, March 2). *End of the line for free Web?* Retrieved March 3, 2002 from http://www.cnn.com/2002/WORLD/europe/03/01/norway.net/index.html.

Evers, J. (2002 February 21). Microsoft media player logging users' DVD picks. *Computerworld*. Retrieved February 22, 2002 from http://www.computerworld.com/storyba.

Finke, M. (2001). An architecture of a personalized, dynamic interactive video system. In R. Earnshaw & J. Vince (Eds.). *Digital content creation*. London: Springer-Verlag.

Halsall, F. (2001). *Multimedia communications: Applications, networks, protocols and standards*. Harlow, England: Addison-Wesley.

McEvoy, S. (2001). *Microsoft Windows Media Player for Windows XP handbook*. Redmond WA: Microsoft Press.

Pew Internet & American Life Project. (2001, November 14). *The dot-com meltdown and the Web*. Retrieved November 20, 2001 from http://www.pewinternet.org/releases/index.asp.

Pew Internet & American Life Project. (2002, March 3). *Getting serious online: As Americans gain experience, they use the Web more at work, write e-mails with more significant content, perform more online transactions, and pursue more serious activities*. Retrieved March 3, 2002 from http://www.pewinternet.org/releases/index.asp.

Pruitt, S. (2001, November 19). IDC: Music service market to reach $1.6B by 2005. *IDG News Service*. Retrieved March 3, 2002 from http://www.idg.net/idgns/2001/11/19/IDCMusicServiceMarketToReach.shtml.

Pruitt, S. (2002a, January 15). Jupiter: Online music hits $5.5B note by 2006. *IDG News Service*. Retrieved March 3, 2002 from http://www.idg.net/idgns/2002/01/15/JupiterOnlineMusicHits55BNote.shtml.

Pruitt, S. (2002b, January 9). Liquid audio spins out new music player line. *IDG News Service*. Retrieved March 3, 2002 from http://www.idg.net/idgns/2002/01/09/LiquidAudioSpinsOutNewMusic.shtml.

RealNetworks.com. (2002, January 29). *RealNetworks surpasses half-million media subscription milestone.* Retrieved March 1, 2002 from http://www.realnetworks.com/company/press/releases/2002/playernumbers250.html.

Schenk, S. (2001). *Digital nonlinear desktop editing.* Hingham, MA: Charles River Media.

Seidner, R. (2001, June 20). *Enterprise streaming: Where ROI means survival.* Keynote speaker at Streaming Media West 2001. Retrieved February 22, 2002 fromhttp://www.streamingmedia.com/video/keynote_west_1.asp.

Terran Interactive. (1995-2000). *Cleaner 5: The camera-to-Web streaming solution* (User manual). Terran Interactive.

Virata, J. (2002, February 26). Streaming the future: An interview with IVT's Mark Lieberman. *Corporate Media News*. Retrieved February 26, 2002 from http://corporatemedianews.com/cgi-bin/getframeletter.cgi.

9

RADIO BROADCASTING

Greg Pitts, Ph.D.[*]

> Consumer demand for improved audio fidelity is undeniable. Access to superior digital audio technologies, such as compact discs and—in the near future—satellite DARS, and the perceived benefits of digitization generally, fuel such demand. We believe that an important benefit of DAB will be enhanced sound quality. DAB technology should permit significant improvements in audio fidelity and robustness over current analog service (FCC, 1999, p. 10).

> The NRSC therefore recommends that the iBiquity FM IBOC [DAB] system ... should be authorized by the FCC as an enhancement to FM broadcasting in the United States, charting the course for an efficient transition to digital broadcasting with minimal impact on existing FM reception and no new spectrum requirements" (NRSC, 2001, p. 9).

The word "radio" fails to deliver the image of sexy technology capable of competing with the latest digital communication technologies. Radio is, however, a big part of the lives of most people. Each week, the more than 13,000 radio stations in the United States are heard by more than 227 million persons, and advertisers spent $18.36 *billion* on radio advertising in 2001, even with a soft economy (FCC, 2001; RAB, 2001; and RAB, 2002). Changes in radio station ownership, digital technology, and marketplace forces are slowly changing the stodgy world of analog radio transmission into a digital radio future. Whether consumers will respond as willingly to the new product offerings will determine the success of new radio technology.

---

[*] Assistant Professor, Department of Communication, Bradley University (Peoria, Illinois).

This chapter examines the factors that have redirected the technological path of radio broadcasting. Among these are radio station ownership consolidation, improved computer automation of stations, delivery of programming through digital networks, competition from two digital satellite radio services, and, most important, technological improvements that will result in a viable digital terrestrial broadcast system.

BACKGROUND

The history of radio is rooted in the earliest wired communications—the telegraph and the telephone—although no single person can be credited with inventing radio. Most of radio's "inventors" refined an idea put forth by someone else (Lewis, 1991). Although the technology may seem mundane today, until radio was invented, it was impossible to simultaneously transmit entertainment or information to millions of people. The radio experimenters of 1900 or 1910 were as enthused about their technology as are the employees of the latest Internet-related start-up. Today, the Internet allows us to travel around the world without leaving our seats. For the listener in the 1920s, 1930s, or 1940s, radio was the only way to hear live reports from around the world.

Probably the most widely known radio inventor-innovator was Italian Guglielmo Marconi, who recognized its commercial value and improved the operation of early wireless equipment. The one person who made the most lasting contributions to radio and electronics technology was Edwin Howard Armstrong. Armstrong discovered regeneration, the principle behind signal amplification. He invented the superheterodyne tuner, leading to a high-performance receiver that could be sold at a moderate price and thus increase home penetration of radios. In 1933, Armstrong was awarded five patents for frequency modulation (FM) technology (Albarran & Pitts, 2000).

The two traditional radio transmission technologies are amplitude modulation (AM) and frequency modulation. AM varies (modulates) *signal* strength (amplitude) and FM varies the *frequency* of the signal.

The oldest commercial radio station began broadcasting in AM in 1920. Though AM technology had the advantage of being able to broadcast over a wide coverage area (an important factor when the number of licensed stations numbered just a few dozen), the AM signal was of low fidelity and subject to electrical interference. FM, which provides superior sound, is of limited range. Commercial FM took nearly 20 years from the first Amstrong patents in the 1930s to begin significant service and did not reach listener parity with AM until 1978 when FM listenership finally exceeded AM listenership.

FM radio's technological add-on of stereo broadcasting, authorized by the Federal Communications Commission (FCC) in 1961, along with an end to program simulcasting (airing the same program on both an AM station and an FM station) in 1964, expanded FM listenership (Sterling & Kittross, 1990). Other attempts, such as Quad-FM (quadraphonic sound), ended with disappointing results. AM stereo, touted in the early 1980s as the savior in AM's competitive battle with FM, languished for lack of a technical standard resulting from the inability of the marketplace and government to quickly adopt an AM stereo system (FCC, n.d.-a; Huff, 1992).

Table 9.1

Radio in the United States at a Glance

Households with Radios	99%
Average Number of Radios per Household	5.6

Source: Statistical Abstract of the United States (2001)

Radio Station Totals

AM Stations	4,727
FM Commercial Stations	6,051
FM Educational Stations	2,234
Total	13,012
FM Translators and Boosters	3,600

Source: FCC (2001)

Radio Audiences

Persons Age 12 and Older Reached by Radio:

Each week:	96% (About 227 million people)
Each day:	78% (About 184 million people)

Persons Age 12 and Older Time Spent Listening to the Radio:

Each week:	20:30 hours
Each weekday:	3:03 hours
Each weekend:	5:15 hours

Where Persons Age 12 and Older Listen to the Radio:

At home:	36.6% of their listening time
In car:	43% of their listening time
At work or other places:	20.4% of their listening time

Daily Share of Time Spent With Various Media:

Radio	44%
TV/Cable	41%
Newspapers	10%
Magazines	5%

Source: Radio Advertising Bureau (2001)

Why have technological improvements in radio been slow in coming? One obvious answer is that the marketplace did not want the improvements. Station owners were unwilling to invest in changes; instead, they shifted music programming from the AM band to the FM band. AM became the home of low-cost talk programming. Listeners were satisfied with the commercially supported and noncommercial radio programming offered by AM and FM stations. Consumers wanting something new bought tape players and then CD players to provide improved audio quality and music choice. Government regulators, primarily the FCC, were unable to devise and institute new radio technology approaches. The consumer electronics industry focused on other technological opportunities, including video recording and computer technology.

THE CHANGING RADIO MARKETPLACE

The FCC elimination of ownership caps mandated by the Telecommunications Act of 1996 is fueling many of the changes taking place in radio broadcasting. Before the ownership limits were eliminated, there were few incentives for broadcasters, equipment manufacturers, or consumer electronics manufacturers to upgrade the technology. Analog radio, within the technical limits of a system developed more than 80 years ago, worked just fine. Station groups did not have the market force to push technological initiatives. (At one time, station groups were limited to seven stations of each service. Later, it was increased to 18 stations of each service, before further deregulation completely removed ownership limits.) The fractured station ownership system ensured station owner opposition to FCC initiatives and manufacturer efforts to pursue technological innovation.

Ownership consolidation, along with station automation and networking, reflect new management and operational philosophies that have enabled radio owners to establish station groups consisting of 100 or more stations. The behemoth of the radio industry is Clear Channel Communications. As of mid-2002, the San Antonio, Texas-based company owned more than 1,200 radio stations and was the highest billing radio group in the United States, with revenues of nearly $3.8 billion in 2000, claiming to reach 120 million radio listeners each week (America's highest, 2001; Mathews, 2002; Clear Channel, n.d.). In terms of station numbers, Clear Channel's nearest competitor is Cumulus Broadcasting, with about 230 stations, mostly in small or medium markets. In terms of revenue, Clear Channel's nearest competitor is Infinity Broadcasting, owned by Viacom, which owns approximately 182 radio stations in some of the largest markets in the United States. Clear Channel's size and Infinity's big-city locations have given both companies the incentive to push for technological improvements and the muscle to see that the ideas move beyond discussion exercises.

RECENT DEVELOPMENTS

There are three areas where technology is affecting radio broadcasting:

(1)   Enhancements to improve the present-day, on-air transmission.

(2)   Supplements to provide new services within the current AM and FM radio systems.

(3)   New transmission delivery modes that are incompatible with existing AM and FM radio.

ENHANCEMENTS—STATIONS INSTALL DIGITAL AUDIO EQUIPMENT

Since the introduction of compact discs and digital recording, most radio stations have upgraded their on-air and production capabilities to meet the new digital standard. Virtually all portions of the audio chain are digital or are capable of handling a digital signal. Music and commercial playback is digital through compact disc, computer hard drive systems, minidisc, or other digital media. Most new audio consoles are digitally capable. The program signal that is delivered to the station's transmitter is digitally processed and travels to a digital exciter in the station's transmitter where the audio is added to the carrier wave. The only part of the process that is still always analog is the final transmission of the over-the-air signal by AM or FM.

Computer systems have also allowed stations to create "walk-away" operational technology, where live announcers and engineers are not needed at the station facility. Announcer comments, along with music and commercials, are stored on a computer hard drive system and played back. Some radio owners, including Clear Channel Communications, have even used the computer system and computer networks to create virtual radio stations where announcers are neither local nor live (Mathews, 2002). The system, called voice-tracking, lets an announcer prepare an on-air show for listeners in another city. The announcer's comments, complete with locale-specific commentary, are sent through a network connection to the receiving station's on-air computer.

Listeners to voice-tracked stations seldom know whether or not their local radio personalities are broadcasting live from the city. Automation technology has been used in radio for more than 25 years, but the newest technology perfectly creates the impression of live broadcasts. Questions surrounding the ethics of creating the live impression through recording technology are sometimes raised. Companies using the systems often describe radio as an "entertainment experience," although not necessarily a "live" experience.

SUPPLEMENTS TO EXISTING SERVICES—RADIO DATA SYSTEM

FM radio broadcasters have, for many years, utilized part of their spectrum to deliver auxiliary services. In the past, subcarrier frequencies (or subsidiary communication authorizations, SCAs) were used to deliver commercial-free background music marketed to businesses, reading services for the visually impaired, or paging. The latest subcarrier use, RBDS/RDS (radio broadcast data system/radio data system) permits specially-equipped radio receivers to display call letters and program information such as station promotional material, music information, or advertising content, and it permits listeners to search for programming based on format (NAB, n.d.-a).

The system was introduced in the United States in 1993. On April 9, 1998, the National Radio Systems Committee (NRSC), a jointly sponsored committee of the National Association of Broadcasters (NAB) and the Consumer Electronics Manufacturers Association (CEMA), approved a revised version of the U.S. Radio Broadcast Data System Standard. The system used by these "smart radios" is a compromise between the previous U.S. system and the European system. The system carries information to FM receivers on a 57 kHz subcarrier utilizing a low bit rate data stream (NAB, n.d.-a).

A 1999 Consumer Electronics Association survey of 603 U.S. radio stations concluded that as many as 20% of all FM stations had the technological capability to broadcast in RDS, but stations cited "a lack of consumer demand" as the primary reason for not transmitting an RDS signal (CEMA, 1999). Generally, the RBDS system has languished since its creation. Neither consumers nor radio station owners have shown much interest. Stations using the system are often only transmitting their call letters. However, efforts to create a digital audio broadcast (DAB) system are expected to make broad use of the RBDS concept.

An NRSC initiative to create a high-speed FM subcarrier (HSSC) was suspended in 1998 after an NRSC subcommittee reached an impasse on establishing a voluntary FM data subcarrier standard (NAB, n.d.-b). Recognizing that the RDS system would not sufficiently handle future data transmission needs, the NRSC began evaluating three competing systems. Test results suggested that a single system was not feasible given the variety of anticipated data transmission uses. Ultimately, the International Telecommunications Union (ITU) approved three systems:

> ➢ The Data Radio Channel (DARC) technology developed by NHK.

> ➢ The Subcarrier Traffic Information Channel (STIC) system developed by MITRE Corporation.

> ➢ The High Speed Data System (HSDS) developed by Seiko Communications Systems.

One factor cited in the NRSC report that inhibited creating a common standard was the consolidation of the radio industry. Broadcasters, especially engineers, were so committed to meeting the engineering needs brought about by consolidation that little attention was available for consideration of the HSSC standard (NAB, n.d.-b).

NEW TRANSMISSION MODES—DIGITAL RADIO

The single biggest change in the free, over-the-air radio broadcasting system in the United States is the anticipated adoption of terrestrial digital audio broadcasting, which is the delivery of today's radio signals as streams of digital information. The digital signal would eliminate many of the external environmental factors that often degrade a conventional AM or FM station's signal (Morgan, 2002). In December 2001, the NRSC submitted an evaluation report to the FCC recommending adoption of a digital system developed by iBiquity Digital (NRSC, 2001). The FCC is expected to approve this digital system during 2002 for use by existing FM stations.

As illustrated in Figure 9.1, this DAB initiative will establish a hybrid in-band, on-channel (IBOC) system that will allow simultaneous broadcast of analog and digital signals by existing FM stations through the use of compression technology without disrupting the existing analog coverage. (A similar system for AM stations is forthcoming, although that system was not as far along in its evaluation as was the FM system when this chapter was being prepared.) The desired IBOC system for FM has been described as being capable of delivering CD-quality audio. The AM IBOC would provide FM stereo quality signals from AM stations.

DAB would also allow wireless data transmission similar to the RBDS technology, including traffic and weather information, programming and promotional material from the station, and, at a future time, the delivery of data to smart telephones or personal digital assistants (PDAs) (iBiquity, n.d.-b). Data streaming or data casting might not only be an in-demand service for consumers, but it could become a significant second revenue stream for radio broadcasters.

iBiquity Digital's IBOC technology consists of audio compression technology called Perceptual Audio Coder (PAC™) that allows the analog and digital content to be combined on existing radio bands and digital broadcast technology that allows transmission of music and text while reducing the noise and static associated with current reception.

One primary impediment to creating a successful system was lifted in August 2000 when the two leading IBOC companies, USA Digital Radio (USADR) and Lucent Digital Radio (LDR), merged to become iBiquity Digital Corporation (iBiquity, n.d.-a). The merger eliminated one of the most compelling issues facing the two IBOC companies, broadcast owners, and regulators—whether the two competitors could have worked together to develop a common system.

Figure 9.1
Hybrid and All-Digital AM & FM IBOC Modes

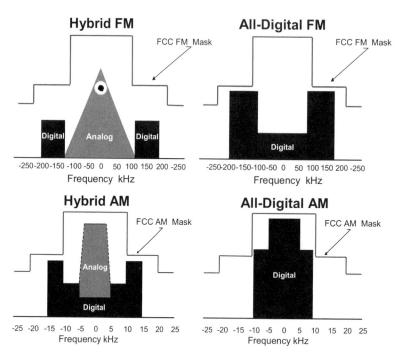

Source: iBiquity

The FCC issued a Notice of Proposed Rulemaking on November 1, 1999 to begin consideration of DAB technology approaches, although the discussion of DAB began in the early 1990s. The technological advances achieved by the merger between USA Digital and Lucent Digital also cannot be overlooked. As recently as mid-2000, experts indicated that IBOC technology looked promising but had not been shown conclusively to work (Rathbun, 2000). The IBOC achievement in the intervening two years involved not only substantial technological progress by iBiquity but also rigorous testing by the NRSC, which recommended the system to the FCC. iBiquity's success reflects the changing radio industry; Clear Channel, Viacom/Infinity Radio, Disney/ABC, Susquehanna Radio, Cox Radio, and Hispanic Broadcasting, some of the largest radio groups, are investors in the company and support the conversion.

Broadcasters are optimistic the system will be approved by the FCC because it is the only terrestrial DAB system under consideration for the United States. After FCC approval, the system would debut in Chicago, Los Angeles, Miami, New York, San Francisco, and Seattle. The first receivers will be presented at the Consumer Electronics Show in January 2003 and would add about $100 to the price of an automotive audio system (Coffey, 2002; iBiquity, n.d.-b).

A different form of DAB service is already in operation in a number of different countries outside the United States. The Eureka 147 system broadcasts digital signals on the L-band (1452-1492 MHz) or a part of the spectrum known as Band III (around 221 MHz) and is in operation or experimental testing in Canada, the United Kingdom, Sweden, Germany, France, and about 50 other countries. Because of differences in the Eureka system's technological approach, it is not designed to work with existing AM and FM frequencies. The World DAB Forum (www.worlddab.org), an international, non-government organization to promote the Eureka 147 DAB system, reports that more than 284 million people can receive DAB signals and that there are more than 400 different DAB services available for listeners (World DAB, n.d.-b). The Eureka numbers seem impressive until put into perspective: 284 million people can *potentially* receive the signal but only if they have purchased one of the required receivers. Eureka 147 receivers have been on the market since the summer of 1998 and should not be viewed as "selling well"; 22 models of commercial receivers are currently available and range in price from around $500 to several thousand (World DAB, n.d.-a).

Germany has been identified as a key country for the success of DAB. The German market contains more than 80 million people, 38 million households, and 42 million cars. Canada is another important DAB market; 57 stations broadcast digital signals in Toronto, Vancouver, Montreal, and Windsor, providing service to 10 million people or 35% of the country's population. General Motors will become the first automaker to provide factory-installed DAB receivers in its cars beginning with the 2003 model year (Bray, n.d.). Eureka DAB in Canada will need a strong lead if it is to succeed against the U.S. IBOC system. Signal spillover from the United States, a flood of receivers from the U.S. marketplace, and the sheer size of the U.S. market could doom Eureka DAB in Canada.

The FCC has emphasized localism as a "touchstone value" of U.S. terrestrial radio service, giving it as one of the reasons for rejecting the Eureka system. The in-band DAB service (IBOC) will enable radio listeners to adopt the technology gradually while continuing to receive radio service through locally programmed stations (FCC, 1999). The saturation of the U.S. radio market would have made it difficult for new local or regional out-of-band services to develop viable programming while trying to establish a critical mass of receivers and listeners.

Digital audio broadcasting, sometimes called digital audio radio (DAR), will require listeners to purchase a new radio capable of receiving the digital programming. There are questions as to whether consumers will want the new service, given the potential expense of new receivers and the possible introduction of other technology including subscriber-based satellite delivered audio services, MP3 audio transfer and playback, and competition from digital television. The availability of reasonably priced, factory-installed DAB automobile receivers will determine the success of the iBiquity system.

OLD TECHNOLOGY, BUT LITTLE COMPETITION—LOW-POWER FM

The FCC approved the creation of a controversial new classification of noncommercial FM stations on January 20, 2000 (Chen, 2000). LPFM, or low-power FM, service will limit stations to a power level of either 100 watts or 10 watts (FCC, 2000). A little more than a year after approving the service, and before any stations were licensed, the commission acquiesced to Congressional pressure on behalf of the broadcast industry and revised the LPFM order. Commercial broadcasters had feared interference with existing stations, added competition in an increasingly competitive radio market, and possible interference with future DAB signals. To prevent encroachment on existing stations' signals, Congress slipped the Radio Broadcasting Preservation Act of 2000 into a broad spending bill which was reluctantly signed by President Clinton (McConnell, 2001; Stavitsky, et al., 2001).

The Congressionally mandated revision required LPFM stations to provide third adjacent channel separation/protection for existing stations. Practically speaking, this would mean that a currently licensed station, operating on 95.5 MHz would not have a LPFM competitor on a frequency any closer than 94.7 MHz or 96.3 MHz. This revision, though minor in appearance, either killed or severely reduced the opportunity for service in most major metropolitan areas (McConnell, 2001). The proliferation of existing FM stations suggests that finding a vacant frequency beyond a station's third adjacent channel is remote except in small communities. One estimate of the drop in service ranged from 75% to 80% (Orange, 2001). Many of the applications on file are competing for the same frequency in a specific community, and it has been estimated that fewer than 80% of the applications for service will ever be established (Labaton, 2000; Stavitsky, et al., 2001). Five "application windows" to apply for LPFM service have been opened and closed by the FCC. As of March 2002, eight stations had been licensed by the FCC but an additional 239 constructions permits have been granted and more than 3,100 applications are scheduled for FCC review (FCC, n.d.-b). Applications for service have been filed by community groups, ranging from church groups to community organizations to schools.

NEW DELIVERY COMPETITION—SATELLITE DIGITAL AUDIO RADIO (SDAR) SERVICE

Subscriber-based satellite radio service, a form of out-of-band digital "radio," launched service in the United States during 2001 and 2002. The service was authorized by the FCC in 1995 and, strictly speaking, is not a radio service. Though reception is over-the-air and electromagnetic spectrum is utilized, the service is national instead of local, it requires users to pay a monthly subscription fee of between $10 and $13, and it requires a proprietary receiver to decode the transmissions. Two publicly traded companies, XM Satellite Radio and Sirius Satellite Radio, offer about 100 channels of music and talk programming, with Sirius charging the higher monthly subscriber fee for its mostly commercial-free product (Sirius Satellite Radio, 2002; XM Satellite Radio, 2002). Both services experienced a number of "false starts" in their efforts to launch satellites, develop receivers, and begin service. Some new model automobiles include factory-installed receivers, although aftermarket receiver sales are the dominant source for receivers and subscriber growth. (As of mid-2002, an automobile receiver and antenna for one of these digital services costs about $300, not including installation.) Sirius and XM have a technology-sharing agreement that allows for production of receivers that work with either service (XM Satellite Radio, 2000).

The question, of course, is whether consumers will pay for an audio product they have traditionally received for free. Both companies are developing program content not typically available from over-the-air broadcasting, including show tunes, blues, folk, bluegrass, American standards, and content from Fox News, BBC, CNBC, CNN, Discovery, Weather Channel, and ESPN. Given the development costs of each service, satellite audio technology faces an uncertain future. XM Satellite (13% is owned by Clear Channel Communications, 23.5% is owned by General Motors, and 17.9% by GM's Hughes Electronics—the parent company of DirecTV) has raised and spent about $1.6 billion since inception, but will likely need another $600 million before achieving positive cash-flow (XM Satellite, 2001; Elstein, 2002). Sirius Radio has raised and spent a similar sum since it began operation. Each service would need between 3.5 million and 4.3 million subscribers to become profitable by 2004 or 2005 (Elstein, 2002; Stimson, 2002). To put their losses of more than $1 billion each and the additional cost of the SDAR technology into perspective, Clear Channel Communications had higher advertising billings in 2000 than the two companies have spent during their entire existence.

The Yankee Group estimates there will be 21 million subscribers by 2006; XM signed 30,000 subscribers in its first 60 days (Coffey, 2002). Another consulting firm, The Carmel Group, expects satellite radio to have 300,000 customers by the end of 2002, with new cars equipped with receivers pushing the subscriber increases (Albiniak, 2001). XM, which had a three-month start-up lead over Sirius, will have more than half of all subscribers. Gross revenue from 300,000 subscribers would range between $3 million and $3.9 million per month. Ad revenue, particularly important for XM, would add a second revenue source.

NEW DELIVERY COMPETITION—INTERNET RADIO

Internet radio and audio streaming were once some of the fastest growing and most popular uses of the Internet. Radio stations first used the Web to distribute promotional information and then added programming. Webcasting became an effective means for radio stations to promote at-work listening and to expand their audience beyond the physical range of the station's broadcast transmitter. Research firm Cyber Dialogue, Inc. reported that 42% of all Internet users have made use of some music-related content and about 37% have visited a radio station Web site (Clark, 1999). Perhaps the most widely known Internet audio service is Broadcast.Com which was purchased by Yahoo! and is now known as Yahoo! Broadcast.

Much of Internet radio's growth, both streaming of radio station audio and custom content, came to a screeching halt when a panel established by the U.S. Copyright Office proposed a copyright payment system for Internet audio providers to the recording industry (Harmon, 2002). Before this decision, radio stations had contended that the copyright fees they paid for over-the-air transmissions covered Internet streaming as well.

The proposed royalty system will create a significant financial burden for radio stations that stream their content over the Internet. Each station must pay a $500 license fee and pay 7 cents for each song played. Put into perspective, a radio simulcast over the Internet of one song heard by 10,000 listeners would cost $7. With 20 songs played each hour each 24-hour day, the copyright fee would be $3,360 per day with an additional 9% fee to cover "ephemeral recordings" (the temporary storage of music files on servers during the streaming process) (Albiniak, 2002). With little tangible revenue coming through Internet transmissions, stations have little incentive to provide the entertainment content. For Internet-only Web sites that do not have simultaneous transmission of content, the fee would be 14 cents per song or nearly $7,000 per day (assuming 10,000 listeners). The new fee structure is not yet in place. Radio broadcasters and Webcasters may appeal the decision in court or ask Congress for help.

NEW COMPETITION—CELL PHONES AND MP3S

Even if the copyright dispute is resolved in favor of radio broadcasters and Webcasters, Internet delivery of over-the-air content or Internet-only content does have limitations—mobile listening, chief among them. Cellular telephones are seen as the leading delivery vehicle in the future for mobile entertainment (Borzo, 2002). Consumers in Europe and Japan already use cell phones to receive music files, video clips, and video games. All of these offerings take consumers away from radio. The iBiquity DAB system might even become a means for radio stations to compete with themselves. DAB broadcasts also might stream information to cell phones.

Downloadable MP3 files that can be saved to portable listening devices also present another technology use that will allow mobile entertainment from a source other than a radio broadcast. Apple's iPod was the first high-capacity, easy-to-use MP3 player that used Firewire technology to reduce transfer time. The device holds about 1,000 songs. MP3 players are also available for home and automobile use (Harris, 2002). (For more on digital audio, see Chapter 17.)

FACTORS TO WATCH

Radio stations have been in the business of delivering music and information to listeners for over 80 years. Listenership, more than any particular technological aspect, has enabled radio to succeed. Stations have been able to sell advertising time based on the number and perceived value of their audience to advertising clients. Technology, when utilized by radio stations, focused on improving the sound of the existing AM or FM signal or reducing operating costs.

DAB technology has the greatest potential to shift the entire radio industry and consumers into a new mode of operation. Digital broadcasting offers the possibility of streaming data content that might hold added value for consumers. Satellite audio itself holds the promise to create multiple revenue streams—the sale of the audio content, sale of commercial content on some programming channels, and possibly delivery of other forms of data. Regulatory barriers for these new technologies are not the issue. Consumer interest in the technologies, perfecting the technology so that it is as easy to use as traditional radio broadcasting has always been, marketing receivers at affordable prices, and delivering content that offers value will determine the success of these new radio technologies.

BIBLIOGRAPHY

Albarran, A., & Pitts, G. (2000). *The radio broadcasting industry*. Boston: Allyn and Bacon.
Albiniak, P. (2002, February 25). Web radio rate set. *Broadcasting & Cable*, 16.
Albiniak, P. (2001, December 31). Target: Radio birds. *Broadcasting & Cable*, 22.
America's highest billing groups. (2001, July 9). *Radio Ink*, 17.
Borzo, J. (March 5, 2002). Phone fun. *Wall Street Journal*, R8.
Bray, D. (n.d.). Have you heard? *The Toronto Auto Show*. Retrieved March 10, 2002 from
    http://www.digitalradio.ca.
Consumer Electronics Manufacturers Association. (1999, February 4). *CEMA Survey shows growing*
    acceptance for radio data system among broadcasters. Retrieved March 11, 2002 from
    http://www.ce.org/newsroom/newsloader.asp?newsfile=5139.
Chen, K. (2000, January 17). FCC is set to open airwaves to low-power radio. *Wall Street Journal*, B12.
Clark, D. (1999, November 15). With Web radio, anyone can be a DJ, but special software confuses users. *Wall*
    Street Journal, B8.
Clear Channel Communication. (n.d.). *Homepage*. Retrieved March 17, 2002 from
    http://www.clearchannel.com.

Coffey, B. (2002, March 18). Big audio dynamite. *Forbes.com*. Retrieved March 17, 2002 from http://www.forbes.com/forbes/2002/0318/166_print.html.

Elstein, A. (2002, March 20). XM Satellite Radio dismisses concerns of its auditor about long-term viability. *Wall Street Journal*, p. B13.

Federal Communications Commission. (2001, September 30). *Broadcast station totals as of September 30, 2001*. Retrieved March 28, 2002 from http://www.fcc.gov/Bureaus/Mass_Media/News_Releases/2001/nrmm0112.txt.

Federal Communications Commission. (2000, January 20). *In the matter of creation of low power FM*. Retrieved February 20, 2002 from http://www.fcc.gov/Bureaus/Mass_Media/Orders/2000/fcc00019.doc.

Federal Communications Commission. (1999, November 1). In the matter of digital audio broadcasting systems and their impact on the terrestrial radio broadcast services. *Notice of proposed rulemaking*, MM Docket No. 99-325. Retrieved March 1, 2002 from http://www.fcc.gov/Bureaus/Mass_Media/Notices/1999/fcc99327.pdf.

Federal Communications Commission. (n.d.-a). *AM stereo broadcasting*. Retrieved March 8, 2002 from http://www.fcc.gov/mmb/asd/bickel/amstereo.html.

Federal Communications Commission. (n.d.-b). *LPFM reports*. Retrieved March 29, 2002 from http://www.fcc.gov/mb/audio/lpfm/index.html#RETURN.

Harmon, A. (2002, February 21). Panel's ruling on royalties is setback for Web radio services. *New York Times*, C11.

Harris, N. (March 5, 2002). Listen up. *Wall Street Journal*, R8.

Huff, K. (1992). AM stereo in the marketplace: The solution still eludes. *Journal of Radio Studies*, *1*, 15-30.

iBiquity Digital Corporation. (n.d.-a). *What is iBiquity Digital?* Retrieved March 10, 2002 from http://www.ibiquity.com/01content.html.

iBiquity Digital Corporation. (n.d.-b). *iBiquity Digital identifies initial markets for conversion to digital AM and FM broadcast technology*. Retrieved March 10, 2002 from http://www.ibiquity.com/news_broadcaster%20support.html.

Labaton, S. (2000, December 19). Congress severely curtails plan for low-power stations. *New York Times*, C1-2.

Lewis, T. (1991). *Empire of the air: The men who made radio*. New York: Harper Collins.

McConnell, B. (2001, January 1). Congress reins in LPFM. *Broadcasting & Cable*, 47.

Mathews, A. W. (2002, February 25). A giant radio chain is perfecting the art of seeming local. *Wall Street Journal*, A1, A10.

Morgan, C. T. (2002, February 1). IBOC: What engineers should know. *Radio World*. Retrieved February 25, 2002 from http://www.radioworld.com/reference-room/iboc/guestmorgan.shtml.

National Association of Broadcasters. (n.d.-a). *NRSC revises U.S. RBDS standard*. Retrieved March 3, 2002 from http://www.nab.org/SciTech/Nrscgeneral/rds.asp.

National Association of Broadcasters. (n.d.-b). *High Speed FM Subcarrier (HSSC) Committee*. Retrieved March 3, 2002 from http://www.nab.org/SciTech/Nrscgeneral/hsscsub.asp.

National Radio Systems Committee. (2001, November 29). *DAB Subcommittee evaluation of the iBiquity Digital Corporation IBOC system, part 1—FM IBOC*. Retrieved March 9, 2002 from http://www.nab.org/SciTech/Fmevaluationreport.asp.

Orange, M. (2001). No power to the people. *Scientific American, 285*, 20.

Radio Advertising Bureau. (2002, February 7). *Radio industry winds up 2001 off 7% in ad sales*. Retrieved March 25, 2002, from http://rab.com/pr/revenue_detail.cfm?id=8.

Radio Advertising Bureau. (2001). *Radio marketing guide & fact book for advertisers 2002-2002 edition*. New York: Radio Advertising Bureau.

Rathbun, E. A. (2000, April 17). Proceeding with digital radio. *Broadcasting & Cable*, 35.

Sirius Satellite Radio Homepage. (2002). Company background. Retrieved March 22, 2002 from http://www.siriusradio.com/main.htm.

Statistical Abstract of the United States. (2001). Washington, DC: U.S. Government Printing Office.

Stavitsky, A. G., Avery, R. K., & Vanhala, H. (2001). From class D to LPFM: The high-powered politics of low-power radio. *Journalism and Mass Communication Quarterly, 78*, 340-354.

Sterling, C. H., & Kittross, J. M. (1990). *Stay tuned: A concise history of American broadcasting.* Belmont, CA: Wadsworth Publishing.

Stimson, L. (2002, February 1). Digital radio makes news at CES. *Radio World.* Retrieved March 22, 2002 from http://www.radioworld.com/reference-room/special-report/ces.shtml.

World DAB: The World Forum for Digital Audio Broadcasting. (n.d.-a). *DAB receiver archive.* Retrieved March 17, 2002 from http://www.worlddab.org/rarchive/rarch_home.htm.

World DAB: The World Forum for Digital Audio Broadcasting. (n.d.-b). *Country updates.* Retrieved March 17, 2002 from http://www.worlddab.org/dabworld/countryupdates.htm.

XM Satellite Radio Homepage. (2002). *Company background.* Retrieved March 22, 2002 from http://www.xmradio.com/js/xmmenu.htm

XM Satellite Radio. (2000, February 16). Sirius Radio and XM Radio form alliance to develop unified standards for satellite radio. *Company press release.* Retrieved March 2, 2000 from http://www.xmradio.com/js/news/pressreleases.asp#.

III

COMPUTERS & CONSUMER ELECTRONICS

If there is one theme underlying the developments discussed in this book, it is "the impact of digital technology." Nowhere is that impact more profound than in the computer industry. This year's computer technology will unquestionably be replaced in less than two years by a technology that has up to twice the performance at almost half the cost (a phenomenon known as "Moore's law"). These advances in computer technology in turn lead to advances in almost every other technology, especially those consumer products incorporating microprocessors or other computer components. The next seven chapters illustrate the speed, direction, and impact of continuous innovation in computer technologies across a wide range of computing and consumer electronics technologies.

The next chapter explores the manner in which computers have moved beyond text to incorporate the mélange of video, audio, text, and data known as "multimedia." The following chapter then explores one of the most pervasive applications of a "dedicated computer:" video game consoles. Chapter 12 addresses the most significant emerging application of personal computers, the Internet and the World Wide Web. The development of Internet-based e-commerce has made such a strong impact on individual businesses and the economy in general to warrant detailed discussion in Chapter 13. The application of computers and other related communication technologies in the office setting (Chapter 14) then provides an illustration of the manner in which these technologies are revolutionizing commerce around the world.

The most forward-looking chapter in this section, Chapter 15, discusses a set of technologies that may have the greatest long-term potential to revolutionize the way we live and work—virtual reality. The production and distribution of video and audio programming is the subject of the last two technologies in this section. The home video chapter reports on the incredible popularity of the existing analog video formats, and on new, digital technology that is beginning to challenge the analog incumbents. Finally, the digital audio chapter reports on the early outcome of the battle between competing analog and digital audio technologies, with digital casualties as well as victors.

In reading these chapters, the most common theme is the systematic obsolescence of the technologies discussed. The manufacturers of computers, video games, etc. continue to develop newer and more powerful hardware with new applications that prompt consumers to continually discard two- and three-year-old devices that work as well as they did the day they were purchased, but not as well as this year's model. Most software distribution, from movies and music to television and video games, has been based upon the continual introduction of new "messages." The adoption of this marketing technique by hardware manufacturers assures these companies of a continuing outlet for their products even when the number of users remains nearly static.

An important consideration in comparing these technologies is how long the cycle of planned obsolescence can continue. Is there a computer or piece of consumer electronics so good that a "better" one will never replace it? Will technology continue to advance at the same rate it has over the past two decades? How important is the equipment (hardware) versus the message communicated over that equipment (software)?

Finally, each of these chapters provides some important statistics, including penetration and market size, which can be used to compare the technologies to each other. For example, there is far more attention paid today to the Internet than to almost any other technology, but only about half of all U.S. households have access to the Internet (with even smaller penetration levels in other countries). On the other hand, the VCR is now found in about 9 out of 10 U.S. homes. In making these comparisons, it is also important to distinguish between *projections* of sales and penetration, and *actual* sales and penetration. There is no shortage of hyperbole for any new technology, as each new product fights for its share of consumer attention.

10

MULTIMEDIA COMPUTERS

Fred Condo, Ph.D.[*]

ew people can deny the impact of multimedia and the personal computer on the world. Multimedia products combine various media, such as text, still and animated images, audio, and video. When these are interactive, they additionally afford the user the opportunity to make navigational choices that determine the sequence of events (Van Buren, 2000). A variety of hardware and software systems exist to deliver this multimedia content. The applications of multimedia computers are greatly varied, ranging from digital video recording (Orlowski, 2002) to medical education (Lynch, et al., 1992). Both authors and users of interactive multimedia can use the same kinds of computer, although the specific hardware and software used may vary somewhat between the two roles. This chapter describes multimedia computers, their background, recent developments, and current status.

BACKGROUND

ROOTS

Although the concept of multimedia far predates the invention of microprocessors (Bush, 1945), multimedia computers today are general-purpose, microprocessor-based computers, as opposed to devices with more focused purposes, such as game consoles or digital video recorders. All such devices are, strictly speaking, computers: they contain memory, input/output interfaces, and a central processing unit (Howe, 1993). General-purpose computers, however, have a broad array of uses, ranging from scientific and financial calculations to office productivity applications such as word

[*] Assistant Professor, Department of Communication Design, California State University, Chico (Chico, California).

processing and electronic mail. They have standard keyboards and a mouse device as typical input devices, and a video screen able to display 800 by 600 pixels or more to create a high-resolution display of text, graphics, and video.

As general-purpose computers, multimedia computers have their origins in the first microprocessor-based computing systems. The first popular personal computer having the ability to integrate text and graphics was the Apple II, introduced in 1977. The IBM PC followed in 1981, as did compatible systems running operating system software by Microsoft (Computer Museum History Center, n.d.).

COMPUTER COMPONENTS

The core or "brain" of every multimedia computer is the central processing unit (CPU). A CPU is an integrated electronic circuit, or "chip," that is able to rapidly execute the instructions in programs or software. A microprocessor alone does not, however, constitute a computer system. The entire system comprises additional hardware, software, and human components. A large printed circuit board holds and electronically interconnects the CPU and other principal hardware components. This board is the mainboard or motherboard. Some other components of the system may be found on daughterboards that plug into matching electronic slots on the motherboard.

Figure 10.1 presents a conceptual model of an entire multimedia computer system. Each layer in the figure provides services that the next layers can use. The hardware is the foundation for the whole system, but cannot do anything useful without the other components of the system. The operating system manages the fundamental operations of the hardware. In turn, application software uses the operating system to provide useful services to the human being operating the system.

Figure 10.1
Conceptual Model of a Computer System

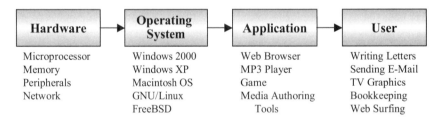

Source: Technology Futures, Inc.

CPU

Most multimedia computers use a microprocessor compatible with the designs of Intel Corporation. Both Intel and Advanced Micro Devices (AMD) manufacture such microprocessors. Intel dominates the microprocessor market with 79% market share, while AMD enjoys a market share of about 22% (Spooner, 2002). Multimedia PCs from Apple Computer use a different type of microprocessor, the PowerPC, manufactured by Motorola. Regardless of the type of microprocessor, its purpose is the same—to execute programs and manipulate information.

Because of the differences between the Intel-compatible microprocessors and the PowerPC, software makers must prepare software specifically for each type of microprocessor. As a result, some commercial software titles may be unavailable for one or the other type of PC. Nonetheless, a great deal of multimedia content is compatible across platforms. This is particularly the case with multimedia delivered on the Internet, such as Web sites, electronic mail, video, and animations.

A CPU's capabilities are often stated as the clock speed in megahertz (MHz) or gigahertz (GHz; 1 GHz = 1,000 MHz). Other measures for CPU performance are millions of instructions per second (MIPS) and millions of floating point operations per second (MFLOPS). For a given processor type, such numbers are comparable, and a larger number means a higher-performing CPU. The interactions of the components of a computer system are complex, however, and it is not generally possible to judge a computer system's capabilities based solely on the CPU's performance ratings.

OTHER HARDWARE

Hardware includes all the electronic and mechanical elements of a computer system. The single most important hardware element is the microprocessor CPU. In addition to the CPU, a computer system contains memory, peripheral interfaces, and peripheral components.

Memory, often called RAM (random access memory), consists of electronic circuits, like the CPU, built on silicon-based chips. The purpose of memory is rapid information storage and retrieval. When software or any kind of information, data, or media content is active in the computer system, it must first be copied from some permanent medium into memory. One of the important characteristics of RAM is that its contents are impermanent. When the system loses power or experiences a software failure or "crash," the RAM loses the information it contained. This is one reason experienced computer users often make it a habit to save their work frequently. The RAM capacity in a current computer is measured in megabytes (millions of letters and numbers), but some systems can accommodate gigabytes of memory (billions of letters or numbers). However, such very large memories are more characteristic of network server computers than of multimedia computers.

Saving information more permanently than does memory is the task of the various kinds of disks that can be part of a computer system. The principal kind of disk in a multimedia computer is a hard disk, which is often in the same case as the motherboard. The disk is permanently sealed in its own metal case, so it does not have the appearance of a disk. The disk itself is made of rigid metal with a magnetic medium on it. The computer system can store information onto the disk or retrieve information from it. Electrical power is not necessary for the disk to retain information, so it is a more permanent medium than memory. Because the disk has delicate moving parts, however, it is both much slower and more susceptible to mechanical damage than RAM. Hard disks commonly have a capacity of tens to hundreds of gigabytes.

Other types of disks may also be present in a computer system. Floppy and Zip® disks, like hard disks, use a magnetic medium for information storage. Floppy disks typically have a capacity of less than two megabytes and are slow and unreliable. As such, they are rapidly becoming an obsolete storage technology. Zip disks can hold 100 or 250 megabytes; they are slower than hard disks, but, because they are removable, users can carry them between computers and maintain a library of disks for various projects.

Still other kinds of disks use an optical medium for information storage and retrieval. These include CDs (compact discs) and DVDs (digital video or versatile disks). Both were initially available only in read-only forms. The computer could retrieve information from the disk, but could not store new information onto it. Such disks are in common use as a publication medium for music (CD) and for movies (DVD–Video). The capacity of a CD is approximately 600 to 700 megabytes. A DVD can hold 8.5 gigabytes (Leyden, 2002; Toshiba, n.d.). Both formats are thin (one millimeter) disks 12 centimeters in diameter.

CDs are now available in CD-R (read) and CD-RW (read/write) formats that allow a computer system with the appropriate CD writer, or "burner," to write new information onto blank disks. CD-R disks can be written once; CD-RW discs can be written, erased, and rewritten many times.

The DVD standard recognizes six formats (DVD Forum, 2000). These vary according to the type of information they store and whether they can be written to or rewritten to (see Table 10.1).

Table 10.1
DVD Formats, Purposes, and Capacities

Format	Purpose	Capacity per Side (Gigabytes)
DVD Video	Storage of movies and other video material for playback.	8.5
DVD Audio	Storage of music and other audio for playback with fidelity better than what CDs offer.	4-8.5
DVD-ROM	Read-only file storage for computers.	8.5
DVD-R	Professional and consumer media authoring. Writable once.	4.7
DVD-RAM	High-performance rewritable data storage, a "virtual hard disk." Rewritable 100,000 times.	2.6-4.7

Source: DVD Forum (2000)

OPERATING SYSTEMS

The operating system is software that manages the fundamental operations of the hardware. It manages files on disk, launches application software, and coordinates the operations of peripherals, the CPU, the memory, and the network. Other software in the system is application software, such as word processors, Web browsers, e-mail programs, games, and media authoring software.

Most multimedia computers use an Intel-compatible CPU, and most such systems use operating system software from Microsoft. Computers from Apple Computer, however, generally use Apple's Macintosh operating system (Mac OS 9 or Mac OS X). Both kinds of computers are capable of running alternative operating systems, such as GNU/Linux and FreeBSD. The great majority of multimedia computers, however, run a Microsoft Windows operating system, such as Windows 98, Windows NT, Windows 2000, or Windows XP.

RECENT DEVELOPMENTS

HARDWARE

In 1965, Gordon Moore, one of the founders of Intel, observed that the number of transistors on a chip (a rough measure of its overall power and speed) roughly doubles every year or two (Interview, 1997). (Moore denies ever using the often-quoted figure of 18 months. Rather, he used one year initially, then revised the figure to two years in 1975.) His prediction has held remarkably true in the 37 years since it first appeared. The trend shows little sign of slowing, despite periodic reports in the popular press to the contrary (Haavind, 2001). Recently, Intel reported development of a microprocessor circuit running at 10 gigahertz (10,000 megahertz). Although it will be years before such laboratory results appear in commercial products, this development proves that microprocessor manufacturers have been able to continue increasing the raw processing power of microprocessors.

A similar, though briefer, trend has occurred in hard disk capacity. From 1997 through 2001, hard disk capacity has doubled, and may quadruple by 2003 (Shim & Spooner, 2001). DVDs, too, have seen capacity increases. Although not yet available on the market, Blu-ray DVD technology, a new standard that takes advantage of the higher resolving power of blue light, allows for storing as much as 27 gigabytes per side (Hitachi, et al., 2002). This represents more than three times the capacity that the current DVD standard affords.

In 2002, Apple introduced a new version of its iMac multimedia computer and iBook multimedia laptop computer. The iMac features a new industrial design, with a hemispherical base and an integrated flat-panel display at the end of an articulated arm. One model of the new iMac computer has both CD and DVD authoring capabilities through a drive able to accommodate both CD-RW and DVD-R disks. At the same time, Apple began offering only flat-panel displays under their brand name—no more monitors based on bulky cathode-ray tubes (CRTs).

SOFTWARE

New operating systems from Microsoft and Apple figured prominently in recent events. In 2001, Microsoft released the most recent edition of its Windows operating system, Windows XP. Microsoft (n.d.) lists multimedia-related features as primary reasons to purchase the new operating system, including DVD and CD playback and recording, instant messaging on the Internet, digital video and still-image editing, and Web browsing. In 2001, Apple released Mac OS X (ten), which represented a major change in the technological and social bases of the Apple operating system. Rejecting completely closed development, Apple (n.d.-b) released the foundation of its operating system as the open-source Darwin project. As such, any software developer who so desires can contribute to the development effort of Darwin, which more closely resembles the GNU/Linux and FreeBSD operating systems than it does either previous Apple or Windows operating systems. Apple (n.d. a), like Microsoft, promotes the multimedia and Internet capabilities of their new operating system.

CURRENT STATUS

The percentage of households having a computer in the United States continues to rise. In August 2000, it stood at 51%, up from 42% at the end of 1998 (Crutsinger, 2000). This continued a trend from 1990 through 1997 (Bureau of Labor Statistics, 1999). By the fall of 2001, 60% of households in the United States had a computer, with even higher penetrations in some cities (Catalano, 2001). This market saturation may explain why computer sales fell in 2001 for the first time since 1985 (Reuters, 2002).

Prices for multimedia computers continue to fall, as raw processing power and other capacities continue to grow. As Table 10.2 shows, in 2001, processor prices have declined by about 40% to 50%, and the price of a CD-RW drive has declined by about 30%. In March 2002, PC manufacturer Dell's (n.d.) Web site advertised multimedia computers for less than $1,300, while Apple (n.d. c) advertised iMac computers for $1,300 to $1,800, depending on processor speed and type of disk drive.

Table 10.2
Price Change for Selected Hardware

Item	Price (USD)	
	January 2001	January 2002
AMD 950 MHz processor	$204	$95
Intel 850 MHz processor	$307	$190
CD-RW drives	$254	$180

Source: PC Today

As of the 3rd quarter 2001, the leading manufacturers of Intel-compatible computers are Dell, Compaq, IBM, Hewlett-Packard, and NEC (Fried, 2001), and 30.5 million units were sold. Apple is the sole producer of iMac and other Macintosh systems, and 850,000 units were sold during the same period. After briefly losing it, Apple regained the market lead in education, with about 3 million installed systems in schools, while Dell's installed base was about half that amount (Haddad, 2001).

FACTORS TO WATCH

PENDING ACTIVITY

Microsoft antitrust case. An antitrust action against Microsoft was continuing as this chapter was written. Nine U.S. states are pursuing aggressive remedies for what they believe are Microsoft's monopolistic practices in the market for operating systems. They are in particular seeking to force Microsoft to separate multimedia capabilities, such as Web browsing, from the core operating system

(Lettice, 2002). It is questionable whether such developments will ever impinge on Windows XP, as the progress of software development tends to be faster than the progress of legal cases.

Open source operating systems. Watch for continued growth in the impact of GNU/Linux, FreeBSD, and Darwin (via Mac OS X) on multimedia technology. Expect GNU/Linux and FreeBSD to gain market share, particularly in server systems that deliver multimedia content via the Internet. Expect Microsoft to maintain its market dominance in the general multimedia operating system market, and Apple to remain important in education and media production.

Government regulation of multimedia computers. Senator Fritz Hollings (D-SC) has held hearings on a bill, the Security Systems Standards and Certification Act, which would enact into law technical standards to protect multimedia content from unauthorized copying (Greene, 2002). Expect the entertainment industry, fearful of the capability of computer users to burn exact copies of music and movies on CD-R and DVD-R media, to continue to militate in favor of such regulations. Expect digital-rights organizations such as the Electronic Frontier Foundation, information-technology makers such as Intel, and independent multimedia authors to oppose such regulations.

LIKELY DEVELOPMENTS

Stratification of the entertainment industry. The decreasing prices of multimedia computers and their components will encourage more individuals and groups to become independent multimedia producers. The impact of this activity on the overall multimedia market (as measured in sales) may be minor. They may, however, be the vanguard of a future multimedia culture that is starkly stratified, with an oligopoly of traditional media producers dominating sales in a mass market in one stratum, and a diverse group of amateur, semi-professional, and niche producers in the other. This latter group would be using modern multimedia computers to compress all the stages of multimedia production to fit into a single office or spare room, while using the Internet to collaborate with other artists all over the planet.

Continued operation of Moore's law. Although Moore's law was a simple observation and never a true law, expect no deceleration in its applicability to processor power. Expect the trend to continue through 2007.

Movement of the locus value from the computer to the network. The Internet affords people the ability to communicate; share files containing music, video, and other media; and thereby collaborate on media production and use. Meanwhile, the cost of computer equipment continually declines. Therefore, people will increasingly derive more value (in terms of knowledge, entertainment, ability, and pleasure) from participation on the Internet than from owning and using a computer in isolation.

Reduction of the digital divide. The inequality in ownership of or access to multimedia computers continues to decline (U.S. Department of Commerce, 2002). As prices also decline, expect the technology to penetrate more and more into lower socio-economic strata. If this trend holds true, the benefits in education and the arts could be substantial, and may well afford new opportunities to persons and groups of limited means. At the same time, expect cost-consciousness to delay the obsolescence of CD-based media production, because CD-RW drives and blank disks will continue to be less expensive than DVD-RW drives and media.

Design rather than performance as distinguishing features. Expect some market actors, particularly Apple, to emphasize industrial design as a distinguishing feature of their products. This is because there is a delay between the development of faster processors and novel ways of using the power they afford. The very next generation of processors will not present a compelling need for users to upgrade, because the capabilities they need and use are already available with current multimedia systems. (See, for example, ZDNet News, 2001.) Expect this trend to accelerate the decreasing price trend for microprocessors.

Where interactive multimedia computers are common, they have the potential to afford the general public opportunities for creative expression that have long been reserved for creative professionals. The declining cost of the technology means that a greater proportion of the populace will enjoy such opportunities. Whether many will seize the opportunity, or remain passive consumers of professionally produced media, remains to be seen.

BIBLIOGRAPHY

Apple Computer, Inc. (n.d.-a). *Apple—Mac OS X*. Retrieved March 15, 2002 from http://www.apple.com/macosx/.

Apple Computer, Inc. (n.d.-b). *Open Source Projects at Apple*. Retrieved March 8, 2002, from http://www.opensource.apple.com/.

Apple Computer, Inc. (n.d.-c). Retrieved March 12, 2002, from http://store.apple.com/1-800-MYAPPLE/WebObjects/AppleStore.woa/213/wo/kRC5V1lTdgOmzkzrg11/0.3.0.3.30.19.0.1.3.1.3.1.1.0?96,54.

Bureau of Labor Statistics. (1999). Computer ownership up sharply in the 1990s. *Issues in Labor Statistics*. Retrieved March 5, 2002 from http://www.bls.gov/opub/ils/pdf/opbils31.pdf.

Bush, V. (1945, July). As we may think. *The Atlantic, 176* (1), 101–108.

Catalano, F. (2001, October 12). Plummeting PC sales worry computer makers. *Puget Sound Business Journal*. Retrieved March 2, 2002 from http://seattle.bizjournals.com/seattle/stories/ 2001/10/15/focus10.html.

Computer Museum History Center. (n.d.). *Timeline of computer history*. Retrieved March 10, 2002 from http://www.computerhistory.org/timeline/1977/index.page.

Crutsinger, M. (2000). Computer ownership up in the U.S. *ABCNews.com*. Retrieved March 10, 2002 from http://abcnews.go.com/sections/tech/DailyNews/computerreport001017.html.

Dell. (n.d.). Retrieved March 10, 2002 from http://www.dell.com/.

DVD Forum. (2000). *DVD primer*. Retrieved March 12, 2002 from http://www.dvdforum.org/tech-dvdprimer.htm.

Fried, I. (2001). PC market hurting, but Dell grabs share. *CNet News*. Retrieved March 14, 2002 from http://news.com.com/2100-1040-274594.html?legacy=cnet

Greene, T. C. (2002). Senator brutalizes Intel rep for resisting CPRM. *The USA Register*. Retrieved March 15, 2002 from http://www.theregus.com/content/54/24195.html.

Haavind, R. (2001). The coming world of nano- and pico-technology. *Solid State Technology, 44* (10), 14.

Haddad, C. (2001, July 11). How Apple is reclaiming the classroom. *BusinessWeek Online*. Retrieved March 8, 2002 from http://www.businessweek.com/bwdaily/dnflash/jul2001/nf20010711_074.htm.

Hitachi, LG Electronics Inc., Matsushita Electric Industrial Co., Pioneer Corporation, Royal Philips Electronics, Samsung Electronics Co., et al. (2002). *Large capacity optical disc video recording format "Blu-ray Disc" established*. Retrieved March 12, 2002 from http://www.matsushita.co.jp/corp/news/official.data/data.dir/en020219-4/en020219-4.html.

Howe, D. (1993). *Free online dictionary of computing*. Retrieved March 10, 2002 from http://foldoc.doc.ic.ac.uk/foldoc/foldoc.cgi?query=computer. Also accessible from http://www.foldoc.org/.

Interview. (1997). *Scientific American*. Retrieved March 14, 2002 from http://www.sciam.com/interview/moore/092297moore2.html.

Lettice, J. (2002). Judge grants states access to Windows source. *The USA Register*. Retrieved March 8, 2002 from http://www.theregus.com/content/archive/24062.html.

Leyden, J. (2002, February 19). Next-gen DVD standard agreed. *The Register*. Retrieved March 9, 2002 from http://www.theregister.co.uk/content/54/24124.html.

Lynch, P. J., Jaffe, C. C., Simon, P. I., & Horton, S. (1992). Multimedia for clinical education in myocardial perfusion imaging. *Journal of Biocommunication 19* (4), 2–8.

Microsoft Corporation. (n.d). *Top 10 reasons to get Windows XP Home Edition*. Retrieved March 15, 2002 from http://www.microsoft.com/windowsxp/home/evaluation/whyupgrade/top10.asp.

Orlowski, A. (2002, February 27). How to TiVO-ize your PC. *The USA Register*. Retrieved March 9, 2002 from http://www.theregus.com/content/54/24167.html.

PC Today. (n.d.). *PC Today® chart selection page*. Retrieved March 15, 2002 from http://www.pctoday.com/input/mwatch/mwchart.asp.

Reuters. (2002). PC sales in Europe could rise in 2002. *ZDNet*. Retrieved March 8, 2002 from http://zdnet.com.com/2100-11-818345.html.

Shim, R., & Spooner, J. G. (2001). IBM "pixie dust" breaks hard drive barrier. *CNet News*. Retrieved March 4, 2002 from http://news.com.com/2100-1001-257943.html?legacy=cnet.

Spooner, J. G. (2002, March 14). AMD grabs a bigger piece of the desktop. *ZDNet*. Retrieved March 14, 2002 from http://zdnet.com.com/2100-1103-860018.html.

Toshiba. (n.d.). *What's DVD?* Retrieved March 12, 2002 from http://www3.toshiba.co.jp/dvd/e/whats/wh05/w0503idx.htm.

U.S. Department of Commerce. (2002). *A nation online: How Americans are expanding their use of the Internet*. Retrieved March 6, 2002 from http://www.stat-usa.gov/stemplate.nsf/Validate?OpenAgent&dbID=pub.nsf&vwID=Abstracts&filekey=nationonline.pdf.

Van Buren, C. (2000). Multimedia computers and video games. In A. Grant & J. Meadows (Eds.), *Communication technology update,* 7th edition. Boston: Focal Press.

ZDNet. (2001, January 22). Is Apple ripe for recovery? *ZDNet News*. Retrieved March 12, 2002 from http://zdnet.com.com/2100-11-527470.html.

11

VIDEO GAMES

Cassandra Van Buren, Ph.D.[*]

Video games are a form of interactive electronic entertainment with a 30-year history as a major economic and cultural force in U.S. and international markets. In 2001, video games generated $9.4 billion in sales in the United States alone (Tran, 2002), with estimates of worldwide sales exceeding $20 billion (Van Grinsven, 2002). Some forecasters predict that, by 2005, as many as 70% of U.S. homes will contain video game consoles, making them almost as prevalent as VCRs (Damuth, 2001). Increasing scholarly, journalistic, and congressional interest in video games demonstrates a burgeoning field of inquiry involving the analysis of the ways in which games shape and recirculate cultural values. This chapter describes the background, recent developments, and current status of video games, including those played on console systems and handheld devices.

BACKGROUND

The history of the video game industry is characterized by distinctive trends.

(1)   Hardware and software development are interdependent, specifically related to the coordinated timing of product release. Hardware systems fail to thrive if compatible and compelling software is not immediately available for the consumer market.

(2)   The industry's rapid technological progression has been driven by the symbiotic relationship between hardware/software innovation and gamers' demand for "realism," meaning increased speed, interactivity, and film-quality imagery and audio.

---

[*] Assistant Professor, Department of Communication, University of Utah (Salt Lake City, Utah).

(3)   The major console manufacturers (Sony, Nintendo, Microsoft, and, until February 2001, Sega) have engaged in intense competition for profits, jockeying for market share and third-party developer relationships.

(4)   Software piracy costs the industry billions in revenue each year.

(5)   The immersive qualities of game technology and narratives have led some politicians and citizens to worry about the social effects of video games, resulting in rounds of congressional hearings, ratings systems, and the removal of certain games from the market.

The work of video game production is a collaborative process requiring groups of people to design, prototype, produce, market, and distribute games. Developers determine the platform, desired user experience, design, and budget. If the title is developed for the console platform, developers target the title for playback on a platform such as Nintendo GameCube, Sony PlayStation 2, or Microsoft Xbox. If the title is developed for playback on a personal computer, developers decide whether to target the Mac platform and/or Windows platform, and the minimum system specifications for RAM and CPU speed. Handheld game developers must consider platforms including Game Boy and Game Boy Advance, personal digital assistants (PDAs) such as Palm and Handspring, and wireless phones.

The video game industry began in the late 1950s as technologists, scattered around the United States, independently began tinkering with existing technology, without much thought about patents or profitability, to create something fun for themselves (Poole, 2000). The first inroad toward successful commercialization began in 1967 when Ralph Baer developed an electronic tennis game for the home. In 1972, as Intel was inventing the tiny microprocessor necessary for small, less expensive computers, Magnavox licensed Baer's game system as *Odyssey*. Simultaneously, after an initial failure with an arcade-based version of *Spacewars*, University of Utah student Nolan Bushnell founded Atari and tried again with *Pong*, a tennis video game. *Pong* proved profitable enough to allow Bushnell to sell out to Warner in 1976 for $28 million (Poole, 2000). By 1977, home-based *Pong* clones saturated the market, causing the first crash in the industry and resulting in the demise of console manufacturers such as Fairchild and RCA (Herz, 1997).

In 1976, Fairchild's Channel F released the first cartridge-based game console that allowed several games to be played on the same hardware. In 1977, Atari released its 2600 programmable video computer system. Sales of the system and games were moderately successful, building through 1979 with the release of the game *Asteroids* and peaking in 1980 with the release of the home console version of *Space Invaders*. Video game sales slowed until 1986, when Nintendo released its 16-bit NES cartridge-based system and *Super Mario Brothers* for home consoles. Sega and Atari attempted to compete with NES by releasing their own systems, but NES outstripped its competitors in U.S. sales by a 10:1 ratio. In 1989, NEC released the first system to use CDs for game storage: the TurboGraphx-16 (Herman, et al., 2002).

In the early 1990s, Sega became a strong force in the video game industry with the 1992 release of the Sega CD system and strong sales of *Sonic the Hedgehog*. After ending a development agreement with Nintendo, Sony announced plans to develop its own 32-bit CD system. In 1993, Nintendo and Sega announced the development of 64-bit and 32/64-bit systems. By 1995, Sega had discontinued production and support of the Sega CD system, while Sony released its PlayStation and Nintendo released Nintendo64 in Japan. In 1996, Nintendo64 made its American debut, selling over 1.7 million units in just three months. Meanwhile, Sony reported PlayStation sales topping $12 million per day during the 1996 holiday season (Herman, et al., 2002). In 1998, console system sales reached between 6.5 million and 7.5 million units in the holiday shopping period (Video game console, 1999), with PlayStation ranking first in sales of console systems to date.

In 1999, Nintendo announced plans to develop a new console, code-named Dolphin. Microsoft threatened competitors when the company announced its entry into the gaming market with the Xbox console. The electronic game market exceeded $7 billion in 1999, marking a 10% gain over the previous year. Part of this gain was due to Sega's 1999 U.S. introduction of Dreamcast, the first 128-bit Internet-ready system. Retailing for $199, Sega reported earnings of $98 million on the first day of U.S. sales (Herman, et al., 2002).

Internet gaming grew in 1999, with casual gamers playing free games on the Web, while hardcore gamers used subscription-based massive multiplayer online game (MMOG) networks. Always hungry for new markets and increased marketing advantages, console manufacturers viewed Internet connectivity as a way to increase sales and open new revenue streams (Berst, 1999). In September 2000, Sega introduced SegaNet, a subscription Internet service for Dreamcast players willing to pay $9.95 per month. The first console-based role-playing game, *Phantasy Star Online,* was released in 2000 for Dreamcast, enabling thousands to play together online in real time (Herman, et al., 2002). Other MMOGs gaining popularity included Everquest, Ultima Online, Majestic, and Anarchy Online.

Sony's PlayStation 2 debuted in Japan in March 2000, breaking industry sales records by selling 980,000 units with retail revenues of more than $3 billion in the first two days of sales (PlayStation sales, n.d.). The console shipped with Internet and DVD/CD capability. While PlayStation 1 sales accounted for an impressive 40% of Sony's profits in 1998, PlayStation 2 was expected to prove even more lucrative for the company (Levy, 2000). U.S. sales, launched in October 2000, were initially slowed by poor availability of units (Pham, 2002).

HANDHELD SYSTEMS

Miniature handheld gaming systems enjoyed steady popularity thanks to their portability and relatively low prices. In 1989, Nintendo released Game Boy, the first black-and-white handheld system, bundled with *Tetris,* for $109. In 1990, NEC released a handheld version of the TurboGrafx-16 called the TurboExpress, marking the first portable system capable of playing games designed for a home console (Herman, et al., 2002). Bandai's November 1996 Japanese release of a tiny "virtual pet" game device called Tamagotchi resulted in a national craze that traveled to the United States by May 1997. One store sold out of its initial stock of 30,000 Tamagotchi in only three days, marking the beginning of a new platform and phase of miniaturization for computer-based games.

The export of *Pokemon* to the United States in 1998, combined with the release of Game Boy Color, spurred sales of Game Boys as *Pokemon* quickly became Nintendo's best-selling game ever. The release of Pocket Pikachu, a miniature game device resembling a Tamagotchi, also raised Nintendo's market share. Meanwhile, several other handheld systems were introduced by companies such as Tiger, Bandai, Hasbro, Cybiko, and SNK whose systems (Pocket Pro, WonderSwan, Pox, Cybiko Xtreme, and NeoGeo Pocket) failed to threaten Game Boy. In 2000, Nintendo continued to expand Game Boy's capabilities with a $49 mobile adapter that allowed players to connect to wireless phones for online game play and e-mail (Herman, et al., 2002).

LOCATION-BASED GAMING

Location-based gaming takes place in business establishments such as video game arcades in movie theaters, malls, and other locations. However, these arcades became youthful all-male bastions, unwelcoming to families, the elderly, girls, and women (LaPlante & Seidner, 1999; Provenzo, 1999), a fact that prevented the industry from tapping into a huge potential market. A notable trend began in 1977 when Atari introduced its venture into restaurant-based locations designed to capitalize on its video game holdings: Chuck E. Cheese Pizza Time Theatre, a location-based entertainment venue combining animatronic characters, video games, food, and beverages (Herman et al., 2002). In 1997, a new location-based gaming business called Gameworks, a joint venture between Sega and Universal Studios, opened its first sites in Seattle, Los Angeles, and Las Vegas. Dave and Buster's, the recognized leader in location-based gaming, opened in 1982 and averages $13 million in revenue per site per year, with 47% of revenue coming from game play. The company clearly hit upon a formula for appealing to previously untapped arcade players: the clientele is, on average, 33 years old with an annual income of $61,000, and 40% of patrons are female (LaPlante & Seidner, 1999).

CONTENT CONTROVERSY

Despite the dearth of evidence causally linking video games to criminal acts (Gunter, 1998), concerns over video games' effects on individuals and society have resulted thus far in four main types of restrictive measures: ratings systems, content-based selection by retailers, developer self-restrictions, and lawsuits blaming games for leading to criminal acts. In 1993, the content of the games *Night Trap* and *Mortal Kombat* prompted U.S. Senators Leiberman and Kohl to launch an investigation of video game violence. As a result, the industry agreed to regulate itself via a voluntary rating system established, implemented, and maintained largely by two organizations: the Entertainment Software Rating Board (ESRB) and the Internet Content Rating Association (ICRA) (Herman, et al., 2002). The ESRB rates games according to content related to violence, sex, crude language, tobacco, alcohol, illegal drugs, and gambling. Rating symbols, such as "M" for mature, are included on the packaging of game boxes and online gaming Web sites. As of February 2002, ESRB reported rating over 7,000 products (ESRBi, n.d.). ICRA staff reported that, as of February 2002, over 195,000 websites were rated using the ICRA self-rating system (personal communication, February 8, 2002).

In 2000, the U.S. Federal Trade Commission reported findings that video game companies targeted kids in ads for M-rated titles. Meanwhile, Indianapolis became the first U.S. city to ban underage children from playing arcade games with violent or sexual content. By March 2001, the ordinance was ruled unconstitutional (Herman et al., 2002). In 2001, the families of some Columbine High School victims filed a $5 billion lawsuit against select game developers, alleging the killers' game-playing habits helped cause the shooting massacre (Associated Press, 2001).

In an effort to respond to consumer pressure, some retailers declined to carry certain video games. In 1998, Wal-Mart refused to carry some 50 arcade games that were deemed "inappropriate" by the company. In 2000, Sears and Montgomery Ward ceased selling M-rated games after a sting operation in Illinois in which underage children were able to purchase forbidden games, and Wal-Mart and Kmart began requiring identification for all M-rated game purchases (Herman, et al., 2002).

RECENT DEVELOPMENTS

Several important developments occurred in 2001. Disappointing sales prompted Sega to discontinue the Dreamcast system in February 2001 (Kent, 2001). Officials announced that the company would instead focus on other business strategies: content development for Dreamcast and other platforms, networked gaming via SegaNet, licensing Dreamcast chip-set technology, and developing game titles for handheld devices (Sega focuses, 2001).

Two console systems and a handheld system were released, renewing cutthroat competition between the big hardware companies. In June 2001, Nintendo released the $99.95 Game Boy Advance and 17 games. The Advance is designed to connect with wireless phones for Internet access and features a 2.9-inch screen, 32-bit ARM CPU, and 240×160 screen resolution (Game Boy Advance, n.d.). Advance sold 500,000 units in its first week, topping one million within six weeks of its release. Four Advance games ranked in the top 10 best-selling games that June (Herman, et al., 2002).

Microsoft's entry into the video console market marked the company's venture into large-scale hardware development and entertainment. The Xbox was unveiled in the United States on November 15, 2001 and, by December 31, had sold 1.5 million units in North America (Tran, 2002). The February 2002 Japanese launch, viewed as crucial for Xbox's worldwide success, proved lackluster, with only 124,000 units sold in the first week (Fox, 2002). The Xbox uses a Pentium III processor, a custom-designed 300 MHz Nvidia X-Chip, and 64MB of RAM (Extended Play Staff, 2001).

Nintendo released its $199.95 Gamecube system in America on November 18, 2001 and amassed over $98 million in console, game and accessories sales in one day. The Japanese release moved 133,000 GameCubes in the first week, outstripping Xbox (Fox, 2002). The Gamecube system is compatible with Nintendo's Game Boy Advance and features a 485 MHz custom-made IBM Gekko processor (GameCube detailed, n.d.).

Sony maintained an undisputed sales lead in 2001 thanks to a one-year head start, reporting U.S. sales figures of 6.6 million PlayStation2 units, with Microsoft and Nintendo competing for second place with North American sales figures of 1.5 million Xboxes and 1.2 million GameCubes (see Figure 11.1). Sony also addressed the low end of the market with a repackaged version of the original PlayStation priced below $100.

After the September 11 terrorist attacks in the United States, several video games featuring terrorist-oriented violence or airplane scenes were altered, delayed, or pulled by developers out of concern that the public would criticize the games for insensitive or inappropriate content. Meanwhile, Sega launched pay-as-you-play games for Japan's i-mode wireless phones based on *Sonic the Hedgehog, Samba de Amigo, Out Run,* and *Fantasy Zone* (Herman, et al., 2002). In Europe, a wireless gaming company called It's Alive touted its two wildly popular pay-as-you-play games called *Botfighters* and *Supafly*, in which players interact with nearby unknown players in real time using wireless phones (Botfighters, 2002; Supafly, 2002).

Figure 11.1
**Console Unit Sales through December 2001,
Worldwide**

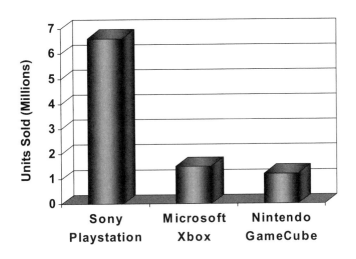

Source: Tran (2002)

CURRENT STATUS

Recent data about video game players reported by the Interactive Digital Software Association refute the notion that game players are adolescent males playing alone in their bedrooms. In 2000, video games were played by 60% of U.S. residents over the age of six. Of frequent players, 59% play with friends and 33% play with siblings. Consoles were found throughout the home (Figure 11.2). The average age of players is 28 years; 43% of players are female, and females influence 25% of console purchase decisions (Console facts, 2002). Strikingly, video games were named "the most fun entertainment activity" by 35% of those surveyed for the third year in a row, with television trailing at 18% and the Internet at 15% (Ten facts, 2002).

The launch of three new systems in 2001 pushed U.S. sales of video games and hardware to a record $9.4 billion, surpassing 2000's figure of $6.6 billion and 1999's record of $6.9 billion. Even Hollywood's 2001 domestic box office receipts were eclipsed at a mere $8.4 billion (Tran, 2002). However, profits would have been higher if not for the estimated $1.9 billion lost to videogame piracy in 2001 (U.S. computer, 2002). Price wars continued as Sega cut the Dreamcast price to $49.95 in late 2001 (Becker, 2001) and Nintendo cut the Game Boy Advance price by 20% in early 2002 (Tran, 2002). Computer and videogame software sales were forecast to reach $16.9 billion by 2003, not including an estimated $1.1 billion from online game revenue (Quick facts, 2002).

Because most PDA and wireless phone games are preinstalled or downloaded as shareware, quantifying sales is difficult (Shim & Fried, 2002). Verizon's 2002 launch of the first 3G wireless phone network in the United States provided a necessary step for the development of networked wireless gaming (Verizon launches, 2002). Steady hardware growth appears certain as 2001 PDA unit sales increased 36% over 2000 figures (Kane, 2002). Color PDA prices are forecast to drop in price by about \$200 in 2002, and development of "smart phones" will mix PDA and wireless phone functions (Kort & Dulaney, 2001). While Palm expects application sales to drive hardware sales (Paul, 2001), the "killer app" that could make PDA sales skyrocket has not yet emerged (Shim & Fried, 2002).

Figure 11.2
Distribution of Video Game Consoles in the Home

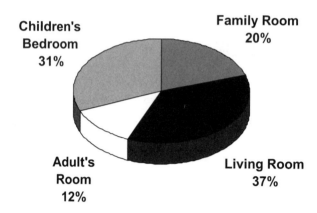

Source: Quick facts (2002)

FACTORS TO WATCH

As the new kid on the block, Microsoft's market position will depend on the availability of titles, the success of the March 2002 European launch, and implementation of the plan to turn the Xbox into a multi-function home entertainment appliance tentatively named HomeStation, designed to handle e-mail, Web access, video recording, and other tasks (Becker, 2002). Industry diversification could be affected by the fate of Botfighters-type pervasive wireless gaming and the continued growth of MMOGs in the United States. As the industry continues to pursue untapped markets, game developers will face pressure to create titles for diverse audiences.

BIBLIOGRAPHY

Associated Press. (2001, April 24). Columbine lawsuit targets video gamemakers. *USA Today Online.* Retrieved February 7, 2002 from http://www.usatoday.com/life/cyber/tech/review/games/2001-04-24-columbine.htm.

Berst, J. (1999, December 10). Five reasons your next PC could be a Nintendo. *ZDNet Anchordesk.* Retrieved February 8, 2002 from http://www.zdnet.com/anchordesk/story/story_4219.html.

Becker, D. (2001, November 20). Sega chops prices to cull Dreamcasts. *ZDNet News.* Retrieved February 8, 2002 from http://zdnet.com.com/2110-11-531132.html.

Becker, D. (2002, January 15). Xbox may spawn entertainment hub. *C|Net Tech News*. Retrieved January 27, 2002 from http://news.com.com/2100-1040-814653.html?legacy=cnet.

Botfighters. (2002). *It's Alive*. Retrieved January 31, 2002 from http://www.itsalive.com/downloads/ botfighters_product_sheet.pdf.

Console facts. (2002). *Interactive Digital Software Association*. Retrieved February 6, 2002 from http://www.idsa.com/consolefacts.html.

Damuth, R. (2001). Economic impacts of the demand for playing interactive entertainment software. *Interactive Digital Software Association*. Retrieved February 6, 2002 from http://www.idsa.org.

ESRBi Online Game & Website Ratings. (n.d.). *Entertainment Software Rating Board*. Retrieved February 6, 2002 from http://www.esrb.org.

Extended Play Staff. (2001, November 2). Microsoft's Xbox specs. *TechTV*. Retrieved February 8, 2002 from http://www.techtv.com/extendedplay/reviews/story/0,24330,3356862,00.html.

Fox, F. (2002, February 28). Japan Xbox launch: Not bad, not incredible. *Linuxworld.com*. Retrieved March 1, 2002 from http://www.linuxworld.com/ic_822477_6996_1-3921.html.

Game Boy Advance detailed specs. (n.d.). *Nintendo.com*. Retrieved February 8, 2002 from http://www.nintendo.com/systems/gba/gba_specs.jsp.

GameCube detailed specs. (n.d.). *Nintendo.com*. Retrieved February 8, 2002 from http://www.nintendo.com/ systems/gcn/gcn_specs.jsp.

Gunter, B. (1998). *The effects of video games on children: The myth unmasked*. Sheffield, England: Sheffield Academic Press.

Herman, L., Horwitz, J., & Kent, S. (2002). The history of video games. *Gamespot*. Retrieved February 7, 2002 from http://gamespot.com/gamespot/features/video/hov/.

Herz, J. C. (1997). *Joystick nation: How videogames ate our quarters, won our hearts, and rewired our minds*. New York: Little, Brown & Co.

Kane, M. (2002, February 8). Palm's grip on PDA market slips. *ZDNet News*. Retrieved February 8, 2002 from http://zdnet.com.com/2100-1103-827360.html.

Kent, S. (2001, January 31). Dreamcast dies. *MSNBC*. Retrieved February 8, 2002 from http://stacks.msnbc.com/news/524046.asp.

Kort, T., & Dulaney, K. (2001, December 20). Handheld PCs in 2002. *ZDNet News*. Retrieved February 8, 2002 from http://techupdate.zdnet.com/techupdate/stories/main/0,14179,2834203,00.html.

LaPlante, A., & Seidner, R. (1999). *Playing for profit: How digital entertainment is making big business out of child's play*. New York: Wiley.

Levy, S. (2000, February 27). Here comes PlayStation 2. *Newsweek*. Retrieved March 6, 2000 from http://www.newsweek.com/nw-srv/printed/us/st/a16816-2000feb27.htm.

Paul, F. (2001, March 9). Palm sees software as key to corporate sales. *ZDNet News*. Retrieved February 8, 2002 from http://www.nintendo.com/news/news_articles.jsp?articleID=5454; http://techupdate.zdnet.com/techupdate/stories/main/0,14179,2694410,00.html.

Pham, A. (2002, January 1). Game sales poised to hit record. *LAtimes.com*. Retrieved January 31, 2002 from http://www.latimes.com/business/la-000000030jan01.story.

PlayStation sales reach 980,000 units during opening weekend in Japan. (n.d.). *Sony.com*. Retrieved March 6, 2000 from http://www.scea.com/news/press_example.asp?ReleaseID=9558.

Poole, S. (2000). *Trigger happy: Videogames and the entertainment revolution*. New York: Arcade.

Provenzo, E. (1991). *Video kids: Making sense of Nintendo*. Cambridge, MA: Harvard University Press.

Quick facts about video game consoles and software. (2002). *Internet Digital Software Association*. Retrieved February 7, 2002 from http://www.idsa.org.

Sega focuses on its content and network strengths. (2001, January). *Sega.com*. Retrieved February 6, 2002 from http://www.sega.com/segascream/corporate/post_pressrelease.jhtml?PressRelID=10020.

Shim, R., & Fried, I. (2002, February 8). Are Palm developers making money? *ZDNet News*. Retrieved February 8, 2002 from http://zdnet.com.com/2100-1103-832609.html.

Supafly. (2002). *It's alive*. Retrieved January 31, 2002 from http://www.itsalive.com/downloads/ supafly_product_sheet.pdf.

Ten facts about the computer and video game industry. (2002). *Interactive Digital Software Association.* Retrieved February 7, 2002 from http://www.idsa.org.

Tran, K. (2002, February 7). Video-game sales top film box office. *Wall Street Journal.* Retrieved February 14, 2002 from http://www.msnbc.com/news/701429.asp.

U.S. computer and video game publishers lose billions worldwide to rampant piracy of entertainment software. (2002, February 14). *Interactive Digital Software Association.* Retrieved February 23, 2002 from http://www.idsa.com/2_14_2002.html.

Van Grinsven, L. (2002, March 13). Microsoft brings Xbox to Europe. *Reuters.* Retrieved March 13, 2002, from http://www.reuters.com/news_article.jhtml?type=technologynews&StoryID=696765.

Verizon launces first U.S. "3G" network. (2002, January 28). *CNN.com.* Retrieved on February 6, 2002 from http://www.cnn.com/2002/TECH/ptech/01/28/verizon.3g/index.html.

Video game console manufacturers to provide ratings information with purchase of all hardware. (1999, September 21). *Interactive Digital Software Association.* Retrieved on February 14, 2002 from http://www.idsa.com/releases/ESRBRatings.html.

12

THE INTERNET & THE WORLD WIDE WEB

Jim Foust, Ph.D.[*]

In less than a decade, the Internet has evolved from a technical curiosity to a major influence on nearly every aspect of life. The Internet has become a social force, influencing how, when, and why people communicate; it has become an economic force, changing the way corporations operate and the way they interact with their customers; and it has become a legal force, compelling reexamination and reinterpretation of the law.

In developed countries, the Internet has truly become a fixture of everyday life. No longer merely the domain of technically advanced "geeks," the Internet has, for many, become an indispensable tool for commerce, research, communication, and leisure. It is estimated that more than half of all Americans use the Internet, and that number is growing at an impressive rate. At the same time, there are concerns about a "digital divide" between those who have access to the Internet and those who do not. Despite the massive proliferation of connectivity in developed countries, significant populations of those countries do not have regular or reliable access to the Internet.

Although the terms "Internet" and "World Wide Web" are often used interchangeably, they have distinct—and different—meanings. The Internet refers to the worldwide connection of computer networks that allows a user to access information located anywhere else on the network. The World Wide Web refers to the set of technologies that places a graphical interface on the Internet, allowing users to interact with their computers using a mouse, icons, and other intuitive elements rather than typing obscure computer commands. The two technologies can be combined to make possible a variety of types of communications, discussed in more detail in the next section.

---

[*] Associate Professor, Department of Journalism, Bowling Green State University (Bowling Green, Ohio).

BACKGROUND

In the 1950s, the U.S. Department of Defense started researching ways to create a decentralized communications system that would allow researchers and government officials to communicate with one another in the aftermath of a nuclear attack. A computer network seemed to be the most logical way to accomplish this, so the military formed the Advanced Projects Research Agency (ARPA) to study ways to connect networks. At the time, there was no reliable way to combine local area networks (LANs), which connected computers in a single location, and wide area networks (WANs), which connected computers across wide geographic areas. ARPA sought to create a combination of LANs and WANs that would be called an "internetwork"; ARPA engineers later shortened the term to Internet (Comer, 1995).

By 1969, ARPA had successfully interconnected four computers in California and Utah, creating what came to be called ARPANET. A key innovation in the development of ARPANET was the use of TCP/IP (transmission control protocol/Internet protocol), a method of data transmission in which information is broken into "packets" that are "addressed" to reach a given destination. Once the data reaches its destination, the packets are reassembled to recreate the original message. TCP/IP allows many different messages to flow through a given network connection at the same time, and also allows for standardization of data transfer among networks. Interest in ARPANET from academia, government agencies, and research organizations fueled rapid growth of the network during the 1970s. By 1975, there were about 100 computers connected to ARPANET, and the number grew to 1,000 by 1984 (Clemente, 1998). In 1983, ARPANET became formally known as the Internet, and the number of computers connected to it continued to grow at a phenomenal rate (see Table 12.1).

Table 12.1
Number of Host Computers Connected to the Internet by Year

Year	# of Host Computers	Year	# of Host Computers
1981	213	1992	727,000
1982	235	1993	1,313,000
1983	562	1994	2,217,000
1984	1,024	1995	4,852,000
1985	1,961	1996	9,472,000
1986	2,308	1997	16,146,000
1987	5,089	1998	29,670,000
1988	28,174	1999	43,230,000
1989	80,000	2000	72,398,092
1990	313,000	2001	109,574,429
1991	535,000	2002	147,344,723

Source: Internet Software Consortium

THE DOMAIN NAME SYSTEM

Each computer on the Internet has a unique Internet protocol (IP) address that allows other computers on the Internet to locate it. The IP address is a series of numbers separated by periods, such as 129.1.2.169 for the computer at Bowling Green State University that contains faculty and student Web pages. However, since these number strings are difficult to remember and have no relation to the kind of information contained on the computers they identify, an alternate addressing method called the domain name system (DNS), which assigns text-based names to the numerical IP addresses, also is used. For example, personal.bgsu.edu is the domain name assigned to the computer at Bowling Green State University referred to above. Domain names are organized in a hierarchical fashion from right to left, with the rightmost portion of the address called the top-level domain (TLD). The TLD identifies the type of information that is contained on the computer. Thus, personal.bgsu.edu is said to be part of the ".edu" domain, which includes other universities and education-related entities. To the immediate left of the TLD is the organizational identifier; in the example above, this is bgsu. The organizational identifier can be a domain as well; the computer called "personal" is thus part of the "bgsu" domain. Table 12.2 lists TLDs in use as of mid-2002.

Table 12.2
Top Level Domain Names

Extension	Definition
.aero	Air-transport industry sites
.arpa	Internet infrastructure sites
.biz	Business sites
.com	Commercial sites
.coop	Cooperative organization sites
.edu	Educational institution sites
.gov	Government sites
.info	General usage sites
.int	International sites
.mil	Military sites
.museum	Museum sites
.name	Individuals' sites
.net	Networking and internet-related sites
.org	Sites for organizations
.pro	Sites for professions

Source: J. Foust

Domain names are administered by a global, nonprofit corporation called the Internet Corporation for Assigned Names and Numbers (ICANN), and the only officially authorized TLDs are those administered by ICANN (http://www.icann.org/). A series of computers called root servers, also known as DNS servers, contain the cross-referencing information between the textual domain names and the numerical IP addresses. The information on these root servers is also copied to many other computers. Thus, when a user types in personal.bgsu.edu, he or she is connected to the computer at 129.1.2.169.

TEXT-BASED INTERNET APPLICATIONS

Several communications applications developed before the rise of the graphical-based World Wide Web. Some of these applications have been largely replaced by graphical-based applications, while others merely have been enhanced by the availability of graphical components. All of these applications, however, rely chiefly on text to communicate among computers.

Electronic mail, or e-mail, allows a user to send a text-based "letter" to another person or computer. E-mail uses the domain name system in conjunction with user names to route mail to the proper location. The convention for doing so is attaching the user's name (which is often shortened to eight or fewer characters) to the domain name with an at sign (@) character. For example, the author's e-mail address is jfoust@bgnet.bgsu.edu. E-mail can be sent to one or many recipients at the same time, either by entering multiple addresses or by using a list processor (listproc), which is an automated list of multiple e-mail addresses. E-mail can also contain computer files, which are called attachments. The rise of graphical-based e-mail programs has also made possible sophisticated text formatting, such as the use of different font styles, colors, and sizes in e-mail communication.

Newsgroups are an outgrowth of early computer-based bulletin board systems (BBSs) that allow users to "post" e-mail messages where others can read them. Literally thousands of newsgroups are available on the Internet, organized according to subject matter. For example, the "alt.video.dvd" newsgroup caters to DVD enthusiasts, while "alt.sports.hockey" caters to hockey fans. One of the advantages of newsgroups is that they allow users to look back through archives of previous postings.

Chat allows real-time text communication between two or more users. One user types information on his or her keyboard, and other users can read it in real time. Chat can be used in either private situations or in open forums where many people can participate at the same time. To use chat, users normally enter a virtual "chat room" and are then able to send and receive messages. A related technology, instant messaging, allows users to exchange real-time text-based messages without having to be logged in to a chat room.

Telnet allows a user to log onto and control a remote computer, while file transfer protocol (FTP) allows a user to exchange files with remote computers. However, since both telnet and FTP are exclusively text-based, both have been, in most cases, supplanted by Web-based applications.

THE WORLD WIDE WEB

By the early 1990s, the physical and virtual structure of the Internet was in place. However, it was still rather difficult to use, requiring knowledge of arcane technical commands and programs such as telnet and FTP. All of that changed with the advent of the World Wide Web, which brought an easy way to link from place to place on the Internet and an easier-to-use graphic interface.

The WWW was the brainchild of Tim Berners-Lee, a researcher at the European Organization for Nuclear Research. He devised a computer language, HTML (hypertext markup language), that allows users with little or no computer skills to explore information on the Internet. The primary innovations of HTML are its graphical-based interface and seamless linking capability. The graphical interface allows text to intermingle with graphics, video, sound clips, and other multimedia elements, while the seamless linking capability allows users to jump from computer to computer on the Internet by simply clicking their mouse on the screen (Conner-Sax & Krol, 1999). WWW documents are accessed using a browser, a computer program that interprets the HTML coding, and displays the appropriate information on the user's computer. To use the Internet, a user simply tells the browser the address of the computer he or she wants to access using a uniform resource locater (URL). URLs are based on domain names; for example, the author's webpage URL is personal.bgsu.edu/~jfoust.

The advent of the World Wide Web was nothing less than a revolution. As illustrated in Table 12.1, the impressive growth rate of the 1970s and 1980s paled in comparison with what has happened since, as users discovered they did not have to have a degree in computer science to use the Internet. Internet service providers such as America Online (AOL) brought telephone line-based Internet access into homes, and businesses increasingly connected employees to the Internet as well. Since HTML is a text-based language, it is also relatively easy to create HTML documents using a word processing program (see Figure 12.1). However, more complex HTML documents are usually created using WYSIWYG (what you see is what you get) programs such as Microsoft Frontpage or Macromedia Dreamweaver. These programs allow users to create Web pages by placing various elements on the screen; the program then creates the HTML coding to display the page on a browser (see Figure 12.2).

Figure 12.1
Simple HTML Coding

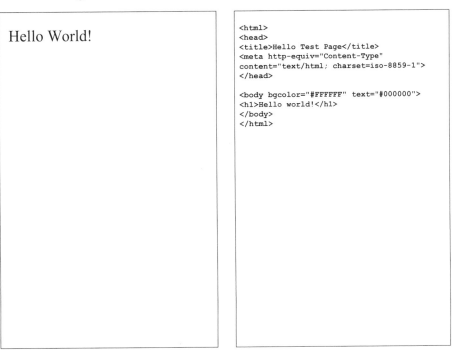

Source: J. Foust

Figure 12.2
Complex HTML Coding

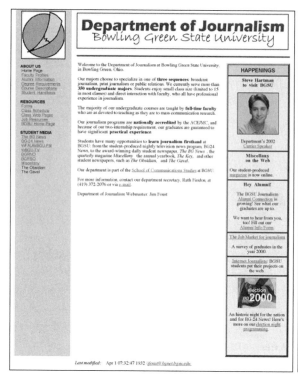

```
<html><!-- #BeginTemplate "/Templates/HomePageTemplate.dwt" -->
<head>
<!-- #BeginEditable "doctitle" -->
<title>BGSU Department of Journalism</title>
<!-- #EndEditable -->
<meta http-equiv="Content-Type" content="text/html; charset=iso-
8859-1">
</head>

<body bgcolor="#FFFFCC" leftmargin="0" topmargin="0"
marginwidth="0" marginheight="0">
<table width="704" border="0" cellspacing="0" cellpadding="0"
align="center" bgcolor="#FFFFFF">
  <tr>
    <td width="150" height="135" bgcolor="#FF9900">
      <div align="center"><img src="images/logo.gif" width="115"
height="120" vspace="0" hspace="0" name="Logo" alt="Journalism
Logo"></div>
    </td>
    <td width="549" height="135">
      <p><img src="images/headertext.gif" width="549" height="89"
alt="Department of Journalism Bowling Green State University"></p>
      <hr>
    </td>
  </tr>
</table>
<table width="704" border="0" cellspacing="0" cellpadding="0"
align="center" bgcolor="#FFFFFF">
  <tr>
    <td width="150" bgcolor="#FF9900" valign="top">
    <table border="0" cellspacing="0" cellpadding="5"
width="97%">
      <tr valign="top">
        <td nowrap>
          <p><b><font face="Arial, Helvetica, sans-serif"
size="2">ABOUT US<br>
             </font></b><font face="Arial, Helvetica, sans-
serif" size="2">Home
            Page </font><b><font face="Arial, Helvetica, sans-
serif" size="2"><br>
             </font></b><font face="Arial, Helvetica, sans-
serif" size="2"><a href="faculty.html">Faculty
            Profiles</a><br>
             <a href="alumni.html">Alumni
Information</a><br>                          (Continued)
```

A small part of the HTML code used to create the Web page on the left is illustrated on the right. Because the Web page is so complex, the actual code is quite extensive; printing it would take many pages.

Source: J. Foust

RECENT DEVELOPMENTS

A 2001 study commissioned by networking company Cisco Systems concluded that the Internet was becoming an inseparable part of the overall economy. "What started out as an alternate marketing channel has quickly turned into a complete economic system ... leading to the day when there will be no separate measure of the Internet economy," the study noted (Measuring the, 2001, p. 1). Similar conclusions could be reached when considering the Internet's effects on legal and social issues. Still, its relative newness and uniqueness as a communications medium often raise issues that have not been dealt with before. Principles, precedents, and even ethical guidelines often need to be reconsidered when applied to an inherently unique medium that has become so ubiquitous so quickly. That very ubiquity has itself changed the nature of the Internet, making it much less a novelty and much more a tool. One observer, in fact, said that 2001 was "the year the Internet stopped being fun" (Pruitt, 2001).

The terrorist attacks of September 11, 2001 had a significant impact on nearly every aspect of life, and the Internet was no exception as it both reflected and affected human existence. On the day of the attacks, Internet-based instant messaging and e-mail allowed people in the vicinity of the attacks to send brief "I'm okay" messages to loved ones while other types of communication were disabled. But popular news sites such as CNN.com and MSNBC.com could not handle the volume of traffic and were thus inaccessible to people looking for the latest updates on the unfolding events (Beacham, 2001). The aftermath of the attacks would also have significant economic, legal, and social repercussions.

ECONOMIC DEVELOPMENTS

For several years, Wall Street investors had been driving up stock prices of Internet-related companies, regardless of whether the companies were making money or even seemed to have the potential for making money. In 2001, however, the so-called "dot.com crash" brought an end to such exuberance, leaving hundreds of Internet-related companies out of business, and thousands of their employees out of work (Regan, 2001). The fall of the dot.coms also had an effect on Internet users, as numerous commercial sites disappeared, and many formerly free services started charging. News sites, for example, increasingly require subscriptions or charge in other ways for the use of their information. Salon.com magazine maintains some free material online, but users must pay $30.00 per year to access everything on the site. Meanwhile, free greeting card services Americangreetings.com and BlueMountain.com began charging a subscription fee for most of their services (Pruitt, 2001). (Internet commerce is discussed in more detail in Chapter 13.)

Economic uncertainty in the aftermath of the September 11 attacks only hastened these trends. However, the Internet and related e-mail, messaging, and virtual conferencing technologies turned out to be a great boon to businesses, as travel became a very uncertain proposition. Instead of flying employees to meetings and conferences, businesses invested in technologies that would allow virtual meetings and increased electronic communication. (For more on virtual meetings, see Chapter 24.)

LEGAL DEVELOPMENTS

In 1997, the U.S. Justice Department sued Microsoft, charging that the company was using monopoly power in bundling its Internet Explorer Web browser program with its Windows operating system. After a trial and several appeals, the Justice Department in 2000 ordered the breakup of the company into two separate entities: one to produce operating systems and the other to produce applications, such as browsers and word processing programs. However, the Justice Department of the incoming Bush administration disagreed with the decision, and late in 2001 announced that it would not seek to break up the company. Instead, the government and Microsoft reached an agreement that will make certain portions of the Windows operating system computer code available to other companies in an effort to prevent Microsoft from having an unfair advantage in producing Windows applications. A number of states including California and New Jersey, however, have refused to accept the settlement and continue their cases against Microsoft. AOL Time Warner, which now owns competing Web browser Netscape, has filed its own lawsuit alleging that Microsoft's anticompetitive practices gave Internet Explorer an unfair advantage over Netscape. As of mid-2002, Internet Explorer was estimated to have more than 90% of the browser market (Olavsrud, 2002).

Another high-profile legal dispute involved Napster, the Internet-based service that allows users to share music files. Napster has lost several court battles with the music industry, which has claimed that Napster infringes on copyrighted musical works. The Web site, which had an estimated 60 million users at its peak, effectively shut down in 2001, but has vowed that it will return using subscription fees to buy licenses for copyrighted material. Many of Napster's former users are now frequenting less centralized and lower profile services for music sharing that have proven more difficult for the music industry to sue, and the music industry is itself launching subscription-based music services of its own (Isenberg, 2001).

The 1998 Digital Millennium Copyright Act (DMCA) was intended, among other things, to prevent the subversion of copyright protection encryption. Such encryption uses sophisticated mathematical algorithms to prevent unauthorized copying of digital material, such as DVDs. In 2001, the DMCA was used to charge a Russian computer programmer with decrypting Adobe e-book files, and to prevent Web sites from posting or linking to files that allow users to copy DVD disks. The DMCA has been criticized by First Amendment advocates, who charge that it threatens free speech on the Internet (Isenberg, 2001).

SOCIAL DEVELOPMENTS

Computer viruses continue to be a danger for Internet users. Increasingly sophisticated viruses are able to circumvent protection software and can not only damage files on individual users' computers, but bog the Internet itself down by infecting Web servers. It is estimated that one in every 300 e-mail messages contains a virus of some sort, and there is increasing fear that viruses may be unleashed by terrorists. While individual users can protect themselves by not opening unfamiliar e-mail attachments and making sure they have the latest updates to their operating system, application, and virus protection programs, the Internet itself remains vulnerable to several types of virus threats (Legard, 2001).

Privacy is also a growing concern, not only from the standpoint of information about users that may be available on the Internet, but also in terms of sites users visit. Internet service provider Comcast alarmed many customers in early 2002 when it was revealed that the company had been tracking the Web usage of nearly a million of its customers without telling them. After the news became public, the company stopped the tracking, but the event only heightened fears of privacy advocates. Such information is potentially very valuable to companies that want to know the habits of customers and potential customers, and Microsoft has been criticized for programs such as Media Player and Passport that track what users have seen and done with them (No tracking, 2002). Privacy advocates are no less alarmed by the USAPATRIOT Act, passed by Congress in the wake of the September 11 attacks. The act gives law enforcement agencies greatly expanded powers to monitor the Internet usage of suspected criminals and to demand information from Internet service providers (Stenger, 2001).

CURRENT STATUS

It is estimated that about half a billion people worldwide have access to the Internet from their homes, and it is a safe assumption that many more who do not have home access can access the Internet from work or school. Forty percent of Internet users are in the United States, but the highest rates of growth are in Asia and Europe (Half a billion, 2002). By 2003, it is projected that the Internet population will be nearly 800 million (Global Internet, 2002). Meanwhile, the number of domains continues to grow, although not at the rate of 1999 and 2000 when the number nearly tripled. It is estimated that there were nearly 30 million domains as of February 2002 (Domain name, 2002).

The most comprehensive study of Internet usage in the United States to date was released in early 2002 by the Department of Commerce. Using census data collected during September 2001, *A Nation Online: How Americans are Expanding their Use of the Internet* revealed that more than half of all Americans are now online, and that number is growing by nearly two million people every month. About 20% of users in the United States were using high-speed data connections for their Internet access, and children and teenagers were found to use computers and the Internet more than any other age group (U.S. Department of Commerce, 2002).

Still, the survey revealed that there remains a "digital divide," especially among low-income people and the less educated:

➢ 75% of Internet non-users reside in households where income is less than $15,000 per year.

➢ 87.2% of people with less than a high school diploma do not use the Internet.

➢ 68.4% of Hispanics and 60.2% of blacks are not online.

The main reason given by those who did not have access was that it was too expensive. There was some good news, however. Users of the Internet in rural areas have nearly closed the gap with their urban counterparts, and access to computers in schools has helped ease the Internet access gap between low- and high-income children and teenagers (U.S. Department of Commerce, 2002).

FACTORS TO WATCH

Despite the dot.com crash on Wall Street, it is unlikely that the growth of Internet use will wane. As more users secure high-speed Internet connections, high-bandwidth applications such as video streaming and virtual conferencing will continue to develop. One area of potentially great growth is in virtual private networks (VPNs), which use shared public networks such as the Internet to create private connections between two or more computers. In a way, VPNs are the Internet come full circle: at first the struggle was to interconnect and make accessible a series of discrete networks, now businesses are working to carve their own private networks out of the vast public network that is the Internet (Tyson, 2002).

Meanwhile, a consortium of nearly 200 universities and 60 corporations is working on the so-called "Internet 2" project. Although it is not, as the name might imply, a replacement for the existing Internet, it seeks to fundamentally change online communication by increasing and then exploiting the speed of the existing Web. Participating universities are investing $80 million per year in the project, and corporations have pledged an additional $30 million over the life of the venture. The major goal of Internet 2 is "to ensure the transfer of new network technology and applications to the broader education and networking communities." The project is simultaneously developing technologies that will make the Internet faster, and applications such as digital libraries, virtual laboratories, and tele-immersion that will take advantage of that increased speed. Internet 2 seeks to do this by "recreat[ing] the partnership of academia, industry and government that helped foster today's Internet in its infancy" (Frequently asked, 2002). Once again, the Internet may be coming full circle.

BIBLIOGRAPHY

Beacham, F. (2001, October 17). 9/11: What worked, what didn't. *TVTechnology.com*. Retrieved March 11, 2002 from http://www.tvtechnology.com/features/Net-soup/f-fb-whatworked.shtml.

Clemente, P. C. (1998). *State of the net: The new frontier*. New York: McGraw-Hill.

Comer, D. E. (1995). *The Internet book*. Englewood Cliffs, NJ: Prentice Hall.

Conner-Sax, K., & Krol, E. (1999). *The whole Internet: The next generation*. Sebastapol, CA: O'Reilly & Associates.

Domain name facts. (2002, February). *netfactual.com*. Retrieved March 7, 2002 from http:///www.netfactual.com.

Frequently asked questions about Internet 2. (2002) Retrieved March 11, 2002 from http://www.internet2.edu/ html/faqs.html#.

Global Internet statistics. (2002). *glreach.com*. Retrieved March 7, 2002 from http://www.glreach.com/ globstats/index.php3.

Half a billion online. (2002, March 8). *NUA.com*. Retrieved March 7, 2002 from http://www.nua.ie/ surveys/?f=VS&art_id=905357729&rel=true.

Isenberg, D. (2001). Internet law 2001: The year in review and a look ahead. *Gigalaw.com*. Retrieved February 22, 2002 from http://www.gigalaw.com/articles/2001/pfv/isenberg-2001-12-pfv.html.

Legard, D. (2001, September 20). Viruses are getting faster, tougher. *CNN.com*. Retrieved March 11, 2002 from http://www.cnn.com/2001/TECH/internet/09/20/faster.virus.idg/index.html.

Measuring the Internet economy. (January 2001). *The Internet Economy Indicators*. Retrieved March 11, 2002 from http://www.internetindicators.com/jan_2001.pdf.

No tracking: Comcast to stop recording customer Web browsing. (2002, March 11). *ABCNews.com*. Retrieved March 11, 2002 from http://abcnews.go.com/sections/scitech/DailyNews/comcast020212.html.

Olavsrud, T. (2002, January 22). Netscape takes aim at Microsoft. (2002, March 11) *Internetnews.com*. Retrieved March 11, 2002, from http://www.internetnews.com/bus-news/article/0,,3_959661,00.html.

Pruitt, S. (2001, December 17). Opinion—2001: The year the Internet stopped being fun. *ITWorld.com*. Retrieved February 22, 2002 from http://www.itworld.com/Tech/2987/IDG011217internetfun/ pfindex.html.

Regan, T. (2001, December 27). After the dot.com crash. *Christian Science Monitor*, p. 13.

Stenger, R. (2001, September 13). Feds enlist ISPs in terrorist probe. *CNN.com*. Retrieved March 11, 2002 from http://www.cnn.com/2001/TECH/internet/09/13/fbi.isps/index.html.

Tyson, J. (2002). How virtual private networks work. *howstuffworks.com*. Accessed February 18, 2002 from http://www.howstuffworks.com/vpn.htm/printable.

U.S. Department of Commerce, National Telecommunications and Information Administration. (2002, February). *A nation online: How Americans are expanding their use of the Internet*. Retrieved March 5, 2002 from http://www.ntia.doc.gov/ntiahome/dn/index.html.

13

INTERNET COMMERCE

Julian A. Kilker, Ph.D.[*]

In the popular media, electronic commerce (e-commerce) is associated, particularly in the United States, with creating a booming "new economy" and a subsequent "dot.com bust" that is based on digital rather than physical commercial transactions. Fueled by a tremendous growth of Internet access and home computer ownership, e-commerce has increased substantially in recent years, and is predicted to continue to do so despite minor economic setbacks. The enthusiasm of recent years has been tempered with the understanding that e-commerce remains at an early stage in which regulations, viable business models, formal and informal standards, and impacts on related business sectors continue to emerge.

In 1999, the U.S. Bureau of the Census released a working definition of e-commerce as "any transaction completed over a computer-mediated network that involves the transfer of ownership or rights to use goods or services" (Mesenbourg, 1999). This definition does not take into account transactions conducted using other communication technologies, "free" transactions (such as downloading trial software), or barter and in-kind transactions (such as exchanging Web advertising placement for services). Most current definitions, and this chapter, emphasize commerce using the Web, although there are electronic predecessors to Web-based commerce. Business-to-customer (B2C) e-commerce has precedents in television shopping (which grossed approximately $4 billion in 1998) and in the use of communication and information technologies for catalog shopping (which grossed over $150 billion in 1998) (Grant & Meadows, 2002). Business-to-business (B2B) e-commerce, which has recently received media and corporate attention, is related to the earlier electronic data interchange (EDI), in which businesses electronically exchanged business documents in standardized formats over private rather than Internet data networks.

---

[*] Assistant Professor, Hank Greenspun School of Communication, University of Nevada Las Vegas (Las Vegas, Nevada).

E-commerce relies on a wide range of hardware, software, and communications technologies. Thus, the major players in e-commerce combine established companies, such as General Electric, IBM, SAP, Oracle, Cisco, and America Online (AOL), with more recent Web-centric companies, such as Network Appliance, Commerce One, Ariba, and Verisign, to name a few. Virtually every business sector is using e-commerce technology. These businesses range from relatively new companies such as Amazon.com (founded in 1996 as an online bookstore, this company has since branched into several product categories) to more established "bricks-and-mortar" companies with a physical presence such as Wal-Mart. The latter have recently realized the value of e-commerce and are establishing "*clicks*-and-mortar" operations that join the efficiencies of e-commerce with traditional services and distribution.

Tracking the rapid pace of e-commerce development requires close monitoring of current business and technology sources. At present, e-commerce is a challenging research topic for several reasons:

(1)   Its rapidly changing nature.

(2)   By the time academic publications on the subject are published, the field has changed.

(3)   Competitive concerns limit detailed or critical coverage (Lohse & Spiller, 1999).

Research is further complicated by the loss of analysis from industry sources such as online metrics firm PC Data Online, which shut down in April 2001, and *The Industry Standard*, a key resource for tracking e-commerce news that failed in late 2001. (At the time of this writing, its archives remain online.)

BACKGROUND

The non-commercial roots of the Internet led many early proponents to extol its virtues as a medium unfettered by commercial content. The same technology that offered early network users an easy and efficient way to share messages, data, and entertainment, however has proven to be just as efficient at connecting businesses with their customers.

Indeed, the commercial online services that many consider to be predecessors to today's Internet, such as Compuserve, Prodigy, and AOL, all considered commercial transactions over their networks to be an important part of their business model. The pioneers who chose to take a chance on the nascent commercialized Internet such as Amazon.com and Travelocity were rewarded for being "first movers" by capturing large market shares.

Because e-commerce replaces physical with digital transactions, it takes advantage of the time—and distance—collapsing nature of the Internet, as well as the data processing capabilities of computers to manage billing and order fulfillment. Ideally, e-commerce provides an efficient means for consumers and merchants to interact because consumers can rapidly locate and compare product specifications and pricing, and merchants can reduce their physical presence, have automated electronic storefronts open around the clock, achieve economies of scale, and reduce payroll expenses.

E-commerce has been influential in three major ways:

> *It adds value to existing services or makes existing services more accessible.* For example, banking and shipping companies provide online account or tracking information.

> *It substantially alters the nature of existing businesses.* For example, online bookstores provide automated searches and reader comments; online brokerages provide rapid, low-priced trades and immediate access to research documents; and online music stores can create custom compact discs and play audio samples.

> *Internet commerce reduces barriers to the implementation of new business models.* For example, eBay.com provides a forum for online trading, Priceline.com provides a "demand collection system" in which individuals make offers that are then aggregated to obtain low prices, and portal sites provide free services subsidized by advertising and tracking customer behaviors.

RECENT DEVELOPMENTS

ECONOMIC

The challenges of defining e-commerce and its rapid development make it difficult—but increasingly important—to quantify its impact in the domestic and global economy. E-commerce baseline and trend estimates should be viewed with particular care. The challenge of determining e-commerce baselines is demonstrated by the eMarketer finding that 1999 estimates for e-commerce varied from a low of $3.9 billion to a high of $36 billion, and that none of these estimates took into account B2B e-commerce. The Department of Commerce's first report of retail e-commerce sales, using techniques similar to its Monthly Retail Trade Survey, estimated sales of $5.3 billion, or about 0.64% of total U.S. retail sales for the fourth quarter 1999. This was in accord with Forrester Research's estimate of $20.3 billion for all of 1999, up from an estimated $400 million for 1997 (Grant & Meadows, 2002).

The Department of Commerce's 4Q 2000 results were $8.9 billion, an increase of 69% over 4Q 1999; and 4Q 2001 results were $10.0 billion, an increase of 13% over 4Q 2000 (see Figure 13.1). All in all, as a percentage of total sales in the United States, e-commerce increased slowly but steadily (taking into account seasonal and other adjustments) from 0.7% in 4Q 1999 to 1.2% in 4Q 2001 (U.S. Department of Commerce, 2002b). It is estimated that U.S. annual online retail revenues will increase to $103.3 billion by 2006 (see Figure 13.2).

Figure 13.1
**Estimated Quarterly U.S. Retail E-commerce
Sales (Unadjusted) (US$ Billions)**

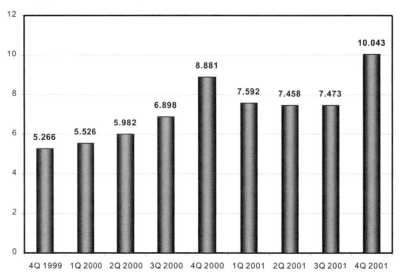

Source: U.S. Department of Commerce

Figure 13.2
**Predicted U.S. Online Retail Revenues
(US$ Billions)**

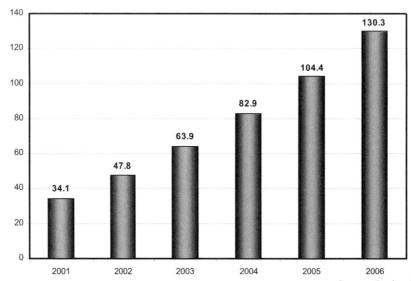

Source: Jupiter Media Metrix, Inc.

Because e-commerce spans geographic, hence regulatory, boundaries, its regulation has received increased U.S. and international attention. The Anticybersquatting Consumer Protection Act was signed into U.S. law in November 1999 in response to corporate concerns about the unauthorized registration of domain names using trademarks, in some cases in order to hold them hostage or resell them for a profit. Fair taxation has also been a serious concern: To support then-fledgling e-commerce, the U.S. Internet Tax Freedom Act of 1998 declared that "the Internet should be a tariff-free zone," and placed a three-year moratorium on "multiple and discriminatory taxes on electronic commerce" (Plain English, n.d.). This moratorium on U.S. Internet-related taxes applies to transactions across the United States, and was renewed for another two years in November 2001. The ongoing controversy is whether a state should be able to tax interstate sales initiated within its boundaries, as well as tax revenues for companies having a physical presence within its boundaries regardless of customer location.

The stakes are high, particularly because B2B sales have the potential to increase rapidly: U.S. state and local sales tax losses for 2001 due to e-commerce were estimated to be \$13.3 billion, rising to \$45.2 billion in 2006 (Bruce & Fox, 2001). In response to concerns about complex and inefficient tax structures, a group of state governors is attempting to simplify tax calculation and collection. Internationally, the European Commission has proposed that companies collect a value-added tax (VAT) on products such as software and music sold and distributed over the Internet. As of this writing, it appears that non-European Union countries will be required to collect a value-added tax for products sold to EU consumers over the Internet starting July 1, 2003.

Developing and maintaining online privacy standards have been ongoing concerns because data gathered online can be misused (at least from the perspective of consumers). In October 1998, the U.S. Congress passed the Children's Online Privacy Protection Act (COPPA), effective April 2000, to protect the collection and use of personal information from children up to the age of 13. In March 2000, the U.S. Department of Commerce and the European Commission agreed on a "safe harbor" arrangement that handles the gap between the two systems for protecting an individual's private information (U.S. Department of Commerce, 2000).

The weaknesses of online security were emphasized by reports of a massive 1999 theft of online credit card records (Brunker, 2000), and the February 2000 malicious "distributed denial of service" attacks that flooded several e-commerce and government Web sites with spurious requests, resulting in outages typically lasting several hours (Abreu, 2000). Shortly thereafter, a U.S. government working group released a report examining whether "existing federal laws are sufficient to address unlawful conduct involving the use of the Internet" (President's Working Group, 2000); the report hints at increased government interest in regulating and protecting e-commerce. This is evident in the technology and practice of digital rights management (DRM) and the associated legal protection of the Digital Millennium Copyright Act (DMCA) of 1998 (U.S. Copyright Office, 1998), both of which have recently received increased attention as content venders and producers cite concerns about the ease with which online e-book, audio, and video content can be copied and protection measures circumvented. In addition, e-commerce technology's ability to act as an intermediary in transactions has hit some established bricks-and-mortar businesses particularly hard (booksellers and travel agents are notable examples), leading them to seek legal protection. Some companies, such as Levi Strauss, have removed e-commerce capabilities from their sites in order to avoid alienating their distributors.

USERS

As Internet usage has grown in recent years, so too has e-commerce. The U.S. Department of Commerce reported, based on census data, that the rate of growth of Internet users in the United States was about two million new users per month by September 2001, and that more than half—54%—of the U.S. population was online for the first time, with Internet use increasing in every demographic category. This is still less than the number of users having access to traditional communication technologies used for commerce—such as the television (approximately 99% of households own at least one) and the telephone (approximately 94% of households have access). The "digital divide," in which access to the Internet is correlated with higher education and income, as well as ethnicity, with implications for their access to e-commerce, appears to be diminishing based on these polls (Hoffman, et al., 2000).

In comparing Internet use between August 2000 and September 2001, the U.S. Department of Commerce found that, while the most common use of the Internet continued to be e-mail (in September 2001, 45.2% of the U.S. population reported using e-mail, up from 35.4% in August 2000), other uses were of particular relevance to e-commerce. Using the Internet to search for product and service information was common (36.2% of the U.S. population in 2001, up from 26.1% in 2000), and 21% reported making online purchases and 8.1% conducted banking online in 2001, as opposed to the 13.3% who reported both activities in the August 2000 survey (U.S. Department of Commerce, 2002a).

As with traditional businesses, key challenges in e-commerce are attracting and retaining customers. The Internet Advertising Bureau reported revenues for online advertising of $1.79 billion in 3Q 2001 (a 9.8% drop from 3Q 2000's result of $1.99 billion because of the soft economy) (Interactive Advertising Bureau, 2001). A key player is the DoubleClick network (founded in 1996), which delivers general and targeted advertisements. DoubleClick reported delivering 172 billion advertising "impressions" on its "Dynamic Advertising, Reporting, and Targeting" e-mail systems during 4Q 2001, an increase of 7% over the previous quarter. Because "click-through" rates for online banner ads are very low—much less than 1%—e-commerce sites relied on expensive traditional mass media such as billboards, radio, and television to attract customers before the U.S. economy softened. More recent, and less expensive, trends involve increasingly obtrusive online advertisements such as "pop-up" and "pop-under" windows and Flash animation with audio effects.

E-commerce sites are especially concerned about alienating customers because of the ease with which they can compare, critique, and discuss products and Web sites. Holiday periods with traditionally high purchasing volumes have previously strained the ability of e-commerce companies to rapidly scale their businesses, maintain customer satisfaction, and retain customers; this was less of a problem during the 2001 holiday season.

TECHNOLOGY

At its simplest, e-commerce relies on the basic client/server model. A customer uses a standard Web browser (the client) to view product information on a merchant's Web server. If the customer decides to purchase a product, encrypted delivery and billing information is entered by the customer and sent to the merchant. After payment verification, the product is then transmitted if it is data (such as software, an electronic report, or a music file), or packed and shipped using delivery services if it is a physical product.

"Fulfillment" technologies, including inventory management, warehousing, logistics, and shipping, are critical but often overlooked components of e-commerce; they proved to be a less serious bottleneck in 2001 than in previous holiday seasons. In 1999, -eToys.com, for example, gave $100 gift certificates to each customer whose promised delivery date was missed because of limitations in fulfillment planning. Just as managing physical inventories is difficult, so is predicting and managing the use of communication resources. This is because of the cost of network technologies, including bandwidth and servers, and because of the challenges of balancing loads across multiple systems and planning for the scale-up of capacity. For example, Stephen King's release of an online novella in March 2000—one of the first such releases by a prominent author—produced 400,000 orders in 24 hours and exceeded the resources of several online booksellers (Carvajal, 2000). Amazon.com has proved to be a leader in inventory management and logistics, and other firms, such as Target and Toys "R" Us, have entered into partnering agreements to benefit from this experience (recent reports suggest, however, that these partners are concerned that Amazon has overextended itself).

Software is crucial to managing e-commerce transactions, maintaining product and customer databases, and serving Web pages. Key challenges for software include linking with legacy systems (older systems with different standards), coordinating multiple communications channels for marketing and customer service (Web, print catalogs, e-mail, and telephone), responding in real time to customer requests, maintaining general system usability (balancing server loads and minimizing downtime), and unobtrusively archiving transaction data. E-commerce packages for small to medium-sized businesses that include server, Web design, and billing and customer service applications are provided by a wide range of companies as different as Earthlink (an Internet service provider), Network Solutions (originally the only domain name registrar), and Broadvision (an e-commerce software developer). Enterprise resource planning (ERP) applications integrate a company's entire range of software tools into a single application that often includes electronic data information communications with other companies (Kumar & Hillegersberg, 2000). Traditional ERP firms such as SAP AG (founded in 1972) and Oracle (founded in 1977) have incorporated Internet e-commerce technologies into their latest products. The recently developed extensible markup language (XML) standard allows new markup elements to be defined and simplifies document viewing on multiple platforms; these characteristics make XML particularly useful for both B2B and B2C e-commerce.

Early e-commerce implementations used simple Web sites, whose design often imitated existing brochures and catalogs with the addition of an online order form. More recent sites use program scripts in CGI, PERL, C++, PHP, and Java, and use specialized production tools (such as Dreamweaver, Flash, and streaming media) to create content. Retail site design has improved since its infancy (Lohse & Spiller, 1999), although key problems remain with downtime, access speed, security, and usability (the last due to confusing navigation and interfaces and slow responses). It is also possible that design innovation has been stunted by patent disputes: To the consternation of e-commerce retailers, Amazon.com patented its "one-click" improvement to the popular shopping cart metaphor, and then took legal action against competitor Barnes & Noble for using a similar system (Gleick, 2000). Priceline.com has been similarly aggressive in patenting e-business concepts.

Online credit card fraud and the theft of billing databases have been serious concerns (Brunker, 2000). One survey found that merchants (rather than customers) must often bear the costs of fraud, and approximately two-thirds of merchants polled viewed fraud as a serious problem (Helft, 2000). Secure Socket Layer (SSL) encryption is used by most e-commerce Web sites to keep transaction information secure, and digital certificates provide additional security. A prominent attempt to balance convenience and security is Microsoft's Passport, which provides a single protected account containing personal and payment information. Despite intensive marketing and the large number of user accounts—Microsoft has leveraged its software and services: using Hotmail and registering software require Passport accounts—security flaws and poor publicity currently hamper this effort. In response to concerns about the online "pirating" of audio, video, and software files, encryption and authentication technologies are also being used by producers to control access to purchases delivered online. Early implementations of such technologies, however, have tended to be clumsy and often overly restrictive, especially when compared by users to the free access provided by earlier peer-to-peer file-sharing tools that demonstrated the technical feasibility of online access to audio and video content.

Online advertising has becoming increasingly sophisticated through the use of "cookie" files that store data on customers' computers, registration databases, click-stream tracking by which e-commerce sites trace the patterns users follow in moving from page to page in a Web site, and personalization tools. The appearance of advertising has also changed: Static banner ads are increasingly being replaced by more intrusive animated and interactive multimedia ads. A key challenge in using technology to track and serve marketing information to customers is no longer technical but social: Customers have become more resistant to marketing as the appeals become more aggressive. E-commerce sites have begun to use e-mail as a sophisticated marketing and customer service tool rather than just for mass mailings. Using appropriate software, e-mail content can be personalized, triggered by customer- or merchant-defined alerts, and programmed to confirm orders. Amazon has proven particularly adept at managing its communication with customers both via personalized e-mail and dynamically generated Web pages.

CURRENT STATUS

Approximately 29 million people in the United States purchased gifts online during the 2001 holiday season, spending an average of $392 per person, up from 20 million people spending an average of $330 during the 2000 holiday season (Pew Internet, 2002). The top 10 shopping sites by visits for the holiday season 2001 are shown in Table 13.1.

The difficult economic conditions that began in 2000 have precipitated a serious shakeout among e-commerce companies. No longer do online retailers spend generous amounts of money, as they did in early 2000, attracting customers through expensive marketing and by routinely providing free shipping, rebates, and substantial discounts on purchases. The Industry Standard reported that a subsection of consumer-oriented Web companies spent an "eye-popping" average of 69% of their revenue on sales and marketing efforts in 1999 (Mowrey, 2000).

Ready access to venture capital had enabled e-commerce "dot.coms" to proliferate, but many of these companies failed or merged in 2001 because of an increasingly tight economy, high start-up expenses (including establishing a brand identity), high ongoing costs (such as delivery, promotions, and technology upgrades), tight margins due to high price sensitivity of online retailing (because customers can easily compare prices), intense competition, and lax spending controls exacerbated by attempts to rapidly achieve an initial public offering (IPO) of stock.

Table 13.1
Top 10 Shopping Sites by Average Daily Visitors, Holiday Season 2001

Company	# of Daily Visitors
Ebay.com	4,515,000
Amazon.com	2,519,000
Mypoints.com	2,016,000
Bizrate.com	683,000
Half.com	660,000
McAfee.com	652,000
Columbia House sites	598,000
Eshop.com	588,000
Americangreetings.com	563,000
Toys "R" Us (Amazon.com)	515,000

Source: Jupiter Media Metrix, Inc.

Such uncertainties have created volatile competitive contexts. In the online toy market, for example, eToys (launched in 1997) bought out its competitor Toys.com in 1998. In contrast, Toysrus.com, affiliated with the traditional retailer, had done poorly since 1998; it closed its own site and entered into a partnership with established online retailer Amazon.com in late 2000. In evaluating e-commerce results, the focus has shifted from number of customers, innovative technologies, and media publicity to the more traditional notion of profitability, although exactly how this is defined remains controversial. For example, Amazon declared its first quarterly profit of $5 million on sales of $1.12 billion in 4Q 2001 under "generally accepted accounting principles," a major landmark for the firm. Industry analysts had been critical of earlier estimates using pro forma accounting (these would have resulted in a "profit" of about $35 million for the same quarter) (Dignan, 2002).

In general, though, the problems listed above were exacerbated by the softening world economy. Online grocer Webvan declared Chapter 11 bankruptcy in July 2001. It had bought out rival Home-Grocer for $1.2 billion in stock and developed sophisticated e-commerce technology, but could not recover the high costs of home delivery) (Farmer & Sandoval, 2001). Online delivery firms Kozmo and Urbanfetch met similar ends. Business-to-business surplus steel site Metalsite went under in September 2001 because it had difficulty locating suppliers, but it has been reborn, perhaps temporarily. Because of the poor advertising market, online publishers such as *Salon* have reluctantly begun to revise their business models and charge for content (Rosenberg, 2001). The soft advertising market has also resulted in a noticeable increase in obtrusive marketing tactics online, including more "unsolicited bulk e-mail" (popularly known as spam); even the venerable *New York Times* site has resorted to the unpopular advertising tactic of popup windows.

Other prominent failures include Biztravel.com (affected by the drop in travel after the September 11 terrorist attack); online pet stores Pets.com, Petstore.com, and Petopia.com; specialty store BBQ.com; community portal theglobe.com (famous for one of the highest IPOs ever—stocks increased over 600% on the first day of trading—it closed in August 2001); and music stores cdworld.com and musicmaker.com (whose customized-CDs business plan was rendered obsolete by consumer CD burners and peer-to-peer file transfers). However, although 12% of U.S. Internet users reported losing a favorite Web site by August/September 2001, and 17% had been asked to pay for previously free content, they easily adjusted, often by finding free alternatives (Pew, 2001). The greatest negative impact on individual users appears to be that many Americans directly or indirectly lost money invested in dot.coms.

Despite the recent negative perceptions of e-commerce, there were positive developments. Holiday sales in 2001 increased over 2000; although spending was weak (U.S. Department of Commerce, 2002b), traffic to shopping sites was up approximately 50% from the previous year (Jupiter Media Metrix, 2002). Several major e-commerce companies were considered successful by standard business metrics: eBay profits increased 8% in 4Q 2001 to $25.9 million (Profit at eBay, 2002), Amazon declared a profit (as noted above), and online payment service PayPal had a successful initial public offering (Internet start-up's, 2002), despite concerns about its customer service and the general challenge for online payment services of balancing usability with security. In addition, existing and new companies benefited from the resources and experiences of collapsed companies. One online business magazine even coined the phrase "first failure" (based on "first mover") to describe the advantages that early failures have in rebuilding, using the example of online grocer Peapod (Moore, 2001).

With rare exceptions, consumers still resist paying for online content. Slate.com and *New York Times on the Web*, having once experimented with access fees, continue without them (the latter charged for non-U.S. Web access). Advertising has become more intrusive, some online publications have closed, and a few have developed tiered access. For example, in April 2001, *Salon* instated a $30 annual fee to access its "premium" content (Rosenberg, 2001), while the *Internet Movie Database*, which has matured into an essential resource for the entertainment industry, provides access to "IMDb-pro" for a $100 annual fee. But other sites continue to provide "free" services such as e-mail (Hotmail.com) and Web site hosting (Tripod.com), and portal sites such as Yahoo.com and Excite.com, among others, maintain a wide array of personalized services (such as tailored weather, news, stock, and sports information). These sites use demographic data, personalization choices, and customer searches to develop marketing profiles and display targeted advertising.

As DSL (digital subscriber line) and cable modems have become more popular, e-commerce sites are beginning to use more bandwidth-consuming multimedia content such as streaming audio and video, interactive views of products, and real-time online customer support. Residential use of broadband service has increased from 11% (August 2000) to 20% (September 2001) of U.S. Internet users (U.S. Department of Commerce, 2002a), although telecom business failures and expensive connections have moderated the adoption rate of these technologies. Early experiments with unique branding features such as Eddie Bauer's brief flirtation with a "virtual dressing room" were not successful in part because such specialized installations require extra effort on the part of potential customers and were slow when using a standard modem. But multimedia tools, including Flash and video players, have become standardized and are widely installed on customer computers.

FACTORS TO WATCH

While mergers and business failures have removed the less-sustainable e-commerce companies, it remains to be seen which companies with which characteristics will thrive rather than merely survive. A key factor to watch is the development and adoption of e-commerce metrics, such as the recently developed guidelines for the measurement of interactive audiences (Interactive Advertising Bureau, 2002). B2B will require a stronger collaboration among suppliers before its revenues increase dramatically. As e-commerce matures and its revenues increase, maintaining a tax moratorium on Internet purchases will become more difficult (the current moratorium is only for two years, less than its proponents requested).

Even though a majority of Americans now use the Internet, differences in access to it—and the reliance on credit cards for B2C billing—continue to suggest that e-commerce growth will vary by demographics and socio-economic status. Consumers are increasingly using the Internet in collaboration with traditional channels (catalogs and retail stores) to make purchases. Forrester Research predicts that the number of new online shopping households in the United States will drop from a high of 11 million in 2000 to 2.6 million in 2004, but that the amount spent by each household will rise as they begin to buy routine or frequent purchases (such as groceries) online. However, routine household purchases have yet to be financially viable for e-commerce marketers, as demonstrated by multiple failures in the online grocery business. The globalization of e-commerce will increase as more international users, particularly those in Europe and Asia, continue to adopt Web technologies. High telecommunications costs and the lower popularity of the credit card as a form of payment remain limiting factors in some countries. Online payment services such as PayPal can facilitate international payments, but security concerns and customer service across international boundaries continue to be problematic.

New e-commerce software and services will likely continue to emphasize:

- ➢ Integration of multiple tools and systems.

- ➢ Real-time analysis of purchasing and browsing behaviors.

- ➢ More sophisticated and personalized customer assistance.

- ➢ Robust scalability and data archiving.

An additional development to watch is the embedding of marketing and monitoring technologies in consumer technologies. Microsoft's Windows XP has been widely criticized for overzealously prompting its users to sign up for Microsoft's Passport service, and companies already pay to place advertising links to their products and services on new PC desktops. The ability to shape user access to e-commerce is technically trivial, but it results in resistance if implemented crudely. This issue will become increasingly critical as new technologies, such as interactive television set-top boxes, provide e-commerce features.

As very high-capacity network bandwidth and storage devices become affordable to consumers, e-commerce retailing is likely to develop into a more personalized and richer multimedia version of television shopping, and more bandwidth-intensive products such as videos will be regularly delivered online. Adequately securing such files is a serious concern for content providers, but recent DMCA-related legal decisions favor the providers and are likely to influence future customer technologies. Low-bandwidth services should also be watched closely: Personal digital assistants and cellular telephones can provide wireless access to Web services, including e-commerce. However, the "killer app" for these tools has yet to emerge. Much depends on whether customers accept the restrictions on the devices: potentially high communication charges, limited bandwidth, and small screen displays. If wireless e-commerce is successful, it may influence the design of e-commerce sites, particularly with respect to low- and high-bandwidth versions.

Recent forecasts and press reports about e-commerce, while not as optimistic as in recent years, tend to emphasize the extent to which e-commerce has become integrated in the daily routines of many people. It is important to keep in mind that the physical nature of many commercial transactions remains a limiting factor. While purchases that involve digital content such as software, music, images, tickets, and e-books can be ordered and delivered online, the physical nature of some purchases makes warehouse and delivery technologies critical to e-commerce success. In addition, some customers and certain product categories will continue to require hands-on interaction. E-commerce will continue to be an interesting topic of study as customers and businesses attempt to identify products and services for this new environment that are useful rather than frivolous, and learn how to integrate e-commerce with traditional commerce channels.

BIBLIOGRAPHY

Abreu, E. (2000, February 21). The hack attack. *The Industry Standard,* 66-67.

Bruce, D., & Fox, W. F. (2001, September 2001). State and local sales tax revenue losses from e-commerce: Updated estimates. *Center for Business and Economic Research, The University of Tennessee.* Retrieved March 14, 2002 from http://cber.bus.utk.edu/ecomm/ecom0901.pdf.

Brunker, M. (2000, March 17). Major online credit card theft exposed. *MSNBC*. Retrieved March 14, 2002 from http://www.zdnet.com/zdnn/stories/news/0,4586,2469820,00.html.

Carvajal, D. (2000, March 16). Long line online for Stephen King e-novella. *New York Times,* A1, C10.

Dignan, L. (2002, January 22). Amazon posts its first net profit. *CNET News*. Retrieved March 14, 2002 from http://news.com.com/2100-1017-819688.html.

Farmer, M. A., & Sandoval, G. (2001, July 9). Webvan delivers its last word: Bankruptcy. *CNET News.* Retrieved March 14, 2002 from http://news.com.com/2100-1017-269594.html.

Gleick, J. (2000, March 12). Patently absurd. *New York Times Magazine,* 44-49.

Grant, A. E., & Meadows, J. H. (2002). Electronic commerce: Going shopping with QVC and AOL. In C. A. Lin & D. J. Atkin (Eds.). *Communication technology and society: Audience adoption and uses.* Cresskill, NJ: Hampton Press.

Helft, M. (2000, March 6). The real victims of fraud. *The Industry Standard,* 152-165.

Hoffman, D. L., Novak, T. P., & Schlosser, A. E. (2000). The evolution of the digital divide: How gaps in Internet access may impact electronic commerce. *Journal of Computer-Mediated Communication, 5* (3). Retrieved March 14, 2002 from http://www.ascusc.org/jcmc/vol5/issue3/hoffman.html.

Interactive Advertising Bureau. (2001, December). *IAB Internet advertising revenue report: 2001 Q3 report.* Retrieved March 14, 2002 from http://www.iab.net/main/IAB_PWC_2001Q3.pdf.

Interactive Advertising Bureau. (2002, January). *IAB issues first-ever guidelines for interactive audience measurement and advertising campaign reporting and audits.* Retrieved March 14, 2002 from http://www.iab.net/news/content/01_15_02.html.

Internet start-up's offering wows investors. (2002, February 16). *New York Times,* B1.

Jupiter Media Metrix. (2002, January 7). Holiday 2001 online shopping results: Traffic up 50 percent compared with last year. *Press Release*. Retrieved March 14, 2002 from http://www.jmm.com/xp/jmm/press/2002/pr_010702.xml.

Kumar, K., & Hillegersberg, J. (2000). Enterprise research planning (ERP): Experiences and evolution [Special Issue]. *Communications of the ACM, 43* (4), 22-69.

Lohse, G. L., & Spiller, P. (1999). Internet retail store design: How the user interface influences traffic and sales. *Journal of Computer-Mediated Communication, 5* (2). Retrieved March 14, 2002 from http://www.ascusc.org/jcmc/vol5/issue2/lohse.htm.

Mesenbourg, T. L. (1999). *Measuring electronic business: Definitions, underlying concepts, and measurement plans*. U.S. Census Bureau. Retrieved March 14, 2002 from http://www.census.gov/epcd/www/ebusines.htm.

Moore, J. F. (2001, August 14). Why Peapod is thriving: First-failure advantage. *Business 2.0.* Retrieved March 14, 2002 from http://www.business2.com/articles/web/0,1653,16795,FF.html.

Mowrey, M. A. (2000, March 6). Financial spotlight: Wall Street impatient with runaway dot-com marketing spending. *The Industry Standard*. Retrieved March 14, 2002 from http://www.thestandard.com/article/0,1902,12553,00.html.

Pew Internet & American Life Project. (2001, November). *The dot-com meltdown and the Web*. Retrieved March 14, 2002 from http://www.pewinternet.org/reports/toc.asp?Report=48.

Pew Internet & American Life Project. (2002, January). *Women surpass men as e-shoppers during the holidays*. Retrieved March 14, 2002 from http://www.pewinternet.org/reports/toc.asp?Report=54.

"Plain English" summary of The Internet Tax Freedom Act. (P.L. 105-277). (n.d.). Retrieved May 3, 2002 from http://cox.house.gov/nettax/lawsums.html.

President's Working Group on Unlawful Conduct on the Internet. (2000). *The electronic frontier: The challenge of unlawful conduct involving the use of the Internet*. Retrieved March 14, 2002 from http://www.usdoj.gov/criminal/cybercrime/unlawful.htm.

Profit at eBay increased 8% last quarter. (2002, January 16). *New York Times*, C5.

Rosenberg, S. (2001, December, 2001). *Inside Salon Premium: Are consumers ready to pay for content?* Retrieved March 14, 2002 from http://www.webtechniques.com/archives/2001/12/rosenberg/.

U.S. Copyright Office. (1998, December, 1998). *The Digital Millennium Copyright Act of 1998* [U.S. Copyright Office Summary]. Retrieved March 14, 2002 from http://www.loc.gov/copyright/ legislation/dmca.pdf.

U.S. Department of Commerce. (2000, March 17). *Documents regarding Department of Commerce's work to develop a "safe harbor" that would help U.S. organizations comply with the European Union's Directive on Data Protection*. Retrieved March 14, 2002, from http://www.ita.doc.gov/td/ecom/.

U.S. Department of Commerce. (2002a, February). *A nation online: How Americans are expanding their use of the Internet*. Retrieved March 14, 2002 from http://www.ntia.doc.gov/ntiahome/dn/.

U.S. Department of Commerce. (2002b, February). *Retail e-commerce sales Q4 2001*. Retrieved March 14, 2002 from http://www.census.gov/mrts/www/ecom.pdf.

14

OFFICE TECHNOLOGIES

Mark J. Banks, Ph.D. & Robert E. Fidoten, Ph.D.[*]

Access cards, metal detectors, electronic search wands, biometric recognition technologies, surveillance cameras, computer surveillance software, and employee identification cards embedded with biometric identification markers. These "new" office technologies are beginning to take a place alongside the traditional technologies of the office, such as the computer and computer networking, the Internet, and multifunction machines. Most of these security-related developments have accelerated in the aftermath of the September 11, 2001 terrorist attacks on the World Trade Center and the Pentagon.

In addition to these technological developments, the attacks have spurred changes in office configurations and work functions—changes that also rely upon the growing use of mobile or portable technologies to accomplish the work usually conducted within the office. Many of the technologies used in office settings and for office functions are described in other chapters. This chapter looks at some specific technologies, but also explores the larger picture of the use and impact of information and communication technologies in the workplace.

Generally, office technologies, both well-established and newer ones, can be grouped into three main categories:

> ➤ *Production technologies,* including such things as desktop publishing (DTP), Web design or multimedia technologies, and electronic editing, allow a single worker to use computers, printers, cameras and media hardware, along with appropriate software, to incorporate research, art, photography, charts, graphs, writing, layout and design, audio and

---

[*] Mark Banks is Professor and Chair and Robert Fidoten is Associate Professor. Both are faculty in the Communication Department, Slippery Rock University (Slippery Rock, Pennsylvania).

video, and printing into documents such as newsletters, presentations, Web sites, notices, and reports at a professional-quality level. These technologies also include typewriters and copiers, and the so-called multi-function products (MFPs) that combine several functions into one unit, such as printing, scanning, fax, word processing, and telephone.

> *Communication technologies* allow workers to send, receive, and otherwise interconnect with other people or with central servers to share or jointly work on information, reports, and databases that are accessible to several people, including both internal and external workers or clients. Among these technologies are the facsimile (fax) machine, telephone, PBX (private branch exchange used to connect telephones within an office or organization), local wired computer networks (LANs), global networks such as the Internet, wide-area networks (WANs), intranets, and, to a growing degree, wireless communication technologies such as cell phones, wireless PDAs (personal digital assistants), wireless LANs, and wireless access to the Internet.

> *Storage technologies* allow for the electronic storage and retrieval of information, and are an important factor in the expansion/creation of the so-called "paperless office." As computer capacity grows, there is a corresponding capability to store and retrieve data in larger bundles and with greater speed, including audio and video files, in addition to the traditional information and data formats.

Some technologies, such as personal digital assistants, desktop, laptop, or notebook computers, accommodate all three functions of production, communication, and storage/retrieval. In fact, the "office" is becoming increasingly more of a function than a singular setting. With telecommuting and mobile offices, teleconferencing, e-mail, and voice mail, the office becomes more where the worker is than a place where the worker goes.

BACKGROUND

The old office technologies often seemed to hang on forever. Although hand writing documents gave way to the typewriter in the early 1900s, that technology and the telephone, dictaphone, and hand-delivered mail dominated the office environment through the first seven decades of the 20th century. The copy machine, which was added in the 1960s, represented one of the first major "modern" additions to the office.

During the late 1970s and throughout the 1980s, a convergence of several technologies including the personal computer led to a mini-revolution in office technologies—innovations that would significantly change the nature of the office and its workers. The change was so significant that Smith-Corona, vender of over 70% of typewriters in the United States, filed for bankruptcy for the second time in late 2000. Its sales fell 95% in the 1990s (Kafka, 2000).

In the late 1970s and 1980s, the nature of sending information from one point to another changed. "Snail mail" from the postal service received competition from overnight delivery services and by the use of better and faster facsimile machines. Over time, electronic mail, aided by the expansion of the Internet, further sped the transmission of information.

Few things were as unchangeable over the decades as plain old telephone service (POTS). Although developments such as direct dialing, easier access to international calling, and switched networks progressed significantly throughout the century, the end user saw little change in the way the telephone was used. In the United States, one telephone company, AT&T, maintained a near-monopoly on telephone service, and it wasn't until the breakup of AT&T's monopoly in the early 1980s that telephone service providers and equipment manufacturers were able to introduce their own equipment and vary the functions of telephone service. This led to a host of add-on technologies, again centered on the desktop or portable computer. Among these technologies were fax machines, private branch exchanges, voice mail, and automated call routing. On the wireless front, portable and cellular telephones proliferated, as well as pagers and personal communication devices. Videoconferencing also grew out of telephony and satellite communication. Laying fiber optic cables increased both the capacity and speed of information that could be carried on the telephone networks.

RECENT DEVELOPMENTS

The technologies named above are not unique to the office environment, and several of them are described elsewhere in this book. The converging application of office technologies in the workplace has led to at least three major developments:

> ➢ The so-called "paperless office."

> ➢ The compression of office activities.

> ➢ The "virtual office."

THE PAPERLESS OFFICE

The "paperless office" is a misnomer. Few offices, if any, will end up paper-free. In fact, the growth of the printer market debunks the notion that offices will become paperless (Printer market, 2002). But many of the functions and activities that relied on the printed form in the past are being replaced by office technologies that allow documents to be put into electronic form. Some companies are making a strong effort to reduce the use of printers and paper (GE embraces, 2001).

An important component of the paperless office is the development of "intranets," private networks that have the look and feel of Internet Web sites. Much of the software used for these intranets is the same as that used for the Internet, and the software allowing users to easily create them continues to improve. A study by International Data Corporation (IDC) estimated more than 133 million intranet users worldwide at the end of 2001 (Fornaserio, 2000).

Intranets can be used for all kinds of information collection, sharing, and even interaction among users. The range of companies and fields of use is myriad, limited by only the imagination of the user. Software development also allows an intranet to have many of the capabilities of the Internet, such as internal searches and fast transfer of information within the intranet itself (Weiss, 2002).

A comparable application is the so-called "extranet," which provides special, secure links between companies and their customers, suppliers, or clients. They are often used in business-to-business applications. Instead of just using the Internet for such links, extranets are more secure, while still allowing external browsers access to internal information systems (Ling & Yen, 2001). Extranets are an extension of virtual private networks (VPNs), which provide "encrypted tunnels across the Internet" to allow secure, remote access to corporate networks (Biggs, 2002).

Indeed, the very definition of "office" is changing rapidly. Because of the proliferation of not only wired technologies such as intranets but also wireless technologies such as cellular, personal communications systems, handheld PDAs, and portable computers, flexibility in the configuration of the office has led to several developments.

COMPRESSION OF ACTIVITIES

Early office computers were used almost exclusively by secretaries, and executives avoided personal computers because of this clerical identity. As the technology evolved through the 1980s and 1990s, however, more and more people at all levels of work, including executives, used computers in the office. This has led to some job compression. For example, memos can be conceived, written, and printed or mailed in one basic operation by the originator. Telephone calls reach the desk of the recipient because automated voice call routing has eliminated most or all of the intervening human steps. The same is true with voice mail messages, which no longer need be written. In some specialties, such as desktop publishing, what used to take several steps in several places for writing, artwork, photography, typesetting, layout, and printing is now compressed into a single workstation where the job can be done by a single person in one place. Employees continue to learn more skills, including communicating through LANs and WANs, and creating information materials that grow ever more versatile, such as audio and video streaming media, multimedia, and presentation materials.

One of the significant office developments in recent years has been the appearance of multifunction peripherals (MFPs). These are automated devices that combine several functions into one unit. Such functions may include printing, scanning, faxing, word processing, telephone answering machines, and other computer functions such as data processing, networking, and CD-ROM. Some researchers predict the gradual replacement of individual office machines through the convergence of several technologies into one MFP. One recently introduced MFP is claimed to have optical character recognition (OCR) that can translate 55 different languages (MFP fax's, 2002). With the proliferation of home or portable offices, MFPs facilitate the portability and compactness that these sites demand.

THE VIRTUAL OFFICE

The "virtual office" has emerged as a feasible solution for contemporary and future work environments. Many types of traditional office work that required a fixed physical setting can be relocated to a wide variety of alternative sites. The employee's home, automobile, client/customer locations, or even temporary hourly/daily space can substitute for traditional centralized office space. Among the advantages of a virtual office are "increased productivity, lower real estate and travel costs, reduced employee absenteeism and turnover, increased work/life balance and improved morale, and access to additional labor pools, including disabled workers, to ease shortage skills" (Hrisak, 1999, p. 54).

Telecommuting technologies also permit freedom of location, instantaneous interaction, and fast response and spontaneity. From a business perspective, the virtual office provides substantial economic benefits. Enterprises are partially relieved of relatively high cost real estate investment or rental.

But the virtual office, or "dispersed collaboration," even in the wake of September 11 when there was a small surge in this development, is not as common or as easy as generally supposed. It requires a commitment across an entire hierarchy of an organization, a safeguard against too much blending of one's work life and personal life, an investment in adequate technology to support telecommuting, and attention to cultural and geographic factors such as work across time zones ("Virtual office," 2002).

A major implication of this approach is the need for homes or telecommuting sites to be equipped with multiple high-speed communication lines, as well as wireless telephone technology. The speed and quality of residential communication lines shifts from primarily providing voice-oriented facilities to one that provides rapid data and image transport as well as videoconferencing capability.

Employers must often provide up-to-date computers with organizationally standardized and compatible software, fast modems, communications services, network access, and other related facilities. Further, it is essential that security be given additional emphasis, since there is markedly increased difficulty in maintaining control and limiting unauthorized access to proprietary information.

Since the office can also move into mobile virtual locations, employers may also provide laptop portable computers, modems, PDAs, and fax facilities so that office workers can have almost infinite flexibility in reaching clients, colleagues, and the home office.

Probably the largest growth in this area will be the increased capability of audio and video "streaming" over the Internet, and the increased capability of wireless systems including network technologies and telephony. Wireless technologies have enjoyed substantial increase in accessibility and bandwidth, including Wi-Fi and 3G (third-generation wireless) developments. The first generation was analog cellular; the second digital cellular. The 3G standards increase the speed of the former by at least 10 times, adding more opportunity for broadband-like data transfer, to make wireless telephony speeds comparable to high-speed digital subscriber line (DSL) that wired systems enjoy (McGrath, 2002; Wireless LANs, 2002; Dickerson, 2002; Making Wi-Fi, 2002). These new technologies will be used not only by cell phones and portable computers, but also by increasingly versatile PDAs (Berger, 2002; Rothenberg, 2002). While there are many wireless services now competing for customers with competing technological developments, consolidation in the wireless sector is likely to follow the model of convergence in other media (Kellner, 2002).

Beyond the issue of meeting and conferencing speed, these mobile communication technologies provide a decided economic advantage relative to the cost of office facilities. Office space selection no longer must be governed by the need for a prestigious address, employee commute time, or client/customer access. Mobile communication technologies permit any required office facilities to be situated in lower-cost locations and be designed for less than the full complement of staff since many staff members will be mobile, working from home office space, a vehicle, a temporary office, or any other nontraditional location.

CURRENT STATUS

As much of the previous discussion shows, office technologies are in a constant state of change. Because so many technologies are involved, they do not change at the same rate, and their changes are seldom coordinated. The changes occur at three levels: internal independent office technologies, wired services, and wireless services.

Independent office technologies—computer workstations, multifunction stations, and desktop publishing—continue to be a large sector of office technology. As the storage capacity and versatility of information management capability of these devices increase, their use has become all the more embedded in the definition of the office of the 21st century. And, of course, all these technologies dovetail more and more with external technologies, the Internet being the most prominent.

The growth of Internet use is probably the largest business application of wired services being used for research, communication, and, increasingly, as a means of promotion and advertising. Faster access through ISDN (Integrated Services Digital Network), T1 and T3 lines, DSL, and other high-speed connections makes this medium even more convenient and versatile enough to also accommodate videoconferencing. The speed and technology of wireless networks is increasing, and is perhaps the area where the most growth will occur in the near future.

FACTORS TO WATCH

The office as a physical place, organizational function, and institution is likely to undergo substantial change as the result of the many dramatic events of 2001. Organizations are reexamining the traditional concept of gathering large segments of staff in one geographic location. Events have demonstrated that traditional concepts of safety and security are no longer valid. Thus, organizations are reconsidering expanding geographically-dispersed office locations and staffing. This concept ranges from smaller units located out of vulnerable central metropolitan areas to having selected members of the organization work from their homes. Thus, the communication linkages and supporting technologies become exponentially more critical to efficient and effective operation.

Geographically-dispersed office locations imply non-stop online communication facilities. Further, the events of 2001 also clearly demonstrated the need for more secure wired and wireless communication facilities. Organizations that are built on services that require real-time availability and response cannot tolerate even brief communication outages, let alone severe disruptions caused by technology failures, accidents, sabotage, or terrorist attacks. This dependency is underscored by the transition of many economies from a manufacturing to a service base. Thus, it is anticipated that the requirements for new and future communication services will have the following attributes: security, redundancy, automatic backup, fail-proof technology, and increased dependence on wireless modes that have broadband capability.

The dispersed office structure will help stimulate the growth of teleconferencing with full visual and interactive capability, which will help bridge the interpersonal gap brought by geographically separated office entities. (For more on teleconferencing, see Chapter 24.)

The events of 2001 related to the disruption of office communication were dominated by external terrorist or sabotage acts. However, with the Enron situation, another classic office communication problem surfaced—the improper destruction of an organization's records. Rules governing the proper maintenance of an organization's paper, film, and electronic documents are governed by law, regulations, tax codes, and have come under renewed scrutiny. In this instance, the sabotage was conducted internally. It is essential that data, whatever the media, have rules governing storage, access, life expectancy, and regulatory and legal requirements. One requirement may be the need for all documents to have digital tags or identifiers that describe all their maintenance terms and conditions.

DISPERSED STAFFING

The events of September 11 have caused organizations to rethink past practices of centralizing office operations. Enabling communication technologies have provided a basis for breaking organizations into smaller units and dispersing personnel geographically. Examples include Empire Health-Choice, which elected to move from a central facility into two offices, with many employees dispersed throughout their business area. Sun Microsystems uses "distributed workplace technologies" so that on any given day 30% of its employees around the world are not working at their desks. At 55 of Sun's field offices, when workers need a desk for a day, they make a reservation (Harmon, 2001). Because communication technologies provide relief from the need to group employees engaged in desk work assignments, some organizations are abandoning "the concentration of offices in an urban corporate campus that was all the rage a few years out. Dispersal is in" (Harmon, 2001, p. C1). "Costly office space in metro areas will be shunned in favor of highly diversified office locations in less expensive areas," says Paul Strassman, a productivity consultant (Bernasek, 2002, p. 112).

A major issue is the anticipated increase in security at major organizational headquarters. One estimate places the added financial burden for security at about $250,000 per year. This would be coupled with a major increase in disaster-recovery or backup computer systems, since many organizations do not have such systems in place. The costs for such systems could be as much as $1 million for a company (Bernasek, 2002). Also, companies will need to install technology that, in a disaster, would ensure continuation of voice and data messaging.

These changes are not solely due to the events of September 11, but to the radically different ways in which work is performed. In the past, work was partially a function of teacher and apprentice working as a team in a shared location. The industrial age brought the assembly line concept, even to the office environment. Communication technologies, however, have obviated the need for physical proximity among workers. And there has been a gradual freeing of workers from locality obligations to the office setting. The terrorist events, however, have spurred these developments, as well as increased use of technology for security purposes.

But the rush to the dispersed office is not a solution without its own set of problems. Worker isolation brings feelings of disconnectedness, and not every worker can function in relative isolation. The work "day" for some has become a 24-hour-a-day, 7-day-a-week obligation as the worker is forever tethered to the so-called "freeing" technologies and as companies become globally dispersed. Moreover, there are increased problems of keeping information secure when it is sent over such public thoroughfares as the Internet and the airwaves.

THE SMART OFFICE

In addition to the movement toward physically dispersed office spaces, there is a parallel movement to develop "smart offices." Combining contemporary office design with a multitude of electronic facilities has gathered momentum. A partnership between IBM and Steelcase has produced a prototype smart office, called BlueSpace, that incorporates many innovative communication features. Upon entering the office, an illuminated sign signals presence or absence of the occupant. "An electronic sensor, picking up signals from a chip embedded in the ID card, adjusts the desk chair, the table, even the temperature.... When the occupant is seated, lighting is adjusted and memos flash on a surface or the tabletop. An electronic sign outside the office door can indicate whether to come in or not to disturb. Projectors can beam memos, notes, and spreadsheets. And if an unauthorized person walks in, screens may switch to less sensitive material" (Deutsch, 2002, p. C4).

Security is a very strong driving influence in the emerging office environment. The traditional safeguards such as locks and shredders are joined by firewalls, virus detectors, encryption technologies, and backup systems. The dispersed office makes security an even more important factor, as proprietary information is sent over systems whose security may be broken.

Physical security is likewise an important factor, as detection devices and even new building architecture become designed to prevent or minimize the danger from threats or terrorism. The new mantra for office environments, regardless of their architecture, is "security, emergency, and disaster recovery programs" (Rivera, 2002, p. 12). The anthrax mail attack of 2001 clearly demonstrated the need to enhance security and basic office communication. Preparation for emergencies need not be based solely on catastrophic events. Events as simple as naturally caused disasters are also a problem, and many organizations that suffer such disasters do not survive.

Of course, one of the prices paid for so much increased attention to security is the erosion of worker privacy. Much has been written about software that tracks employee computer productivity and activity, systems that monitor where in the building an employee is at any time, and cameras that watch over people, not only in the workplace but even on the street.

The streaming of information is also an important development. It is no longer necessary to organize information by calendar or clock constraints. Financial information, statistics, events, news, or any continuous stream may have a substantial impact on an organization's operations, tactics, and strategies. Traditionally, such information was reviewed and assessed on a periodic basis, and only after such a process were decisions generated that would result in significant changes in operations. The continuous streaming of information, perhaps targeted at computer-based decision support systems or artificial intelligence models, permits organizations to initiate change in a more dynamic mode. As conditions change, analysis and recommended changes in operations can be initiated. Weekly or daily review meetings can literally be compressed into automated decision making—an almost minute-by-minute change process.

Continuous streaming of information plays a major role in changing the manner in which office work is managed. It is no longer necessary to have structured layers of employees that traditionally have accessed and interpreted information. This traditional process of decision-making required a hierarchical structure, many discussions and conferences, and delays. Streaming brings the information directly from its source to the decision maker, obviating the need for intermediate layers. Thus, information availability and accessibility become the criteria for organizational structure.

Until recently, our society maintained an allegiance to the traditional five-day, eight hours per day work schedule. Although this schedule may have been nominal for many office situations, it nevertheless has been the formal work system. The explosion of communication technologies has all but eliminated that work system. Online accessibility, 24-hour communication availability from anywhere, information streaming, and related technologies have created a new office work culture. Further, business structures continue to grow on a global basis with little or no recognition of national boundaries or time zones. Thus, the workday becomes a manifestation of the global demands of the business, and, in turn, its office functions. Office workers become available wherever and whenever needed. Weekends also are no longer privacy-protected. Since various cultures have varying workweek structures, it is essential to many organizations that employee availability not be bound by local national traditions.

The outsourcing of work has become an exponentially growing practice. Work can be shipped via telecommunications to any corner of the earth where the requisite skills are available at the right price. This practice allows organizations to ferret out the required skills, employ personnel on a contract basis without longer-term commitment, handle overload situations, and shop for the most advantageous labor rates. Tasks can be assigned, reviewed, and submitted regardless of the workers' location. The economy has grown at such an accelerated basis that labor shortages have become commonplace. Recruiting, employing, and training a sufficient staff to meet workload requirements has become a major organizational problem. Outsourcing alleviates this problem since work can be shifted to underutilized workers in areas with less booming economies. Communication technologies provide the engine to facilitate this manner of outsourcing work.

The ability of office organizations to manage their information resources and sources has reached a critical juncture. In the very recent past, the typical office organization could access manageable chunks of information on demand, essentially as required to perform required work. Now, there is a glut of information available, making it not only more difficult to manage, but also more difficult to determine its reliability. Data flood into organizations in an exponentially increasing amount, and it is an almost insurmountable task to filter it and convert it into useful information. The task of making decisions is hampered by this "information overload." But there is the potential for the development of artificial intelligence technologies to assist with that filtering process. There may also be an increased role for the so-called "information manager" whose job it is to find, organize, evaluate, and synthesize information for decision-makers.

This discussion ends with some attention to paper trails and information security. The communication revolution of the past decades, and the changes anticipated for the next decades, demand that the traditional "paper trail" concept be rethought and redefined. As office records continuously shift from paper to electronic media, the nature of how to organize, maintain, and secure an information trail is being redesigned.

Traditionally, paper trails were specified and defined in organizational manuals, government and tax regulations, and operational procedures. Today's movement of information has accelerated with respect to time and technologies. No longer can a reviewer anticipate locating a clear, clean, logical trail. Office information may be located in computer files, e-mail messages, paper memos and letters, voice mail, file cabinets—in effect, ever-widening geographic domains. The problem calls for a new set of generally accepted procedures that can be incorporated into office practice, particularly as electronic forms of information carry increasing status as legal documents.

Related to the electronic paper trail is the issue of security. As records and information move more and more through cyberspace, the future will bring more challenges and solutions to the need for security, confidentiality, and protection of sensitive or proprietary information. While modern communication methods have brought enormous flexibility in how information is generated, stored, and distributed, it has brought a corresponding need for ways to protect that information.

BIBLIOGRAPHY

Berger, M. (2002, April 3). Corporate PDA buyers wait for wireless. *Network World.*

Bernasek, A. (2002, February 18). The friction economy. *Fortune, 145* (4), 104-112.

Biggs, M. (2002, March 11). Narrowing the tunnel—Flatrock's instant extranet goes beyond VPNs to provide secure, Internet-based access to specific applications. *InfoWorld, 24* (10), 30.

Deutsch, C. (2002, January 14). New economy. *New York Times*, C4.

Dickerson, C. (2002, April 8). CTO connection: 802.11 is revolutionary—The wireless technology is a strategic choice that can help save money and speed responsiveness. *InfoWorld, 24* (14), 54.

Fornaserio, R. (2000). Extending information throughout the enterprise: Portals, browsers, and Intranet adoption trends. *International Data Corporation*. Retrieved April 12, 2002 from http://www.idc.com/getdoc.jhtml?containerId=23207.

GE embraces the paperless office. (2001, June 25). *Business Week*, 10.

Harmon, A. (2001, October 29). Breaking up the central office. *New York Times*, C1, C12.

Hrisak, D. M. (December, 1999). Millions move to the home office. *Strategic Finance, 81* (6), 54-57.

Kafka, P. (2000, November 13). Tapped out. *Forbes, 166* (13), 234.

Kellner, M. (2002, April 1). Wireless choices grow: Wi-Fi products are set to take off, while faster standards are on the way. *Government Computer News, 21* (7), 52.

Ling, R., & Yen, D. (2001, Spring). Extranet: a new wave of Internet. *SAM Advanced Management Journal, 66* (2), 39.

Making Wi-Fi pay: Wireless networking. (2002, April 6). *The Economist.*

McGrath, P. (2002, March 18). 3G phone home! Wireless: High speed phone service is coming. Will it work? Will it sell? Will you be able to translate all the acronyms? *Newsweek*, 38.

MFP fax's OCR translates 55 languages. (2002, March). *Office Solutions, 19* (3), 38.

Printer market growth debunks paperless office myth, says BMI. (2002, March 12). *Africa News Service,* 1008071.

Rivera, C. (2002, January). Expecting the unexpected. *Office Solutions, 19,* 12-18.

Rothenberg, R. (2002, April 1). Forward-looking marketers will lead the way on wireless. *Advertising Age, 73,* 19.

"Virtual office" not yet common. (2002, March). *Financial Executive, 18* (2), 10.

Weiss, T. (2002, February 18). Google targets intranet data searches with new devices. *Computerworld*, 20.

Wireless LANs to make mark on wired world; IDC predicts revenue for WLAN equipment worldwide will reach $1.45 billion in 2001, growing to $3.72 billion by 2006. (2002, April 10). *InternetWeek.*

VIRTUAL & AUGMENTED REALITY

Karen Gustafson, M.A.[*]

Virtual reality (VR) describes a diverse collection of technologies used to create computer-generated artificial environments. Although popular images of virtual reality are often highly immersive, such as in the environments depicted in Neal Stephenson's 1992 bestseller *Snow Crash* or in the 1999 blockbuster film *The Matrix*, practical implementations of VR often require a variety of hardware interfaces, such as head-mounted displays, treadmills, and force-feedback devices. Still, "virtual reality is all about illusion. It is about the use of high technology to convince yourself that you are in another reality, experiencing some event that doesn't physically exist in the world in front of you" (Pimental & Teixeira, 1995, p. 7). Virtual reality, an "emerging communication system" continues to develop technologically and industrially, with more applications appearing every year (Biocca & Levy, 1995, p. 15). Virtual reality technology is currently utilized in medicine, aviation, computer-aided design (CAD), entertainment, and defense, among other fields. With VR, surgical students can learn how to create a realistic incision in a "virtual" body, and soldiers can safely train in simulated combat environments (McLaughlin, et al., 2002, p. 3; Hafner, 2001).

Virtual reality, also referred to as virtual environments, consists of three basic principles: interactivity, real-time three-dimensional graphics, and varying levels of user immersion dependent upon the VR interface (Pimental & Teixeira, 1995, p. 11). In the virtual environment (VE), graphics change according to the user's apparent point of view and movement, so that the user feels encompassed within a dynamic, interactive virtual space (Heudin, 1999). Many popular applications of VR focus on the senses of vision and hearing, and some systems have even incorporated the user's senses of touch, temperature, and smell into the virtual experience.

---

[*] Doctoral Student, Department of Radio, Television, Film, University of Texas at Austin (Austin, Texas).

One growing field of development within VR research is the study of haptics—techniques for producing tactile sensations through force feedback. Haptics allow a user to feel the weight and texture of virtual objects, and can be used in a wide variety of applications, from scientific visualization to medicine (McLaughlin, et al., 2002). Virtual reality interfaces range from desktop VR, in which the visual display is limited to a standard computer screen, to more immersive formats using head-mounted displays or wide-screen graphics. Engineering teams can experience immersive VR as a group in a CAVE (Cave Automated Virtual Environment). With this technology, groups of people can collaborate on virtual automobile or building designs without having to build costly models.

Augmented reality (AR) straddles these extremes, superimposing virtual images upon the user's view of the real world, while not occluding the user's field of vision (Pierarski & Thomas, 2002). Augmented reality's flexibility makes it a desirable technology in a variety of fields including medicine, industrial manufacturing, and commerce (Mahoney, 1999).

BACKGROUND

Virtual reality technologies can be traced back to early vehicle simulators (Hillis, 1999). As early as 1929, the Link trainer was used to educate prospective pilots. Consisting of a motion platform and a simulated fighter cockpit, the Link trainer proved important to the development of military training, and was followed by tank, helicopter, and ship simulators (Pimental & Teixeira, 1995). Although these simulators did not initially include visual feedback devices, the military's demand for visual interactivity later drove the development of powerful computer graphic systems (Kalawsky, 1993). In an entirely different field, Morton Heilig saw the potential for simulated environments in the entertainment industry, creating the Sensorama in 1956. This device offered visual, auditory, and scent cues, but was not interactive and did not do well commercially (Pimental & Teixeira, 1995; Kalawsky, 1993).

By the 1960s, the primary elements behind virtual reality systems were coming into place. Douglas Engelbart, a pioneer in human/computer interface design, had explored the potential of computer screens as input and output devices, and had introduced a prototype of the mouse (Pimental & Teixeira, 1995). Visual displays made the manipulation of information much more accessible, and, in 1965, Ivan Sutherland, a researcher at MIT and later at Utah University, theorized an "ultimate display," a visual interface which would produce images indistinguishable from real objects (Kalawsky, 1993).

Sutherland's prescient paper, "The Ultimate Display," described an ideal concept of virtual reality in which users could be surrounded with 3D displays of information, and today Sutherland is considered one of the pioneers of VR (Vince, 1998; Larijani, 1993; Hillis, 1999). Sutherland focused on VR's development in flight simulation, and he created an early head-mounted display (HMD) in 1969 (Hillis, 1999). Although the HMD was notorious for its bulk and weight, it provided the visually-coupled imagery that has become a hallmark of VR technology. Also credited as the "father of computer graphics," Sutherland developed early image generators that were forerunners of contemporary graphics accelerators (Pimental & Teixeira, 1995). These generators could produce very simple scenes at a rate of 20 frames per second (fps), but more complex animation suffered at these speeds (Burdea & Coiffet, 1994).

By 1979, the military produced HMDs capable of real-time visual feedback, using technology similar to that manufactured by Sutherland (Pimental & Teixeira, 1995). The developing computer graphics industry became a powerful partner in VR development by the 1980s, producing the visually-coupled, stereoscopic imagery necessary for emerging VR applications. Virtual reality systems demand the highly efficient production of complex computer graphics. Because of this need for speed, images are constructed from polygons, a flat shape frequently used in 3D modeling. Polygons are very simple shapes and are an ideal building block for more complex 3D images, such as desks, cars, or houses, one might encounter in VR (Vince, 1998). Also, these shapes can be generated quickly, an important factor in truly interactive VR applications. To create a convincing sense of real-time interactivity, a VR system must have a response time below 100 milliseconds. In this fraction of a second, the computer must access the user's input, calculate changes in the visual display, and render new images using powerful graphics rendering hardware (Pimental & Teixeira, 1995).

Ever since the early days of vehicle simulators, the military has been a primary source for VR development, investing funds and research in the development of interface hardware and computer graphics (Burdea & Coiffet, 1994; Pimental & Teixeira, 1995). During the 1970s, research interests in VR broadened, although the National Aeronautical and Space Administration (NASA) continues to be a driving force (Hillis, 1999). The United States was the primary nation involved in VR research until the 1980s, when researchers and engineers in Japan, Germany, Canada, and France began to show serious interest (Larijani, 1994; Burdea & Coiffet, 1994). Although different components of VR research have been in development for decades, it was not until 1989 that another VR pioneer, Jaron Lanier, coined the term "virtual reality" (Heudin, 1999).

THE IMPORTANCE OF INTERFACE: VR HARDWARE

In 1991, cyberspace pundit Howard Rheingold predicted "in the coming years, we will be able to put on a headset, or walk into a media room, and surround ourselves in a responsive simulation of startling verisimilitude" (Rheingold, 1991, p. 388). Two years later, researcher Roy Kalawsky cynically noted, "current users of cyberspace are generally legally blind, headed for a stereoscopic headache, about to be motion sick, and soon to be struck by a pain in the neck due to the helmet weight" (Kalawsky, 1993, p. v). Although Rheingold's hopes have yet to be achieved, VR experiences are improving dramatically, and advancing beyond these early obstacles. As a potentially immersive technology, VR depends upon the transparency of its interfaces, with the goal of producing a sense of unmediated experience.

Today, virtual reality configurations consist of three primary hardware elements. CRT displays, or increasingly, liquid crystal displays, are used to provide the visual component of a virtual environment. A tracking system monitors the user's head position and orientation and communicates this information to the computer, which alters the displayed image to reflect changes in the user's perspective. This continuous, high-speed operation is accomplished through a variety of tracking technologies, including mechanical, electromagnetic, inertial, optical, or acoustic systems (Mahoney, 2000).

Finally, high-speed image rendering systems generate graphics in close to real time to produce a convincing experience of virtuality (Larijani, 1994). If image processor speeds are unable to produce graphics at a speed of at least 20 fps to 30 fps, the system latency disrupts the immersion of the user. Latency refers to the delay between user movement and the readjustment of the graphics system

(Loeffler & Anderson, 1994). Ideally, the computer running the graphics would adjust to the user's movement in real time. If the delay between the tracking system and image generation is too long, the user may experience motion sickness (Kennedy, et al., 2000).

As mentioned earlier, there are varying levels of immersion in virtual reality, and these correspond to the mode of presentation. High-immersion modes of VR presentation include:

➢ CAVEs (Cave Automated Virtual Environments) or dome displays.

➢ Cylindrical displays.

➢ Head-mounted displays.

Less immersive modes of VR include:

➢ Panoramic displays.

➢ Workbenches.

➢ BOOMs (Binocular Omniorientation Monitors).

➢ Traditional computer screens.

Group viewing technologies such as CAVEs and dome displays give a surrounding view of the virtual environment, and several people can experience it at once, making these viewing technologies useful for groups of industrial designers. However, these multiple image-generation systems can be quite costly. Panoramic displays are also useful for groups, but these do not surround the users. Workbenches allow one or two viewers to be individually tracked so that each receives customized point-of-view images; this hardware can also offer high-resolution stereoscopic imaging, allowing users to interact with three-dimensional data (Input/output, 2001).

Individual VR technologies include the HMD and the BOOM, an alternative to HMDs, which consists of a viewing box suspended on a rod. The BOOM, which provides stereoscopic viewing but restricts users' mobility, has been employed extensively in aerospace engineering and scientific visualization projects (Mayer, 2000). The HMD is one of the oldest and most common components of VR and AR systems, and, in the past several years, a variety of HMDs has appeared on the market. HMD hardware ranges from football helmet-sized devices to unobtrusive lightweight glasses, offering varying degrees of resolution and angles of view to the user. Prices range from a few hundred dollars up to ten thousand dollars (Mayer, 2000).

Other interface devices for virtual and augmented reality track the user's hands and, in some cases, can even provide the sensations of tactile and force feedback. The DataGlove, which was originally developed in 1982 by VPL Research Incorporated, consists of an exoskeleton device that tracks the movement of the user's hands and individual fingers (Kalawsky, 1993). With this interface, the user can manipulate virtual 3D objects, rearranging a room or redesigning an automobile engine. Haptic, or force feedback VR technologies, can give virtual objects a sense of weight and solidity, using a variety of mechanical actuation systems for force generation (McLaughlin, et al., 2002). Fred Brooks, a pioneer in VR haptics at the University of North Carolina, created the Grope system in the late 1960s as a feedback tool to aid in scientific visualization projects, such as the virtual manipulation of molecular models (Kalawsky, 1993). Other force feedback devices include SensAble's Phantom, a desk-grounded robotic arm, and certain joysticks.

CURRENT USES OF VIRTUAL REALITY TECHNOLOGY

Entertainment. The entertainment industry has embraced VR in a multitude of applications, with varying rates of success. Virtual reality technologies enhance many amusement park rides, and some theme parks and arcades, such as DisneyQuest in Chicago and Orlando, have been constructed completely around VR. While Disney's Chicago venue was forced to close in 2001, its Florida park continues to attract visitors. Another entertainment company utilizing VR technologies, GameWorks, was co-founded by film director Steven Spielberg and originally planned to have 100 locations across the country. Like Disney, GameWorks has had difficulty in drawing clientele to its chain of arcades, and is not reaching original market expectations (Walt Disney, 2001). Another entrepreneur, Xulu Entertainment, is expected to enter the field by the middle of 2002 (Yim, 2001) (see Figure 15.1). Based in San Francisco, Xulu intends to create an immersive entertainment environment that will allow patrons to travel to virtual solar systems and participate in alien adventures and nightlife (Brill, et al., 2002).

Figure 15.1
Xulu Photo

Source: Xulu Entertainment

While the future of VR entertainment centers remains uncertain, VR is being enthusiastically adapted to popular online games, such as *Quake*. At the 2001 Ars Electronica Festival in Linz, Austria, participants had the opportunity to play *Quake*, a well-known "first-person-shooter" game, while fully immersed in a CAVE (Robertson, 2001). Meanwhile in the United States, researchers have adapted *Quake* to an untethered AR format, allowing players to move about freely while wearing a transparent HMD and shooting with a haptic feedback gun (Piekarski & Thomas, 2002).

Education. Virtual reality technologies are being utilized in a variety of educational environments from elementary to college level. PaulingWorld, a section of Project ScienceSpace, uses immersive distributed virtual environments for science education. Geographically distant students share a virtual environment for exploring and manipulating molecular structures in a variety of formats. Tracking systems on the users' heads and hands give a sense of direct contact within the shared virtual space, allowing participants to interact with molecular models (Su & Loftin, 2001).

Another educational project uses VR to bring ancient civilizations back to life. The CLIOH (Cultural Digital Library Indexing Our Heritage) program at Indiana University recreates the sights and sounds of ancient civilizations such as at Chichen Itza. The project, a collaboration between IBM and the School of Informatics, will be connected to the Internet to allow distant users to virtually walk through ancient cities (Ancient civilizations, 2001). CARTE (Center for Advanced Research in Technology for Education) at USC Information Sciences Institute has produced several applications using virtual reality technologies for education development. Many of these projects involve the use of guidebots, or animated pedagogical agents, in three-dimensional immersive and networked virtual environments. STEVE (Soar Training Expert for Virtual Environments) functions as a virtual tutor, teaching students physical skills, such as equipment maintenance, in a 3D virtual work environment. Students use an HMD, data gloves, and head tracking equipment to interact with STEVE as he guides them through various tasks and encourages positive learning habits (Mahoney, 2001). Another of STEVE's cohorts, ADELE (Agent for Distributed Learning Environments) guides medical students through diagnostic scenarios in a Web-based, virtual environment. One of the principal challenges for CARTE has been the simulation of natural dialogue, and this has been partially solved through using speech recognition technology (Johnson, 2001). Much of this work at CARTE is funded by the Office of Naval Research and the U.S. Army Research Office, which hope to utilize these virtual teaching agents in training soldiers to perform various missions (Mahoney, 2001).

Aerospace and Defense. VR and AR technologies are being used in a variety of ways by government and military institutions. VR has been used for years to simulate a variety of battlefield situations, and networked VR has allowed soldiers in different locations to take part in the same virtual exercise. With advanced animation, a wrap-around screen, and voice recognition technologies, soldiers can be put in increasingly complex situations beyond the battlefield. In one exercise developed at USC at the Institute for Creative Technologies, the soldier is confronted with the tensions of peacekeeping duties in a Bosnian village, and, in another stressful scenario, the soldier must deal with a bus containing children and would-be terrorists. These complex exercises help the participants and researchers better study the effects of fear, anxiety, and stress on military personnel, while training soldiers for the complexities of contemporary military situations (Hafner, 2002).

Virtual reality technologies have also been useful in the simulation of non-terrestrial experiences, and have been implemented by NASA. At Glenn's Reconfigurable User-Interface and Virtual Exploration (GRUVE) lab in Cleveland, applications for flight simulation and for the study and improvement of jet engines are being used. The GRUVE environment allows collaborative stereoscopic viewing of virtual spaces and data visualization and can be networked to allow communication between geographically distant research teams (Gunter, 2001). Meanwhile, at the NASA Ames Research Center in California, researchers are using virtual reality applications to experiment with neuroengineering technologies; in one instance, a pilot was able to land a virtual jet using muscle signals from his forearm. Eventually, this remote control technology could be applied to microsurgery, uninhabited flight vehicles, or space shuttle repair, and could eventually replace hardware such as the data glove (Mecham, 2001).

Medicine. Surgical simulation is one of the most common uses of virtual reality in the medical field. Medical simulators are currently in use around the world, helping to train surgeons in a variety of medical techniques from simple blood draws to neurosurgery. Virtual reality is also utilized in actual surgical procedures, especially in the case of microsurgery and the use of teleoperation systems. Master/slave robotic procedures facilitate precise, minimally invasive operations such as the grafting of artery bypasses (McCloy & Stone, 2001).

Beyond helping surgeons, VR technologies are also directly assisting patients in a variety of ways. These technologies have been used for acute pain management, stroke therapy, and the management of phobias (Harrar, 2001; Bennett, 2001; Blough, 2002). In clinical tests, virtual reality is reported to have reduced anxiety and perceptions of pain in patients undergoing surgery, and fully immersive environments have proved effective in distracting burn patients from painful treatments (Harrar, 2001; Hoffman, et al., 2001). Finally, VR has proved useful in the treatment of phobias, such as fear of public speaking or fear of heights, as many clinicians turn to inexpensive VR therapy systems to help patients address their anxieties (Hodges, et al., 2001).

RECENT DEVELOPMENTS

Although virtual and augmented reality tools are clearly useful to a variety of industries, use of these emerging technologies is still generally limited to very large organizations, such as automobile and aerospace manufacturers. While it will take time for system costs to drop, VR and AR technologies are becoming more powerful, precise, and flexible. Powerful graphics generators are becoming more widely available on the consumer market, while HMDs may eventually be displaced by direct retinal displays. Tracking technology over wide area environments is growing in flexibility, and VR touch technologies can convey feelings of detailed patterns or even the changing temperatures of virtual objects.

IMAGE GENERATION

Virtual environment generators are large, high-performance computer systems with powerful databases and a considerable memory capacity, so that the computer can keep up with the actions and changing perspectives of the VR user (Kalawsky, 1993). Visualization engines and graphics processing continue to improve at both industrial and consumer levels. SGI recently introduced a new generation of supercomputers, the Onyx 3000 series, which offer advanced visualization capabilities and have been adopted by the Japan Atomic Energy Research Institute and Lockheed-Martin Aeronautics Company, among others (SGI posted, 2002). The SGI Onyx 3000 series is based on new modular computing architecture that enables the system to better deliver scalable shared-memory computing, bandwidth, and connectivity. Nvidia Corporation, a leading producer of graphics processing units, recently made graphics cards based on the new GeForce 4 unit available in retail centers (Nvidia, 2002). The graphics processing units offer real-time 3D image generation and special effects, and are an example of faster, cheaper computing technology diffusing into consumer markets.

DISPLAYS

There has been considerable improvement in HMD displays as well, and liquid crystal displays (LCDs) are becoming increasingly lightweight, although resolution and field-of-view remains a tradeoff for most models (Mayer, 2000). Eventually, the virtual retinal display (VRD) may obviate the need for head-mounted screens. Researchers at the University of Washington's Human Interface Technology laboratory (HITlab), in cooperation with Microvision Corporation, have designed a visual technology that beams images directly into the viewer's eyes. The VRDs still require a visor, but are more lightweight than traditional HMDs and can provide image resolution equal to high-definition television (HDTV) standards. The technology uses a very low-power light source to convey a single pixel at a time to the retina, and, at such low intensities, should pose no danger to the eye (Siuru, 2001).

TRACKING

Tracking technologies face multiple challenges: They must accurately track and relay the user's movements and orientation to the visualization engine, while not overly constraining the user. Ideally, these tasks must be performed in close to real time, so as not to disorient the user. The HiBall Tracking System, developed at the Tracker Research Group at the University of North Carolina, utilizes an electro-optical tracking technology that can be deployed over large areas. This technology lets users move freely through virtual environments, such as architectural plans, while allowing real-time graphics processing (Mahoney, 2000). Using a grid of LED ceiling panels, the HiBall system can provide up to 2,000 pose estimates per second with very low latency (Welch, et al., 2001).

HAPTICS AND TACTILE DEVICES

Haptic technologies provide a sense of physical force in VR, offering resistance in joysticks, levers, or other tactile interfaces. These interfaces are important for VR deployment in medical training, surgery, and interactive museum displays. Current force feedback technologies include the Cybergrasp (an exoskeletal device produced by Immersion Corporation) and the HapticMaster (from the University of Tsukuba in Japan), which can be used in conjunction with surgical tools. Recent research has focused on delivering more finely detailed touch sensations and avoiding the problem of system latency. Vibrotactile systems developed at Carnegie Mellon and MIT use tiny actuator arrays to deliver more detailed sensations of texture, and the Thermopad, developed at the Fraunhofer Institute for Computer Graphics in Darmstadt (Germany), can transmit sensations of temperature change to the VR user (McLaughlin, et al., 2002).

Haptic technologies are currently used in a variety of fields and research environments:

➢ Surgical simulation.

➢ Minimally invasive surgery.

➢ Microscopic telerobotics.

➢ Dental training.

➢ Bio-molecular research.

➢ Interactive museum displays.

➢ Gaming.

➢ Military training.

➢ Scientific visualization.

➢ Computer-assisted design.

➢ Assistive technologies for visually or neurologically impaired.

VRML AND THE WEB3D CONSORTIUM

The VRML Consortium set standards in the 1990s for versions of virtual reality modeling language (VRML), which is used to enable Internet browsers to interact with 3D virtual worlds (Vince, 1998). Currently known as the Web3D Consortium, the non-profit organization's membership includes many major institutions, such as Nvidia, Sun Microsystems, and the National Institute for Standards and Technology. In August 2001, the Consortium introduced Extensible 3D, or X3D™, the successor to the previous VRML standard. The development and diffusion process provides interested companies with open access to the new standard, and is expected to provide greater portability and flexibility to Web-based 3D applications (Web3D consortium, 2001).

CURRENT STATUS

Although much of the research on virtual reality continues to be funded by government and military interests, current development of VR and AR technologies includes a variety of private corporations, many working in cooperation with one another. Also, several major academic centers of research work with private and government projects. The HITlab at the University of Washington is the center of various research initiatives on VR and medical applications, including pain management (Hoffman, et al., 2001). The Electronic Visualization Laboratory at the University of Illinois is a focal point for research and applications of CAVE technologies, and the University of Southern California hosts several units pursuing the use of VR in education (Johnson, 2001). Other research centers include the Massachusetts Institute of Technology, MINDlab at Michigan State University, UC

Berkeley, Stanford, and Carnegie-Mellon University. Aerospace research in teleoperations and neural interfaces continues at NASA's Ames Research Center in California, while NASA's GRUVE in Cleveland is creating new opportunities for collaborative research and aerospace design using virtual environments (Mecham, 2001; Gunter, 2001).

MARKET AND CURRENT USES

The primary fields employing VR and AR technologies include higher education, medicine, and virtual prototyping. While these emerging technologies are still largely used in a research-oriented paradigm, the past decade has revealed a variety of practical applications. Automotive manufacturing and aerospace design are some of the fastest growing areas of VR use. Military operations continue to be a primary driver for ongoing VR research and development, but VR technologies are increasingly moving into everyday applications in museums and trade show exhibits, as this previously esoteric collection of technologies moves more and more into regular public use (Delaney, 2001). At the same time, due to issues of cost and maintenance, VR systems are still mainly found within large institutions with considerable capital resources (Hodges, et al., 2001).

In the past few years, the virtual reality industry has undergone a process of consolidation. The casualties of the recent economic downturn have mainly been smaller and newer companies, while larger and more established companies such as FakeSpace, SGI, and Evans & Sutherland persevere, sometimes absorbing smaller firms (Delaney, 2001). Because of this fallout, it is likely that the sizable surviving companies will be among those most likely to take advantage of virtual reality applications coupled with Internet connectivity (Delaney, 2001).

Despite recent setbacks, such as the slowing of technology markets, the market for VR and AR technologies continues to grow. Although VR systems sales and service revenue declined between 2000 and 2001, the number of VR and AR system units sold actually increased nearly 20% (Delaney, 2001). CyberEdge Information Services, a VR industry analyst, measured the total value of the VR and visual simulation market at $13.6 billion in 1998 and $17.7 billion in 1999, with an estimated growth rate of 35%. Although the industry exceeded expectations in 2000, climbing to $27 billion, this growth was followed by a drop back to $22 billion in 2001, coinciding with the general technology slump (Delaney, 2001).

COMPANIES IN VR

While there are several companies currently in the VR and AR markets, major commercial virtual reality developers and vendors include SGI (formerly known as Silicon Graphics), FakeSpace, Elumens Corporation, SensAble Technologies, and Ascension Technologies Corporation. SGI's Onyx™ graphics engines are made to respond to the huge demand real-time immersive visualization puts on a computer system, and come in several varieties, including the Onyx2™, Onyx300™, and Onyx3000™ with InfiniteReality3™ graphics. The recently introduced Onyx3000™ supercomputer systems are designed to simultaneously process 3D graphics, 2D images, and video information, and pricing starts at about $118,000 (SGI announces, 2002).

FakeSpace is best known for display systems that facilitate interactive viewing of computer-generated simulations and development of virtual prototypes (Schmitz, 2001). Recently, the company announced a collaborative project with graphics developer Alias/Wavefront. At the 2001 SIG-GRAPH, a leading annual convention in the visualization industry, the companies presented a "Portfolio Wall" which will allow industrial designers to create full-scale 3D renderings of large manufactured items, such as automobiles (Elliott, 2001). In late 2001, FakeSpace commercially introduced a digital version of its CAVE immersive visualization product, developed in conjunction with researchers at the Electronic Visualization Lab at the University of Illinois at Chicago (Input/output, 2001). Like FakeSpace, Elumens Corporation also develops screens for the creation of immersive virtual environments. Formerly known as Alternative Realities, Elumens offers the VisionStation, a dish shaped screen that costs about $20,000 and offers an180-degree view (Ditlea, 2000).

One of the main suppliers of tracking and motion capture tools, Ascension Technologies, works with a variety of other companies in the VR industry to provide head, hand, and body tracking for users of VR applications. Finally, SensAble provides a variety of haptic technologies, including the FreeForm modeling system. Used with SensAble's Phantom force-feedback hardware, FreeForm creates a direct touch interface between the user and a virtual 3D object, so precise that users can feel the texture of a virtual surface (Feel the texture, 2002). These technologies may eventually be used to provide museum patrons with the opportunity to touch virtual copies of rare artifacts (Ignatius, 2001).

FACTORS TO WATCH

Virtual reality is increasingly implemented as a collaborative tool for both remote users and groups interacting in the same space, but VR is still largely used by specialized industries. As the demand for collaborative applications increases, more VR applications are taking to business intranets or even the Internet. Augmented reality may be closer to consumer-level diffusion. Devices are getting less expensive and more wearable, so that users may eventually put on a pair of eyeglasses and walk down the street, viewing commercial sales or their e-mail superimposed over their normal field of vision.

COLLABORATIVE VIRTUAL ENVIRONMENTS

Collaboration continues to be a key application of VR and AR technologies. Workbenches and cylindrical "video tunnels" facilitating efficient multiple use are crucial to the many design industries that employ VR (Lyman, 2000). The first generation of these collaborative technologies include HMDs, BOOMs, and stereographic monitor displays; more recent interfaces include the CAVE and wall-sized 3D projection screens, such as FakeSpace's WorkWalls™ (Ragusa & Bochenek, 2001). Several engineers can assemble in front of large stereoscopic projection screens, and view full-sized models of their prototypes. Also, avatars—projections of remote workgroup members—can represent collaborators who are geographically distant (Smith, 2001). In this way, design crews in Australia and North America can simultaneously examine the same full-sized prototype of a car, and even interact with one another. Such virtual prototyping increases organizational efficiency, since most development costs are established in the early stages of product design (Ragusa & Bochenek, 2001). However, technology does not always bring greater communication and productivity; studies have shown that the success of virtual collaboration depends largely on how the systems are actually used, and how collaborative teams are assembled (Qureshi & Zigurs, 2001; Maybury, et al., 2001).

AUGMENTED REALITY

Although there is no mass market for virtual reality yet, augmented reality consumer items are already on the market, with more products to follow. Augmented reality allows the user to see their actual surroundings, but adds information to the view, so that a surgeon might see the tumor below a patient's skin or an auto mechanic might see a malfunctioning engine part, conveniently highlighted. In the past few years, certain luxury cars have been equipped with heads-up displays that use infrared technology to detect nighttime roadway obstacles and then project these images onto the windshield (Eisenberg, 2000). Other applications of AR are in development for use in everything from medicine to marketing. AR's lower costs and increasing system portability make it possible that, in the next decade, tourists will see maps and historical information as they walk through a strange city's streets, without ever having to consult their guidebooks.

There are two primary types of AR systems, optical see-through and video see-through. While the optical technology uses a half-silvered mirror to project a computer display over the user's regular field of view, the video system is opaque, sending images from a camera to stereoscopic screens in front of the user's eyes. Optical see-through AR allows perfect resolution, while the video images are inferior to the human eye. At the same time, optical systems may project information that is transparent and difficult for the user to read (Feiner, 2002). Also, tracking technologies must be improved before the user can take to the street with an AR headset. Although global positioning systems (GPS) can provide some information on a person's whereabouts, GPS tracking is not finely grained enough to constantly follow a user's every move (Feiner, 2002).

VR ON THE NET

Greater bandwidth and more efficient compression technologies are allowing VR and other 3D applications to take to the Internet (Mirapaul, 2000). Internet 2, a system of high-speed data connections between research institutions, government, and corporate users, was arranged in 1996 to foster such experimental applications as VR and teleoperations. Average use of this network doubled in 2001 as more institutions network high-bandwidth projects (Hafner, 2002). Many of the corporations taking advantage of this connectivity are mid- to top-range industry players, and the VR systems most likely to use the Internet for content delivery cost $50,000 and above (Delaney, 2001).

Virtual reality has transformed the industrial design industry and has been employed in many aspects of military, defense, and aerospace research for decades. Recently, these technologies have diffused into other key areas, from medicine and education to psychotherapy and online gaming. The future of VR and AR technologies often sounds like science fiction, but may eventually become everyday reality. In the future, neural impulses may direct the computer, creating a seamless interface between user and system (Mecham, 2001). Augmented reality glasses may eventually replace personal digital assistants, supplying accurate information in front of every building and storefront we see (Feiner, 2002). As broadband networks spread, processor capacity increases, and compression technology advances, VR and AR technologies will enter and transform everyday life.

BIBLIOGRAPHY

Ancient civilizations come alive online. (2001, October). *THE Journal, 29* (3), 14.

Bennett, R. P. (2001, June/July). Virtual reality: Could the future be now? *Wemedia, 5* (3), 104-106.

Biocca, F., & Levy, M., Eds. (1995). *Communication in the age of virtual reality*. Hillsdale, NJ: Lawrence Erlbaum Associates.

Blough, K. (2002, February 18). Virtual reality to aid stroke therapy. *Information Week, 876*, 20.

Brill, L., Coledan, S., Cottrill, K., Dunne, J., & Lowe, M. V. (2002, March). Alien games for cybertourists. *Popular Mechanics, 179* (3), 26.

Burdea, G., & Coiffet, P. (1994). *Virtual reality technology*. New York: Wiley.

Delaney, B. (2001, October). Moving to the mainstream. *Computer Graphics World, 24* (10) 18.

Ditlea. S. (2000, November/December). Rounding out displays. *Technology Review, 103* (6), 29.

Eisenberg, A. (2000, March 16). Seeing the skull beneath the skin, on the skin. *New York Times*.

Elliott, L. (2001, December 3). "Digital corkboard" creates a 45-ft blank slate. *Design News, 56* (23), 39.

Feel the texture of virtual clay. (2002, January 7). *Design News, 57* (1), 51.

Feiner, S. K. (2002, April). Augmented reality: A new way of seeing. *Scientific American, 286* (4), 48-56.

Gunter, H. (2001, May). Engineers get into the GRUVE. *Computer-Aided Engineering, 20* (5), 12.

Hafner, K. (2001, June 21). Get hold of yourself, Lieutenant. *New York Times*.

Hafner, K. (2002, January 10). Internet 2; An incubator for technology. *New York Times*.

Harrar, S. (2001, April). Virtual reality eases surgery pain. *Prevention, 53* (4), 75.

Heudin, J. (Ed.). (1999) *Virtual worlds: Synthetic universes, digital life, and complexity*. Reading, MA: Perseus Books.

Hillis, K. (1999). *Digital sensations*. Minneapolis: University of Minnesota Press.

Hodges, L. F., Anderson, P., Burdea, G. C., Hoffman, H. G., & Rothbaum, B. O. (2001, November/December). Treating psychological and physical disorders with VR. *IEEE Computer Graphics and Applications*.

Hoffman, H. G., Patterson, D. R., Carrougher, G. J., Nakamura, D., Moore, M., Garcia-Palacios, A., & Furness, T. A. (2001). The effectiveness of virtual reality pain control with multiple treatments of longer durations: A case study. *International Journal of Human Computer Interaction, 13* (1), 1-12.

Ignatius, A. (2001, June 4). Hands on. *Time South Pacific, 22*, 51.

Input/output. (2001, December). *Computer Graphics World, 24* (12), 54-56.

Johnson, W. L. (2001, Winter). Pedagogical agent research at CARTE. *AI Magazine*, 85-94.

Kalawsky, R. S. (1993). *The science of virtual reality and virtual environments*. Reading, MA: Addison-Wesley.

Kennedy, R. S., Stanney, K. M., & Dunlap, W. P. (2000, October). Exposure to virtual environments: Sickness curves during and across sessions. *Presence: Teleoperators & Virtual Environments, 9* (5), 463-473.

Larijani, L. C. (1994). *The virtual reality primer*. New York: McGraw-Hill.

Loeffler, T., & Anderson, T. (Eds.). (1994) *The virtual reality casebook*. New York: Van Nostrand Reinhold.

Lyman, R. (2000, July 31). Virtual reality comes back in new guise: Collaboration. *New York Times*.

Mahoney, D. P. (2001, April). Virtual teacher. *Computer Graphics World, 24* (4), 15-17.

Mahoney, D. P. (2000, April). On the right track. *Computer Graphics World, 23* (4), 16-18.

Mahoney, D. P. (1999, February 1). Better than real. *Computer Graphics World, 22* (2).

Maybury, M., D'Amore, R., & House, J. (2001, December). Expert finding for collaborative environments. *Communications of the ACM, 44* (12), 55-57.

Mayer, J. H. (2000, July). It's all in your head. *Government Computer News, 19* (17) 11.

McCloy, R., & Stone, R. (2001, October 20). Virtual reality in surgery. *British Medical Journal, 323* (7,318), 912-916.

McLaughlin, M. L., Hespanha, J. P., & Sukhatme, G. S. (Eds.). (2002) *Touch in virtual environments: Haptics and the design of interactive systems*. Upper Saddle River, NJ: Prentice-Hall.

Mecham, M. (2001, March 5). Ames researchers "land" 757 with nothing but muscle. *Aviation Week and Space Technology*, 154 (10) 51.

Mirapaul, M, (2000, October 5). Three-dimensional space is the next frontier for the Internet. *New York Times*.

Nvidia GeForce4 GPUs hit retail shelves. (2002, March 18). *PR Newswire*.

Pierarski, W., & Thomas, B. (2002, January). ARQuake: the outdoor reality gaming system. *Communications of the ACM*, *45* (1), 36-38.

Pimental, K, & Teixeira, K. (1995). *Virtual reality: Through the looking glass*, *Second Edition*. New York: McGraw-Hill.

Ragusa, J. M., & Bochenek, G. M. (2001, December). Collaborative virtual design environments: Computer-generated animated and enhanced movies and training simulations are now commonplace because of an exponential increase in cost-effective computer power and software robustness. *Communications of the ACM, 44* (12), 40-44.

Rheingold, H. (1991). *Virtual reality*. New York: Touchstone.

Robertson, B. (2001, November). Immersed in art. *Computer Graphics World*, *24* (11), 24-30.

Qureshi, S., & Zigurs, I. (2001, December). Paradoxes and prerogatives in global virtual collaboration. *Communications of the ACM*, *44* (12), 85-89.

Schmitz, B. M. (2001, January). Visualization center reduces time-to-market for jet designs. *Computer-Aided Engineering*, *20* (1), 28.

SGI announces the world's fastest graphics—Infinite performance. (2002, January 29). *PR Newswire*.

SGI posted significant sales wins in December quarter. (2002, March 13). *PR Newswire*.

Siuru, B. (2001, January). Virtual-retinal display. *Poptronics*, *2* (1), 31-34.

Smith, R. (2001, December). Shared vision: General Motors proved fertile ground for implementing collaborative technologies for designing cars. *Communications of the ACM, 44* (12), 45-49.

Stephenson, N. (1992). *Snowcrash*. New York: Bantam Books.

Su, S., & Loftin, R. B. (2001, December). A shared virtual environment for exploring and designing molecules. *Communications of the ACM, 44* (12) 57-59.

Sutherland, I. E. (1965). The ultimate display. *Proceedings of the IFIP Congress 2*, 506-508.

Vince, J. (1998). *Essential virtual reality fast*. New York: Springer-Verlag.

Walt Disney to close Chicago DisneyQuest. (2001, July 7). *Los Angeles Times*.

Web3D Consortium starts work on successor to VRML. (2001, August 9). *New Media Age*.

Welch, G., Bishop, G., Vicci, L., Brumback, S., Keller, K., & Colucci, D. (2001, February). High-performance wide-area optical tracking. *Presence: Teleoperators & Virtual Environments*, *10* (1), 1-21.

Yim, R. (2001, March 12). Xulu universe gets closer to lift-off. *San Francisco Chronicle*, E1.

16

HOME VIDEO TECHNOLOGY

Bruce Klopfenstein, Ph.D. [*]

This chapter reviews what was arguably the most significant home communication technology in the 1980s in the United States—home video. Today, a number of technologies make up "home video" and, trite as it may sound, revolution seems an apt descriptor for the near future of home video. Home video includes the venerable videocassette recorder (VCR), digital personal video recorders (PVRs), DVD (formerly digital video or versatile disc), and the rapidly emerging recordable DVD formats. Sony's PlayStation2 videogame doubles as a CD and DVD player, and baseline home PCs are now including a DVD player as standard equipment, so the DVD market actually goes well beyond the traditional home video living room. While video game consoles may include DVD players, video games are explored in Chapter 11.

Home video recording dates back to the earliest days of magnetic recording technology, while VCRs can trace their heritage to the professional Sony U-Matic, the first videotape *cassette* recorder (Klopfenstein, 1985). The first consumer Betamax was introduced in 1975 at a list price of $2,295 (nearly $8,000 in 2001 dollars when accounting for inflation, the same price for some digital high-definition television [HDTV] sets in 2001). Consumer electronics products, of course, have a history of rapid price declines.

A recurring and continuing theme in the history of home video is that of format standards battles. The most famous is the Sony Betamax versus JVC VHS battle of the 1970s that was won handily by VHS. Since then, there has been a similar battle on the camcorder front where the tide was turned: Sony's 8mm camcorders had been displacing VHS and VHS-C camcorders. Now, digital camcorders including DVD camcorders are replacing their analog predecessors. Even the DVD found itself in a

---

[*] Associate Professor and Director, Dowden Center for New Media Studies, University of Georgia (Athens, Georgia).

brief format "war" with the aborted DivX format pushed by retailer Circuit City in a bid to use the discs to compete in the VHS home video rental business. DivX was "reborn" (in name only) as a new format for digital video that enables the distribution of DVD-quality video over Internet protocol (IP) networks at small file sizes, much like MP3 does for digital music (Wagner, 2002).

Many lament the lack of standards when they occur in home video because of the belief that lack of standards slows market adoption (Klopfenstein, 1985). There is evidence on both sides of this debate. Many thought that the original DivX DVD format would hurt DVD adoption and diffusion (DVD players could not play DivX discs). The standards battle between Betamax and VHS led both sides to push technological innovations such as longer recording times, high-fidelity sound, lower prices, and others that probably accelerated VCR adoptions (Klopfenstein, 1985). (The same drive to innovation was evidenced in the early battle between the Netscape and Microsoft Internet browsers.) It is reasonable to assume that DVD advocates may have increased their innovative efforts to differentiate DVD features and hold prices as low as possible to prevent DivX from upending DVD.

Another serious issue that comes up continuously with home video recording involves concerns over copyright issues (see the pro-consumer electronics manufacturers' "Home Recording Rights Coalition" Web site for a history of events in recording rights at http://www.hrrc.org). The Consumer Electronics Manufacturers Association (CEMA at http://www.ce.org) also argues for recording rights because its constituency wishes their equipment to be as functional as possible for the consuming public. On the other side representing anti-piracy/pro-copy protection views are the Motion Picture Association of America (MPAA at http://www.mpaa.org) and the Recording Industry Association of America (RIAA at http://www.riaa.org). Look to these organizations' Web sites to see their latest stances on critical recording and anti-piracy technologies and related issues.

The relevance of concerns over copyright is dramatized by the plight of the digital audiotape (DAT) format. Challenges by recorded music copyright holders slowed the diffusion of this audio technology and may have effectively killed it as a consumer audio format. Clearly, prudent consumer electronics manufacturers will take piracy seriously if they wish to avoid litigation. Such legal challenges pose one of the only threats to the continued rapid diffusion of home recording technologies.

RECENT DEVELOPMENTS

As is or has been the case with all media technologies, home video is progressing in a transitional stage from the past's analog systems to today and tomorrow's digital technologies. The sale of digital television sets, for example, exceeded 1 million sets in 2001 (Shapiro, 2001a). The VCR is in decline. In number of units sold, VCR sales fell 35% from 2000 to 2001, according to Consumer Electronics Association (CEA) figures. CEA conservatively projects that the average price of a VCR in 2005 will be $50. Month-by-month sales figures showed DVD players passed VCRs in September 2001 and were forecast to outsell VCR decks for the first time in 2002: 16.25 million to about 14.5 million. By buying just over 13 million DVD players in 2001, consumers made the DVD player the fastest product in consumer electronics history to reach a 25% household penetration rate (Ward, 2002; Zorn, 2002). In the first quarter of 2002, DVD disc sales were up nearly 75% from the same period in 2001 (DVD News, 2002).

Table 16.1
CEA DVD Player Sales

| | \multicolumn{6}{c}{*DVD Player Sales History (Since Format Introduction)*} |
	2002	**2001**	**2000**	**1999**	**1998**	**1997**
January	542,698	572,031	370,031	125,536	34,027	--
February	736,118	555,856	401,035	109,399	34,236	--
March	1,162,568	1,207,489	412,559	123,466	38,336	Launch
April	N/A	631,353	409,192	269,107	42,889	34,601
May	N/A	523,225	453,435	279,756	47,805	27,051
June	N/A	920,839	654,687	326,668	79,044	29,037
July	N/A	693,013	537,453	325,151	84,709	19,416
August	N/A	673,926	557,617	260,225	81,170	34,021
September	N/A	1,768,821	1,296,280	501,501	113,558	34,371
October	N/A	1,516,211	1,236,658	603,048	163,074	56,407
November	N/A	1,781,048	866,507	449,242	136,908	37,657
December	N/A	1,862,772	1,303,091	646,290	233,505	42,575
Yearly Total	2,441,384	12,706,584	8,498,545	4,019,389	1,089,261	315,136

Source: CEA (2002)

DVD-R, DVD record and playback devices, fell dramatically in price by early 2002. While the average recording model was priced above $1,500 in 2001, retailers advertised DVD-R machines for less than $500 by 2002. One company, CenDyne introduced a DVD/CD burner for recording any type of video, photography, music, or data onto DVD-R, DVD-RW, CD-R, and CD-RW media at a list price of $450 (see www.cendyne.com). (Note: These "list" prices are not street prices. Manufacturers set list prices at the highest point a retailer might sell them, but it's not all that unlike the "sticker price" for a new car.) Without the recording functionality in the lower price ranges, DVD does not replace the VHS but merely supplements it for the time being. With DVD hardware and blank disc prices falling, the story will probably change by 2003. More than 20% of U.S. households now have standalone DVD players; that figure does not count videogame consoles or home computers equipped with DVD players. CD players took eight years to reach that percentage of the population, while VCRs took 12 years (Arensman, 2002).

The VHS format continued to be the rental choice of consumers by a margin of almost three-to-one. While DVD rapidly is increasing its share of the home video market, the total home video market is also growing. According to the Video Software Dealer Association's VidTrac program, the 2001 year-end rental revenue market share for VHS was 83.4%, while DVD accounted for 16.6%. This variance shows that VHS rental spending outpaced DVDs by $5.6 billion. The first quarter 2002 video rental market, however, shows a significant gain for the DVD rental revenue market-share, rising 12 percentage points to 28.9%. The 2002 year-to-date numbers showed consumers had spent $542 million renting DVDs (172 million rentals), but still spent $1.32 billion on VHS (500 million rentals) (VHS is not dead, 2002).

Current VCR user behavioral research is surprisingly difficult to locate in the public domain. U.S. VCR penetration is estimated to be about 95% of all television households in 2002. Multiple VCR households have also increased dramatically in the last five years. According to one source, 47.4% of all households had at least two VCRs as of June 2001, up from 44.2% in June 2000 (Centris, 2001b). VCRs are even showing up in family automobiles.

As of the late 1990s, average VCR use in U.S. households totaled about six hours per week. This was divided between 1.5 hours for recording programming and 3.5 to 4 hours for playing both home-recorded and rented or purchased tapes. One-third of program recordings were never being played back. Over half of taping (55% to 60%) is done while the TV set is off, 25% with the set tuned to the same channel taped, and 15% to a different channel. Most recording occurs in prime time (30%), the weekday daytime hours (30% to 35%), and fringe hours (13%). About 60% of recordings are of shows aired by the big three networks or their affiliates. Serials are the most commonly taped genre, accounting for half of all shows taped (Everything about, 1998). Another source indicated that, in an average month between January and June 2001, 48% of all VCR households rented cassettes, versus 39% who attended a movie theater (Centris, 2001a). One source estimated that 47% of all VCR households rented an average of 6.6 tapes per month. Thirty-four percent of all VCR households purchased an average of 4.0 tapes per month (Centris, 2001c).

Ernst & Young reported that 138 million DVDs shipped in the fourth quarter of 2001, up 73% from the 80 million units shipped in fourth-quarter 2000. Consumers spent $4.6 billion on DVD software in 2001, up from $1.9 billion in 2000, marking the first time that DVD purchases pulled ahead of VHS purchases. The enthusiastic DVD Entertainment Group (DEG) estimated that consumers spent $6 billion on DVD purchases and rentals combined in 2001, up from $2.5 billion in 2000. Such robust growth led to total U.S. home-video revenue (including purchases and rentals of both DVD and VHS) of an estimated $16.8 billion in 2001, a 21% increase over 2000. Consumers thus spent more than twice as much on home video as they did on movie tickets ($8.1 billion) in 2001. Total home video numbers were nearly triple 2001's videogame sales ($6.4 billion) and higher than CD sales ($11.2 billion) and book sales ($16.5 billion). Adams Media Research predicted that home video consumer spending is on a trajectory toward $30 billion by 2006, according to AMR's newest home video projections (see www.adamsmediaresearch.com; Rivero, 2002).

DIGITAL PERSONAL HOME VIDEO RECORDERS

The latest home video technology in position to "revolutionize" television is called the personal video recorder. The term "revolutionize" is not used lightly. As discussed below, PVR users are, in fact, significantly changing their television viewing. The PVR is a digital recording technology that allows users to record television programs, but goes beyond VCRs by including intelligent features that allow recording based on viewer program interests. PVRs are basically stripped-down PCs with large hard drives and a modem that dials in to download program listings on a regular basis.

PVRs are at the forefront of a set of devices that allow users to actually control live programming, with these capabilities generating a lot of interest among both consumers and manufacturers. One example of the control is the capability PVRs offer to enable users to pause live programming or skip through commercials. In a late 2001 survey by NextResearch, Inc., 69% of PVR users said they almost always fast-forward through commercials. Also in late 2001, ReplayTV, a subsidiary of SONICBlue, Inc., upset the broadcasting and advertising industries by launching a PVR that can be programmed to edit commercials out altogether (Ranch, 2002).

The inner workings of a standalone PVR system are fairly simple. Similar to cable set-top boxes, PVRs also contain several key components: MPEG decoder and encoder chips that digitize and play back the video and a hard drive that stores the video and provides a buffer so users can pause live programming without missing anything. Satellite customers now can opt to rent or buy a PVR-functional set-top box instead of the standard satellite set-top box. While satellite customers pay a premium for the combination, they also save money because they only need one box instead of two (Ranch, 2002). It is easy to imagine rapid diffusion if and when satellite and digital cable providers include PVR offerings as standard fare once the incremental cost of the additional PVR electronics declines to become effectively negligible.

Cable television companies also see the potential for building PVRs into their set-top boxes and charging subscribers an extra monthly fee for the service. One product for the cable market is Scientific-Atlanta's Explorer® 8000 Home Entertainment Server, a digital cable box that includes full PVR functionality, announced in April 2002 (Scientific-Atlanta, 2002).

Camcorders

Jennifer Meadows

Camcorders are becoming as ubiquitous in the home as a regular still camera. A study by the Consumer Electronics Association found that, as of January 2001, 39% of households in the United States had a camcorder (CEA, 2001). The consumer in 2002 has a plethora of analog and digital formats and cameras to choose from. These formats include VHS, VHS-C, SVHS, SVHS-C, 8mm, Hi-8, Digital8, MiniDV, and DVD-CAM. While analog formats (VHS, VHS-C, SVHS, SVHS-C, 8mm, and Hi-8) remain in the majority of homes, digital formats are growing the fastest. From 1998 to 2000, digital camcorders experienced 860% growth (CEA, 2001).

Digital camcorders often serve as digital still cameras as well. The user can choose to take a still image or record moving video. Users record images onto tape or memory cards. The cameras have analog video outputs to allow the user to view footage on a television or transfer it to videotape using a VCR. These cameras usually have a Firewire and/or a USB connection to allow users to import video into their personal computer for editing or transfer.

Digital8 is a Sony format that uses regular 8mm or Hi-8 tapes. Strictly for consumers, these cameras offer backward-compatibility so users that previously owned an 8mm or Hi-8 camera can still view and use their tapes. There are several different models available, starting with the DCR-RTV140 which retails at $499.99 (Digital8, n.d.).

MiniDV camcorders are available from a variety of manufacturers including Sony, Canon, and JVC. These cameras record a digital video image onto a DV tape. Prices for these cameras range widely, from $799.99 for an entry-level camera like the Sony DCR-TRV18 to the high-end SONY DCR-VX2000 for $2,999.95 (MiniDV, n.d.; Canon DV, n.d.).

DVD-CAM is a new format. Hitachi debuted DVD-CAM camcorders that record video or still images onto small (three-inch) DVD-R or DVD-RAM discs in late April 2002. DVD-R discs can then be played back on most DVD players and drives, and the DVD-RAM discs can be played back on DVD players meeting the DVD-RAM book 2 standards. In late April 2002, two cameras were available: the DZ-MV200A (680K pixel CCD) which retails for $899.95 and the DZ-MV230A (1.1 Mega pixel CCD) for $999.95 (Summary of, 2002).

Regardless of the technology, users are fairly united in their reasons for purchasing a camcorder. The Consumer Electronics Association study found that emotions top the list. "Recording events for keepsake" was the number one reason people bought a camcorder. Interestingly, while consumers are adopting this technology at a steady pace, actual use is quite low. Most respondents (two-thirds) reported using their camcorder one hour or less a month (CEA, 2001).

CURRENT STATUS

One difference among PVR systems is the presence or absence of subscription fees. On one hand, SONICBlue's ReplayTV does not require any monthly fee, with a portion of the upfront cost of the unit reserved for paying for the continuing expense of the program guide that the device needs to function. TiVo and UltimateTV, on the other hand, charge a monthly subscription fee or a one-time upfront fee for the service. (For example, TiVo's fees as of mid-2002 were $12.95 per month or an upfront fee of $249, in addition to the price of the hardware.)

In the previous edition of this book, this author questioned the program subscription approach unless the PVR is virtually given away. "If it's eventually 'built-in' as an added cable or ISP service, it might be more plausible" (Klopfenstein, 2000, p. 177). Ironically, the press in 2002 continually reported extremely high satisfaction ratings for PVRs, yet their overall penetration remained very low. It seems likely that PVRs will diffuse widely and rapidly when they (or their equivalent functionality) are simply included in the electronics of the future set-top box or the television set itself.

One company, NextResearch, Inc., has released some research on PVR use. These results come from a panel of about 350 PVR adopters and more than 1,000 non-adopters. According to the company, the findings showed that in 2001, PVR adopters were disproportionately likely to subscribe to a satellite television service, as well as the following:

➢ 92% of PVR adopters said it is easier to find the shows they want to watch.

➢ 51% of PVR adopters said they rent fewer movies now.

➢ 32% PVR adopters saved money by not renting.

As for the utility of their PVRs, adopters responded this way:

➢ 82% indicated one of the most appealing features is the ability to record two shows simultaneously (an advantage not available on all older models).

➢ 69% say they find the PVR's ability to keep a list of favorite shows appealing.

The greatest barriers to potential adoption cited included users being restricted to a specific service in order to use the box and the necessity of subscribing to an additional service, with the additional monthly fee, in order to access many of the recording features. Seventy percent cited an inability to watch one channel while recording another as a barrier to adoption (The personal, 2002).

FACTORS TO WATCH

DVD and DVD-R will replace VHS (or, even more certainly stated, some digital disk technology will replace the analog VHS tape format) as surely as the CD replaced the LP. VCR sales in the first quarter of 2002 were down 25% from a year earlier (Vital statistics, 2002). A full 90% of current DVD owners, however, said they still used their VCR, and over a million VHS units were still selling every month through the end of 2001. According to research by the Consumer Electronics Association, the average VHS household had 39 tapes compared to 20 DVD discs in DVD households. In total, this represents a $59 billion investment in VHS tapes, compared with $10 billion in DVD software. VHS titles also outnumbered DVD in 2001 by 10 to 1. VHS wins out in sheer numbers over DVD with its 95+ million households to 20 million for DVD. The number of multiple VCR households also significantly outnumbers those households with more than one DVD player (Wargo, 2001), although it is intriguing to ponder how many recent PC adopters can or do watch DVD movies on their computers. The low costs of both DVD players and VCRs suggest they may coexist as long as manufacturers continue to produce VCRs. A visit to a Blockbuster video store will continue to serve as a barometer of home video activity as they increase their DVD-to-videocassette ratio.

On the PVR front, Moore's law seems likely to continue upping the ante on low-cost digital storage. It is not clear that the 2002 market leaders (TiVo and ReplayTV have few customers) will be the ones that bring personal video recording into the home. Do not be surprised to see combination TV/PVRs eventually lead to an explosion in the PVR market when the cost of the PVR is relatively insignificant to the overall cost of the television receiver itself. Given the high initial prices for digital television sets, this scenario seems especially possible. For the time being, PVRs appear to be slightly ahead of their time.

The overall economic health is presumably always a factor in the adoption of new media, although surveys show many people would give up *other* devices rather than their home video equipment. In fact, as of mid-2002, economic conditions *appear* to already be on the rebound from a recession that was punctuated by the attacks on the United States on September 11, 2001. Nevertheless, we are living in interesting times, and the world's economic and political outlooks are far from certain, so this chapter concludes with observations that may take a backseat to unforeseen world events.

The most obvious factor to watch is the inevitable decline and probable extinction of the VCR, despite the introduction of sexy new VCRs including some with a PVR-like recordable hard drive (Preston, 2001). While this may seem a bold forecast, the CD has already shown how a new medium can force its predecessor off the stage in short order. Just as LP turntables can still be found today, "vintage" VCRs (or VCPs, videocassette players) will probably still be available in 2010 somewhere for those who really want them.

DVD player and disc sales should continue their dramatic growth as we have yet to reach the diffusion curve inflection point in player sales. The 2002 Christmas season will be very telling for the next technology likely to be welcomed with open arms into the U.S. consumer household, the recordable DVD. Initial high prices for recordable disks will have to come down, and they clearly will.

Less certain is the future of the current iteration of PVR technology. The DVD has kept the spotlight away from the PVR, and consumers are clearly confused by the need for the TiVo subscription charge. In addition, higher-end PVRs with more recording capacity are far more costly than sophisticated VCRs for a technology that, on the surface, promises the same recording ability *without* the prerecorded libraries most users associate with VCRs. In other words, most households have yet to embrace time shifting. As noted earlier, PVR owners express high satisfaction with the machines, but their experience is not easily demonstrated to potential adopters at a retail store. The battle brewing on the horizon would appear to be one between functional and less expensive video on demand services and the next generation of PVRs.

TiVo appears to be in financial peril as of mid-2002, while ReplayTV's parent SONICBlue is not. Worth noting is Microsoft's apparent abandonment of its UltimateTV PVR, although it reportedly is being used by DirecTV's DBS service (Shim, 2002). Moxie, the all-in-one "black box" for satellite, cable, home video, and audio, is another wildcard if and when it is placed on the consumer market. Paul Allen's Digeo announced it had purchased Moxie in spring 2002.

As of mid-2002, another factor to watch appears to be home networking. There is presumably a conceptual barrier for many potential adopters that may have to be crossed here, as "networking" carries with it the connotation of "computing," a term that may still intimidate more than "cable-ready," for example. This author has watched with near amazement as new homebuilders have by and large opted out of including wiring for home networking as part of their default new house specifications. An Ethernet port in the same faceplate as a telephone jack would not be difficult to include in new homes, but this has not happened. (For more on home networking, see Chapter 20.)

One legal obstacle to be addressed is the question about PVRs automatically eliminating commercials. Why would a consumer want to watch a commercial? That is the question at hand. Two solutions are based on value to the viewer: first, advertisers may have to produce targeted ads to those individuals who actually are in the market for the product advertised; second, advertisers and programmers may have to develop economic incentives for viewers to watch the commercials (e.g., watch a video-on-demand program at no charge in return for allowing the commercials to run).

Will copyright issues stymie the adoption and diffusion of home recording technologies? This question was asked 20 and 30 years ago, and it was not copyright that prevented adopters from recording programs off the air, it was the complexity of how to program one's VCR to tape off the air or cable. IBM is among the many companies working on copy protection solutions for video recording today. VCRs also became a video playback device, and so there may have been too little incentive for VCR manufacturers to get serious about usability and produce a truly simple interface between the user and the program stream. Still, PVR copyright issues have yet to be resolved as of this writing.

If, instead, there simply is no interest in time shifting, then home video recording technologies will remain in the realm of home video enthusiasts. Perhaps there should be more audience research that investigates the reasons why many viewers are perfectly content to watch video programs at a time decided upon by the corporate programmer. There is no question that off-the-shelf technology, technology near market introduction, and new technologies in the lab will make recording video easier, "intelligent," and more limitless in length. The most poignant question is where will home video fit into our daily lives in the future. The simple answer is, if anything, more prominently than it already is.

BIBLIOGRAPHY

Arensman, R. (2002, March). DVD sales soar, while prices plunge. *Electronic Business, 28* (3), 34.
Canon DV: Digital video camcorders: ZR50MC. (n.d.). Canon U.S.A., Inc. Retrieved on April 28, 2002 from http://www.canondv.com/zr50mc/index/html.
Centris Insights. (2001a, December). *Content/software activity incidence.* Retrieved April 6, 2002 from http://www.centris.com/sostwareactivity.htm.

Centris Insights. (2001b, December). *Home video penetration, installed base*. Retrieved April 6, 2002 from http://www.centris.com/data/insights_1201_pg1subviewdata.html.

Centris Insights. (2001c, December). *Insights—Monthly activity*. Retrieved April 6, 2002 from http://www.centris.com/data/insights_1201_pg2subviewdata.html.

Consumer Electronics Association. (2001, March 7). Camcorder popularity on the rise. *Press Release*. Retrieved on April 28, 2002 from http://www.ce.org/press_room/press_release_detail.asp?id=7282.

Consumer Electronics Association. (2002). *CEA DVD player sales*. Retrieved April 29, 2002 from http://www.thedigitalbits.com/articles/cemadvsales.html.

Digital8: Connect and create. (n.d.). *Sony Style*. Retrieved on April 28, 2002 from http://www.sonystyle.com/digitalimaging/F_D8.shtml.

Everything about television is more. (1998, March 6). *Research Alert, 16* (5), 1.

Klopfenstein, B. C. (1985). Forecasting the market for home video players: A retrospective analysis. *Dissertation Abstracts International*, 46-03, 0546A. (University Microfilms No. AAI8510588).

Klopfenstein, B. (2000). Home video. In A. Grant and J. Meadows (Eds.). *Communication technology update*, 7th edition. Boston: Focal Press.

MiniDV: Life in the digital domain. (n.d.). *Sony Style*. Retrieved on April 28, 2002 from http://www.sonystyle.com/digitalimaging/f_DV.shtml.

The personal video recorder: What is it? Is it the next VCR or the rirst VOD? How do *you* brand and profit from it? (2001, September). *NextResearch, Inc*. Retrieved April 20, 2002 from http://www.nextresearch.com/custom/pvr.html.

Preston, M. (2001, November 29). The future of home video recording? Retrieved April 20, 2002 from http://it.mycareer.com.au/news/2001/11/29/FFXNKERELUC.html.

Ranch, H. (2002, March). PVR popularity on the rise. *Electronic Business, 28* (3), 38.

Rivero, E. (2002, March 4). Report: Studio video revenues surged 20 percent to $13.2 billion in 2001. *Video Store*, 24 (11), 1.

Scientific-Atlanta ships first Explorer® 8000 home entertainment servers for customer trials. (2002, April 18). *Scientific-Atlanta*. Retrieved April 20, 2002 from http://www.sciatl.com/news/02apr18-2.htm.

Shapiro, G. (2001, December). Devoted to DTV. *Broadcast Engineering, 43* (14)**,** 56-57.

Shapiro, M. (2001, December). Going wireless. *Camcorder & Computer Video, 17* (12), 82.

Shim, R. (2002, April 4). Microsoft upgrades UltimateTV service. *CNet News*. Retrieved April 20, 2002 from http://news.com.com/2110-1040-876120.html.

Summary of DVD-CAM FAQ. (2002, March). Hitachi Home Electronics.

VHS is not dead; Format remains dominant in video rentals and will co-exist with DVD for years to come according to the VSDA. (2002, March 27). *Video Software Dealers Association*. Retrieved May 12, 2002 from http://www.vsda.org.

Vital statistics. (2002, April 2). *DVD insider*. Retrieved April 3, 2002 from http://www.dvdinsider.com/news/view.asp?ID=3690.

Wagner, H. J. (2002, March 10). DivX dives for mainstream VOD with version 5.0. *Video Store, 24* (11), 15.

Ward, D. (2002, January 13). CEA: DVD hardware sales to top VCRs in 2002. *Video Store, 24* (3), 1, 6.

Wargo, S. (2001, November). VHS is dead. Long live VHS. *Dealerscope, 43* (11), 18.

Zorn, E. (2002, January 29). VCR may be doomed, but it's not toast yet. *Chicago Tribune*. Retrieved April 20, 2002 from http://www.chicagotribune.com/news/columnists/chi-0201290038jan29.column?null.

17

DIGITAL AUDIO

Ted Carlin, Ph.D.[*]

D igital diversity is driving today's audio industry, and it is giving consumers more ways than ever to enjoy digital audio. The industry has released a number of digital audio options designed to enhance the realism and impact of prerecorded music, deliver new digital audio recording options to suit diverse lifestyles, harness the diversity of Web content, and provide Internet-delivered music without users being tied to a computer.

The audio industry, like many business sectors, has been concerned about the impact of the 2001 U.S. economic recession and the events of September 11. However, the audio industry has been able to benefit from a refocusing of consumer spending *toward* home entertainment and *away* from travel and going out, in addition to the innovations in digital technology (eBrain Consulting Services, 2002).

The audio industry did modestly well in 2001, as sales of home and portable audio components grew 2% over 2000 to $6.1 billion (CEA, 2002). Demand was driven by the increasing affordability and performance of products, as well as by growing demand for home theater-in-a-box systems and MP3 units.

Advances in home theater and MP3 technology are not the only developments pointing to renewed audio industry vibrancy. The last few years have included technological developments that have given consumers compelling reasons to buy products that will rekindle their passion for listening to music—the passion that launched the hi-fi industry in the 1950s. Key manufacturers have worked assiduously to develop DVD-based multichannel replacements for the two-channel compact disc

---

[*] Associate Professor and Chair, Department of Communication and Journalism, Shippensburg University (Shippensburg, Pennsylvania).

(CD). Combination audio and video players incorporating one or both of the two standards, DVD-Audio and Super Audio CD, are being sold.

Audio suppliers, however, do not appear to be banking solely on technological advances to make their profits: 2000 and 2001 saw continuing changes in tradition-defying product designs that simplify purchase and hookup, fit unobtrusively in a home's décor, or fill a need that reflects new music-listening habits at home. Whereas the core audio customer was once a serious music listener who assembled a complex system of standard-size components in a room set aside for serious music listening, audio consumers today are listening to music in more than one room in a house—often as background to other activities. In addition, hundreds of electronic games exist in surround- and high-quality sound (CEA, 2002).

As a result, traditional suppliers of home audio components have developed multimedia PC speakers, tabletop radio/CD players, under-cabinet CD/radios for the kitchen, and a broader selection of stylish "microsized" stereo systems for use in offices and home offices. Microsized systems are a key driver of audio sales since the stereo can be placed anywhere from a dorm room to a bedroom, office or den. Suppliers also have improved the price/performance ratio of custom-installed whole-house audio systems that distribute music throughout the house to in-wall speakers powered from a single rack of audio components (CEA, 2002).

Convenience, portability, and sound quality have, therefore, re-energized the audio industry for the 21st century. Competition, regulation, innovation, and marketing will continue to shape and define this exciting area of communication technology.

BACKGROUND

ANALOG VERSUS DIGITAL

Analog means "similar" or "a copy." An analog audio signal is an electronic copy of an original audio signal as found in nature. For instance, an analog audio signal follows the same pattern as the vibration in air pressure caused by the original sound, or a copy of the vibration of your eardrum as the air pressure hits it. Microphones turn audible sounds into electronic analogs of those sounds through mechanical reproduction and transduction. For example, an original sound wave travels through a dynamic microphone's port and causes an internal wire to vibrate. This vibration is transduced (changed) into electronic pulses that are sent through the audio system to a recorder, loudspeaker, etc., where they are stored or reproduced as analog sound waves. As this analog copy is created, some unwanted system distortion (noise) is also being recorded or broadcast. This is due primarily to the amount of electrical impedance present in the system components and cables. In an analog audio system, this distortion can never be totally separated from the original signal. Subsequent copies of the original sound suffer further signal degradation, called generational loss, as signal strength lessens and noise increases for each successive copy. However, in the digital domain, this noise and signal degradation can be eliminated (Watkinson, 1988).

In a digital audio system, the original sound is encoded in binary form as a series of 0 and 1 "words" called bits. The process of encoding different portions of the original sound wave by digital words of a given number of bits is called "pulse code modulation" (PCM). This means that the original sound wave (the modulating signal, i.e., the music) is represented by a set of discrete values. In the case of music CDs using 16-bit words, there are 2^{16} word possibilities (65,536). The PCM tracks in CDs are represented by 2^{16} values, and hence, are digital. First, 16 bits are read for one channel, and then 16 bits are read for the other channel. The rest are used for data management. The order of the bits in terms of whether each bit is on (1) or off (0) is a code for one tiny spot on the musical sound wave (Watkinson, 1988). For example, a word might be represented by the sequence 1001101000101001. In a way, it is like Morse code, where each unique series of dots and dashes is a code for a letter of the alphabet (see Figure 17.1).

Figure 17.1
Analog Versus Digital Recording

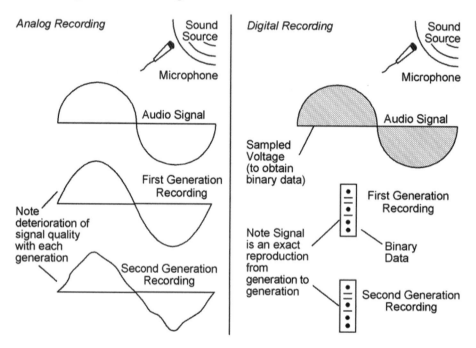

Source: Focal Press

Digital audio systems do not create bit word copies for the entire original sound wave. Instead, various samples of the sound wave are taken at given intervals using a specified sampling rate. Three basic sampling rates have been established for digital audio: 32 kHz for broadcast digital audio, 44.1 kHz for CDs, and 48 kHz for professional digital audiotape (DAT) and videotape recording.

This digitalization process, then, creates an important advantage for digital audio versus analog audio:

A digital recording is no more than a series of numbers, and hence can be copied through an indefinite number of generations without degradation. This implies that the life of a digital recording can be truly indefinite, because even if the medium (CD, DAT, etc.) begins to decay physically, the sample values can be copied to a new medium with no loss of information (Watkinson, 1988, p. 4).

But with this ability to make an indefinite number of exact copies of an original sound wave through digital reproduction comes the incumbent responsibility to prevent unauthorized copies of copyrighted audio productions in an effort to safeguard the earnings of performers and producers. Before taking a closer look at the various types of digital audio systems in use, a brief examination of the important legislative efforts and resulting industry initiative involving this issue of digital audio reproduction is warranted.

AUDIO HOME RECORDING ACT OF 1992

This 1992 legislation exempts consumers from lawsuits for copyright violations when they record music for private, noncommercial use and eases access to advanced digital audio recording technologies. The law also provides for the payment of modest royalties to music creators and copyright owners, and mandates the inclusion of the Serial Copying Management Systems (SCMS) in all consumer digital audio recorders to limit multigenerational audio copying (i.e., making copies of copies). This legislation also applies to all future digital recording technologies, so Congress is not forced to revisit the issue as each new product becomes available (HRRC, 2000).

Multipurpose devices, such as a general computer or a CD-ROM drive, are not covered by the AHRA. This means that they are not required to pay royalties or incorporate SCMS protections. It also means, however, that neither manufacturers of the devices, nor the consumers who use them, receive immunity from suit for copyright infringement (RIAA, 2002).

THE DIGITAL PERFORMANCE RIGHT IN SOUND RECORDINGS ACT OF 1995

For more than 20 years, the Recording Industry Association of America (RIAA) has been fighting to give copyright owners of sound recordings the right to authorize digital transmissions of their work. Before the passage of the Digital Performance Right in Sound Recordings Act of 1995, sound recordings were the only U.S. copyrighted work denied the right of public performance.

This law allows copyright owners of sound recordings the right to authorize certain digital transmissions of their works, including interactive digital audio transmissions, and to be compensated for others. This right covers, for example, interactive services, digital cable audio services, satellite music services, commercial online music providers, and future forms of electronic delivery. Most non-interactive transmissions are subject to statutory licensing at rates to be negotiated or, if necessary, arbitrated.

Exempt from this law are traditional radio and television broadcasts and subscription transmissions to businesses. The bill also confirms that existing mechanical rights apply to digital transmissions that result in a specifically identifiable reproduction by or for the transmission recipient, much as they apply to record sales.

NO ELECTRONIC THEFT LAW (NET ACT) OF 1997

The No Electronic Theft law (the NET Act) states that sound recording infringements (including by digital means) can be criminally prosecuted even where no monetary profit or commercial gain is derived from the infringing activity. Punishment in such instances includes up to three years in prison and/or $250,000 in fines. The NET Act also extends the criminal statute of limitations for copyright infringement from three to five years.

Additionally, the NET Act amended the definition of "commercial advantage or private financial gain" to include the receipt (or expectation of receipt) of anything of value, including receipt of other copyrighted works (as in MP3 trading). Punishment in such instances includes up to five years in prison and/or $250,000 in fines. Individuals may also be civilly liable, regardless of whether the activity is for profit, for actual damages or lost profits, or for statutory damages up to $150,000 per work infringed (U.S. Copyright Office, 2002a).

DIGITAL MILLENNIUM COPYRIGHT ACT OF 1998

On October 28, 1998, the Digital Millennium Copyright Act (DMCA) became law. The main goal of the DMCA was to make the necessary changes in U.S. copyright law to allow the United States to join two new World Intellectual Property Organization (WIPO) treaties that update international copyright standards for the Internet era. Among its provisions, the DMCA specifically addresses the licensing for Webcasting.

The DMCA amends copyright law to provide for the efficient licensing of sound recordings for Webcasters and other digital audio services. In this regard, the DMCA does the following:

> ➢ Makes it a crime to circumvent anti-piracy measures built into most commercial software.

> ➢ Outlaws the manufacture, sale, or distribution of code-cracking devices used to illegally copy software.

> ➢ Permits the cracking of copyright protection devices, however, to conduct encryption research, assess product interoperability, and test computer security systems.

> ➢ Provides exemptions from anti-circumvention provisions for nonprofit libraries, archives, and educational institutions under certain circumstances.

> ➢ In general, limits Internet service providers from copyright infringement liability for simply transmitting information over the Internet.

> ➢ Service providers, however, are expected to remove material from users' Web sites that appear to constitute copyright infringement.

> ➢ Limits liability of nonprofit institutions of higher education—when they serve as online service providers and under certain circumstances—for copyright infringement by faculty members or students.

> ➢ Requires that Webcasters pay licensing fees to record companies.

> ➢ Calls for the U.S. Copyright Office to determine the appropriate performance royalty, retroactive to October 1998.

> ➢ Requires that the Register of Copyrights, after consultation with relevant parties, submit to Congress recommendations regarding how to promote distance education through digital technologies while maintaining an appropriate balance between the rights of copyright owners and the needs of users (U.S. Copyright Office, 2002b).

The DMCA contains the key agreement reached between the RIAA and a coalition of Webcasters and satellite audio delivery services. It provides for a simplified licensing system for digital performances of sound recordings, such as those on the Internet and through satellite delivery. This part of the DMCA provides a statutory license for non-interactive, non-subscription digital audio services with the primary purpose of entertainment, if terms of the license are met. Such a statutory licensing scheme guarantees Webcasters and satellite services access to music without obtaining permission from each and every sound recording copyright owner individually and assures record companies an efficient means to receive compensation for sound recordings.

The RIAA has established the Performance Right Program (the Soundexchange) to administer the performance right royalties arising from Webcasting, on behalf of the sound recording copyright owners (usually record companies). The Soundexchange involves both licensing and royalty administration and distribution efforts (RIAA, 2002).

SECURE DIGITAL MUSIC INITIATIVE (SDMI)

The Secure Digital Music Initiative is a forum of more than 160 companies and organizations representing a broad spectrum of information technology and consumer electronics businesses, Internet service providers, security technology companies, and members of the worldwide recording industry working to develop voluntary, open standards for digital music. A DMAT® mark is the trademark and logo specifying that products bearing this mark (including hardware, software, and content) meet SDMI guidelines. A list of SDMI members is available on the SDMI Web site (http://www.sdmi.org).

These guidelines will permit artists to distribute their music in both unprotected and protected formats—the choice is up to them. The SDMI guidelines permit consumers to copy their CDs onto their computers and digital recorders for personal use. In fact, the guidelines enable consumers to do so as many times as they wish—as long as they have the original disc.

Professor Edward Felten's research team at Princeton University is one of two groups that claimed to have successfully broken four schemes of SDMI's digital watermark technology in a highly-publicized SDMI hacking challenge in 2000. But when the team was scheduled to present its findings at a conference in April 2001, Felten came under pressure from SDMI and RIAA. The organizations threatened to sue him for violating the DMCA's provisions against distributing information on how to bypass copyright-protection measures. Felten relented to the pressure. Since then, the SDMI coalition has been slow to create a viable standard, citing disagreements over priorities between music and consumer-electronics companies. Digimarc, Microsoft, Real Networks, DivX

Networks, and the Fraunhofer Institute (originator of the MP3 format) are all developing digital watermark technologies on their own.

RECENT DEVELOPMENTS

DIGITAL AUDIOTAPE (DAT)

Digital audiotape is a recording medium that spans the technology gulf between analog and digital. On one hand, it uses tape as the recording medium; on the other, it stores the signal as digital data in the form of numbers to represent the audio signals. A DAT cassette is about half the size of a standard analog cassette. Current DAT recorders support all three digital sampling rates and do not use compression, allowing them to deliver excellent sound reproduction. However, as a tape format, DAT does not offer the random access capability of a disc-based medium such as a CD or MiniDisc (MD) (TargetTech, 2000).

COMPACT DISC (CD)

In the audio industry, nothing has revolutionized the way we listen to recorded music like the compact disc. Originally, engineers developed the CD solely for its improvement in sound quality over LPs and analog cassettes. After the introduction of the CD player into the market, consumers became aware of the quick random-access characteristic of the optical disc system. In addition, the size of the 12-cm (about 5-inch) disc was easy to handle compared with the LP. The longer lifetime for both the media and the player strongly supported the acceptance of the CD format. The next target of development was to be the rewritable CD. Sony and Philips jointly developed this system and made it a technical reality in 1989. Two different recordable CD systems were established. One is the write-once CD named CD-R, and the other is the re-writable CD named CD-RW (Disctronics, 2002a).

MINIDISC (MD)

MiniDiscs were announced in 1991 by Sony as a portable disc-based digital medium for recording and distributing "near CD" quality audio. There are two physically distinct types of discs: premastered MDs, similar to CDs in operation and manufacture, and recordable MDs, which can be recorded on repeatedly using a magneto-optical recording technology (Yoshida, 2000).

Magneto-optical disc recording technology has been used for computer data storage for several years. Sony updated the technology, developed a shock-resistant memory control for portable use, and applied a high-quality digital audio compression system called ATRAC (Adaptive Transform Acoustic Coding) to create the MiniDisc. With a diameter of 64mm, smaller than a CD, a MiniDisc can hold only one-fifth of the data stored by a CD. Therefore, ATRAC data compression of 5:1 is necessary in order to offer a CD-comparable 74 minutes of playback time (Sony Electronics, 2000). However, the use of compression also results in a slight reduction in sound quality.

MiniDisc's advantages over a tape format such as DAT include its editing capabilities and quick random access. MiniDisc's main advantage over the MP3 format is that MP3's sound quality is inferior to MD. MP3 compresses sound data in a 10:1 ratio, which results in a greater loss of sound quality. (MD FAQ, 2002)

HOME THEATER

Since 1998, key innovations in digital audio processing and digital audio distribution have helped to create a vibrant consumer marketplace for audio products. The primary home entertainment experience today is delivered by a quality home theater system. Home theater has grown increasingly affordable over the years, and a respectable system with all the necessary audio and video gear costs as little as $600. According to the Consumer Electronics Association (CEA), a home theater system is a TV set with a diagonal screen size of at least 25 inches, a video source such as a hi-fi/stereo VCR or DVD player, a surround-sound-equipped stereo receiver or compact shelf system, and four or more speakers.

According to a 2001 CEA survey, 23% of American households own a home theater system meeting this definition. At the very least, surround-sound puts the viewer in the center of the action, enveloping the viewer with the ambient background sounds (i.e., a driving rainstorm, a thunderous explosion, a fast-paced chase scene). It also enhances dialogue intelligibility and realism by channeling dialogue to a TV-top center-channel speaker, making the voices of on-screen actors come from the same direction as their images (CEA, 2002).

Three companies have taken the lead in developing digital audio processing technologies, like surround-sound and 3D audio, which are being used to enhance the media experience of the consumer: Q-Sound, Dolby, and DTS.

Q-Sound Digital Audio. Providing "audio processing technologies to meet your needs," Q-Sound's key innovations are QSurround, Q3D, and QXpander (QSound Labs, 2000). By creating virtual speakers, QSurround reproduces spatially correct, multichannel output on regular two-channel equipment and enhances sound reproduction on five- and six-channel systems. Designed for professional audio recording and video game applications, Q3D places multiple individual sounds in specific locations outside the bounds of conventional stereo reproduction to provide a 3D listening environment. QXpander is a stereo-to-3D enhancement process available for both headphones and speakers. A robust algorithm allows it to process any stereo signal, making QXpander suitable for a broad range of consumer electronic applications (QSound Labs, 2000).

Dolby Digital 5.1 Audio. Dolby revolutionized tape recording in the late 1960s and early 1970s with Dolby A-type (for professional applications) and Dolby B-type (for consumer applications) noise reduction. In the late 1980s and early 1990s, Dolby Surround and Dolby Pro Logic home theater systems entered the marketplace. These systems allowed home viewers to create the same four-channel theater-type setup in the home (Dolby Labs, 2000).

Today's Dolby Digital 5.1 audio system takes the next step, providing six channels of digital surround-sound. Dolby Digital 5.1 delivers surround-sound with five discrete full-range channels—left, center, right, left surround, and right surround—plus a sixth channel for those powerful low-frequency effects (LFEs) that are felt more than heard in movie theaters. As it needs only about one-tenth the bandwidth of the others, the LFE channel is referred to as a ".1" channel (and more commonly as the "subwoofer"channel) (Dolby Labs, 2000).

Dolby Digital 5.1 is the audio standard for the movie industry and for new digital media applications. Also, for DVD-video, Dolby Digital is called a "mandatory" audio coding format, meaning that a Dolby Digital soundtrack can be the only one on a disc. Discussed below, DTS, by comparison, is an "optional" coding format, meaning that the disc must have a mandatory-format soundtrack as well.

DTS Digital 5.1 Audio. Another new technology for surround-sound entertainment, DTS Digital Surround, is an encode/decode system that delivers six channels (5.1) of 20-bit digital audio. In the encoding process, the DTS algorithm encrypts six channels of 20-bit digital audio information in the space previously allotted for only two channels of 16-bit linear PCM. Then, during playback, the DTS decoder reconstructs the original six channels of 20-bit digital audio. Each of these six channels is audibly superior to the 16-bit linear PCM audio found on CDs, with the ability to provide effective channel separation and a wide dynamic range of sound (DTS, 2000).

CURRENT STATUS

CD TECHNOLOGY

High-Density Compatible Digital (HDCD), developed by Pacific Microsonics, is a recording process that enhances the quality of audio from compact discs, giving an end result more acceptable to audiophiles than standard CD audio. HDCD discs use the least significant bit of the 16 bits per channel to encode additional information to enhance the audio signal in a way that does not affect the playback of HDCD discs on normal CD audio players. The result is a 20-bit per channel encoding system. HDCD is claimed to provide more dynamic range and a very natural sound. Many HDCD titles are available and can be recognized by the presence of the HDCD logo (Disctronics, 2002a). HDCD recordings will always sound somewhat better than conventional CDs when played on any CD player. A fuller, richer sound is available on all types of players, from portables to high-end systems. To bring out the full bandwidth and higher fidelity of HDCD recordings, a player with HDCD decoding should be used.

Super Audio CD (SACD). When the CD was developed by Philips and Sony in the 1980s, the PCM format was the best technology available for recording. Nearly two decades later, huge strides in professional recording capabilities have outgrown the limitations inherent in PCM's 16-bit quantization and 44.1 kHz sampling rate. Philips and Sony have produced an alternative specification called Super Audio CD that uses a different audio coding method, Direct Stream Digital (DSD), and a hybrid disc format. Like PCM digital audio, DSD is inherently resistant to the distortion, noise, wow, and flutter of analog recording media and transmission channels. DSD samples music at 64 times the rate of compact disc (64×44.1 kHz) for a 2.8224 MHz sampling rate. As a result, music companies can use DSD for both archiving and mastering (Super Audio CD, 2000).

The result is that a single Super Audio CD can contain three versions of the music, stored on two separate layers, that sounds noticeably better than conventional CDs when played on a conventional CD player. The Super Audio CD makes better use of the full 16 bits of resolution that the CD format can deliver, and is backward-compatible with existing CD formats (Sony, 2002).

There are more than 200 SACDs available from Sony, Virgin, and 15 "audiophile" labels, such as Telarc and DMP. An average of 20 SACDs were released per month in 2001, with artists such as Aerosmith, Miles Davis, Allison Kraus, and Destiny's Child using the format (DVD Audio, 2002).

Double Density CD (DDCD). In 2001, Sony and Phillips launched the Double Density CD-R/RW. It offers 1.3 GB of capacity per DD disc, and it reads and writes older single-density media as well. The DD media is about $1.99 per CD-R and $2.99 per CD-RW—a dollar or so more than its single-density counterpart, but very close in price per megabyte. The current real competitors are DVD-RAM, which is slower and suffers a high cost per megabyte, and DVD-R/RW, which costs about the same per megabyte but requires paying $700 to $800 for the drive.

The Sony Double Density drive is being marketed as a solution for those who like the reliability of CD media but find its capacity limiting. The $249 retail price is somewhat steep for a 12X/8X/32X drive, but the backup software and doubled capacity make the Sony an interesting option. Unfortunately, double the density does not mean twice as fast. The Sony's write performance has been average for a drive of its class, and while it runs slightly faster writing to DD media than to CD, it is by only a slight margin (Jacobi, 2001).

MultiLevel CD (MLCD). In 2002, TDK launched a CD drive which incorporates a MultiLevel read/write chip that adds a 2 GB recording mode to its standard CD-R/RW capability. And, equally important for recording large files, the ML encode/decode chip triples the drive's native recording speed to 36X (over the entire disc) when recording in ML mode. Though recordable DVDs offer more than twice the capacity of ML media, TDK's ML-enhanced CD-RW drive records at about twice the speed of a DVD burner. The integration of the ML technology into the CD-R/RW drive involved no change to standard CD burner optics, mechanics, or manufacturing systems (TDK introduces, 2002).

ML's ability to triple the drive's data capacity and transfer speed potentially means very little price differential over conventional CD-R/RW drives and recording media. For many value-conscious PC users, who will have to pay about $200 more for a DVD drive and twice as much for DVD recordable discs, the cost and speed advantages of TDK's ML CD-RW drive and media make it a compelling option for disc-based backup as well as many other archiving applications. (TDK, 2002).

MINIDISC TECHNOLOGY

MiniDisc Long-Play (MDLP) is a new encoding method for audio on MiniDisc that offers two long-play modes: 160 minutes of stereo audio (LP2) and 320 minutes of stereo audio (LP4). MDLP uses a new encoding technique called ATRAC3, which is also used in Sony's MemoryStick Walkman, Vaio Music Clip, and Network Walkman. Sound quality in LP2 is equal to standard MD, although LP4 quality may sometimes be inhibited by digital artifacts (The future, 2002).

NetMD. The one feature that has been limiting the popularity of the MiniDisc format is the necessity of recording music in real time. Rival MP3 products allow the user to download an hour's worth of songs in just a few minutes. With the December 2001 release of the Sony MZ-N1 NetMD recorder, MD promoters finally have an answer to the MP3 speed advantage. Sharp has also launched its own version of the NetMD recorder, with prices for units of both brands around $350.

By employing a standard USB interface, music data can be transferred from a PC to a NetMD product at high speed. In addition, the Net MD protocol will enable a PC to control a NetMD product while editing music that has been recorded on a MiniDisc.

To ensure backward compatibility with existing MD products, NetMD will support both ATRAC and ATRAC3 audio compression technologies. NetMD can securely transfer music data from an authenticated PC hard drive to an authenticated Net MD player. The NetMD interface is also designed to support Serial Copy Management System to prevent second-generation music data digital copying (NetMD info, 2002).

DVD-Audio is the latest member of the DVD family of prerecorded optical disc formats. As of mid-2002, DVD players were in one of every four U.S. homes (31 million units sold), reaching this penetration level faster than any other consumer electronics product (DVD Entertainment Group, 2002) (see Table 17.1). According to figures compiled by the DVD Entertainment Group (based on data from the Consumer Electronics Association, retailers, and manufacturers), 16.7 million DVD players were sold to consumers in 2001 representing 1.7 times more units sold than in 2000. In addition, there are currently more than 54 million DVD playback devices including set-top players, DVD-ROM drives, and game machines. There are now more than 125 DVD player models marketed under 50 different consumer electronics brands.

Table 17.1
Years To Reach 30 Million Players Sold

Format	Years
VCR	13 Years
CD	8 Years
DVD	5 Years

Source: DVD Entertainment Group

The initial version of DVD-Audio was released in April 1999, with discs and players appearing in the second half of 1999. DVD-Audio can provide higher-quality stereo than CD with a sampling rate of up to 192 kHz (compared to 44.1 kHz for CD). The standard DVD-Audio specification makes use of one or two layers (channel groups) of PCM multichannel stereo encoding. However, by using Meridian Lossless Packing, a digital compression codec, DVD-Audio can deliver even more data: up to six channels of 96 kHz/24-bit surround-sound (compared to a standard CD's two channels of 44.1 kHz/16-bit data) (Disctronics, 2002b). The improved audio quality, however, is only available through DVD players that are designed to play DVD-Audio.

Both write-once and re-writable DVD discs have been developed and are now available. There are four different formats, all with a capacity of 4.7 GB per side: DVD-R, DVD-RAM, DVD-RW, DVD+RW. As of April 2002, specifications have been approved by the DVD Forum for version 1.0

of DVD-Multi, a new set of hardware specifications to enable disc compatibility for all formats officially created by the DVD Forum. Formats created outside the DVD Forum (currently only DVD+RW) will not be covered by the DVD-Multi specifications (DVD Forum, 2002).

MP3. Since long before MP3 came onto the scene, computer users were recording, downloading, and playing high-quality sound files using a format called .WAV. The trouble with .WAV files, however, is their enormous size. A two-minute song recorded in CD-quality sound would eat up about 20 MB of a hard drive in the .WAV format. That means a 10-song CD would take up more than 200 MB of disk space.

The file-size problem for music downloads has changed, thanks to the efforts of the Moving Picture Experts Group (MPEG), a consortium that develops open standards for digital audio and video compression. Its most popular standard, MPEG, produces high-quality audio (and full-motion video) files in far smaller packages than those produced by .WAV. MPEG filters out superfluous information from the original audio source, resulting in smaller audio files with no perceptible loss in quality. On the other hand, .WAV spends just as much data on superfluous noise as it does on the far more critical dynamic sounds, resulting in huge files (MPEG, 1999).

Since the development of MPEG, engineers have been refining the standard to squeeze high-quality audio into ever-smaller packages. *MP3*—short for MPEG 1 Audio Layer 3—is the latest of three progressively more advanced coding schemes, and it adds a number of advanced features to the original MPEG process. Among other features, Layer 3 uses entropy encoding to reduce to a minimum the number of redundant sounds in an audio signal. Thanks to these features, the MP3 standard can take music from a CD and shrink it by a factor of 12, with no perceptible loss of quality (MPEG, 1999).

To play MP3s, a computer-based or portable MP3 player is needed. Dozens of computer-based MP3 players are available for download. Winamp, the most popular of these MP3 players, sports a simple, compact user interface that contains such items as a digital readout for track information and a sound level display.

This user interface can be customized by using "skins." Skins are small computer files that let the user change the appearance of the MP3 player's user interface. Another intriguing part of the MP3 world is CD rippers. These are programs that extract—or rip—music tracks from a CD and save them onto a computer's hard drive as .WAV files. This is legal as long as the MP3s created are used solely for personal use, and the CDs are owned by the user. Once the CD tracks have been ripped to the hard drive, the next step is to convert them to the MP3 format. An MP3 encoder is used to turn these .WAVs into MP3s.

All the copyright laws that apply to vinyl records, tapes, and CDs also apply to MP3. Just because a person is downloading an MP3 of a song on a computer rather than copying it from someone else's CD does not mean he or she is not breaking the law. The Secure Digital Music Initiative is the recording industry's main effort to prevent unauthorized duplication of digital audio using MP3 technology.

HOME THEATER TECHNOLOGY

The audio industry is trying to advance the state of the art in the Dolby and DTS digital surround formats by debuting two new formats on DVD discs. Dolby Digital Surround EX (extended) and DTS ES (extended surround) are 6.1 formats that add a back center channel, which allows DVD makers to accurately place sound effects behind the listener, not just to the sides and front. Side-surround speakers still reproduce left and right surround information, but the new combination allows for effects that are supposed to be seamless (CEA, 2002).

HARD-DRIVE JUKEBOX/RECORDER

With computer hard drive prices falling rapidly in the last five years, and with computer technology becoming more crash-resistant, more suppliers saw potential for audio-component-style digital recorders that use hard drives as their recording media. The number of companies offering such products rose to 10 in 2000, with the leaders being MusicMatch, Nomad, and XPlay (Eckhouse, 2001). Typically, storing music in MP3 or Windows Media Audio, these devices archive hundreds of hours of songs, which can be accessed by album title, song title, artist name, or music genre.

Some of the recorders feature built-in modems to download music from Web sites without the assistance of a PC. Some deliver streaming audio content from the Web to connected AV systems. For superior sound quality, the devices can be connected to cable modems or digital subscriber line (DSL) modems. Some hard-drive recorders also come with a built-in CD player, making it possible to rip songs from discs for transfer to the hard drive.

FACTORS TO WATCH

By the end of 2002, the two most important factors in digital audio will be (1) which digital audio format(s) consumers will adopt and (2) the protection of copyrighted material. With a myriad of digital audio choices making it to the marketplace, consumers will be evaluating products on ease-of-use, audio quality, storage capacity, and price. The successful worldwide adoption of the CD, with over 500 million existing CD players and 10 billion existing CDs, will be a challenge for any new digital audio format to top. Will consumers replace a perfectly good CD audio system with a new technology? Will they supplement their CD system with additional new technologies that fill specific needs such as Internet audio or home theater systems? Are there just too many new technologies for consumers to evaluate?

For example, two companies, Liquid Audio and AT&T's A2b, are already offering MP3-like Internet audio formats. They are *not* compatible with MP3 players as they use different digital compression techniques (i.e., MPEG-2 AAC Low Complexity Profile Audio Coding for A2b), forcing users to try yet another format. However, they are promising quicker downloads that use a smaller amount of disk space while maintaining sound quality that is virtually indistinguishable from the CD (i.e., a three-minute song downloaded from A2b would take up only 2.25 MB of disk space) (A2b, 2000).

Listen.com, an independent Internet audio subscription service, has created a nonexclusive licensing agreement with AOL Time Warner's Warner Music Group. Subscribers to Rhapsody, Listen.com's online music service, will have access to Warner catalog recordings. With this deal, Rhapsody has four of the five major recording companies on board—its existing agreements are with Sony Music Entertainment, Bertelsmann's BMG Entertainment, and EMI Recorded Music. This puts Rhapsody in the lead for the race to land all five companies, an accomplishment that appears to be absolutely crucial for fee-based online music subscription services to be viable. Described below, the other two online music competitors, MusicNet and Pressplay, both offer recordings from only three of the major record labels (Swanson, 2002).

A firestorm of controversy involving the Internet and music has been centered on the issue of unauthorized downloading of copyrighted music from MP3 hosting sites (i.e., Napster, Aimster) and person-to-person (P2P) networks, (i.e., Gnutella, Streamcast, KaZaA, AudioGalaxy). MP3 hosting sites maintain a central directory server on their Web site to connect "peer-to-peer" MP3 file swappers. This server enables users to find uploaded music from other users—legal and pirated—and download it without paying the artists or the record companies. It also allows the MP3 hosting site to monitor which users are online and what files they are sharing via the server.

In 1999, the RIAA filed and eventually won a lawsuit against Napster in the Ninth U.S. Circuit Court, alleging that Napster, as a central MP3 hosting site, encouraged piracy by enabling and allowing its users to trade copyrighted songs through its servers (RIAA, 2001). Napster, which has been shut down by the Court since the RIAA decision, will remain inoperable until it can show the Court proof that all copyrighted music has been removed from its service. Napster recently postponed the launch of a new secure service that would pay royalties to artists and recording companies. Napster has been unable to obtain performance licenses from the five major music companies (Warner Music Group, EMI Group, Bertelsmann Music Group, Universal Music Group, and Sony Music Entertainment), which are reluctant to deal with Napster while their own copyright infringement lawsuit against Napster is still pending in the courts (Hiatt, 2001).

While Bertelsmann is still a party to this lawsuit, it broke ranks with the other companies in October 2000 to buy a stake in Napster and help finance Napster's transformation from a renegade, free service into the (postponed) secure, membership-charging service. In March 2002, Bertelsmann announced it had tabled plans to purchase the remaining portion of Napster until Napster's legal affairs are decided (Reuters, 2002).

During this time of uncertainty for Napster, these major music labels launched two online subscription services of their own, Pressplay and MusicNet, in an effort to tap the 60 million users who were originally drawn to Napster. With Napster charged with copyright infringement by the record labels, Napster has counter-argued that the record labels have sought to impose anti-competitive licensing terms even as they have created Pressplay and MusicNet. The Court has ordered an investigation into the record labels' business practices to determine if Napster's claims have merit (Borland, 2002a). Rather than face a potentially "dirty" investigation, this may finally force the record labels to expedite an out-of-court settlement with Napster.

Pressplay is an equally-held venture of Sony Music Entertainment and Universal Music Group. It offers consumers on-demand access to a wide variety of music from the world's three largest record companies—Sony, Universal, and EMI—as well as many independent labels (i.e., Jive, Madacy,

Sanctuary, TVT). Music can be streamed, downloaded, or burned onto a CD, through affiliates that currently include MSN Music, Yahoo!, MP3.com, and Roxio. However, burning the downloads to a CD is severely restricted by Pressplay subscription rules (see http:// pressplay.comfaq.html#burns).

MusicNet, which is currently available through Real One Music and AOL Music, combines RealNetworks' Internet media delivery technology with the music catalogs of Warner Music Group, BMG Entertainment, EMI Recorded Music, and Zomba. RealOne Music clips can only be used on the computer they were downloaded to and can only be used on a personal computer—clips cannot be transferred to CD or MP3 players (see http://service.real.com/realone/music.html).

With Napster sidelined, P2P networks have evolved to keep free Internet downloading a reality. Unlike users of MP3 hosting sites, persons using P2P networks can download the P2P networking software (such as Morpheus or FastTrack) and then trade files without going through a storage center. This is accomplished because any computer running P2P network software can search all of the other computers connected to the Internet that are also running the software and retrieve information that each user makes publicly available. This makes it virtually impossible to shut down the network without unplugging every computer on the network. It also makes it difficult to control by legislation because there is no central authority to restrict (Cha, 2000).

The record labels are not standing idly by, but are seeking to prevent P2P networks from replacing Napster. The RIAA and the Motion Picture Association of America (MPAA) have jointly filed suit against KaZaA and licensees of its FastTrack file-trading technology. Licensees include West Indies-based Grokster and Nashville-based Streamcast, which has distributed more than 60 million copies of its Morpheus file-sharing software.

The RIAA lawsuit claims that, like Napster, these P2P companies "have created a 21st century piratical bazaar," providing a tool to quickly and easily exchange copyrighted audio and video (RIAA, 2002). The plaintiffs are seeking a permanent injunction to shut down the services, as well as penalties of $150,000 for each work infringed through copying on the network (RIAA, 2002).

The defendants are using many of the same arguments that Napster used in its defense. They plan to argue that their services have perfectly legal uses (this is similar to the 1984 Betamax defense strategy discussed in Chapter 16). Consumers can use them to trade copies of public-domain books, personal documents, even recipes. Moreover, the companies argue that they cannot be held responsible for the illegal actions of their users. Unlike Napster, which maintained a central directory to connect file swappers, the P2P networks maintain no such central control point. The software lets computers talk directly to one another.

And in that crucial respect, this case differs greatly from the Napster lawsuit. Napster was not shut down by the circuit court because it gave away file-trading software, but because its server enabled users to find pirated music and download it without paying the artists or the record companies. However, these P2P networks float anonymously on the Internet, without any set physical location. The software's architecture makes it impossible for any of these services or their users to know who is doing what and when they're doing it (Schonfeld, 2002).

Congress is also proposing action in the digital anti-piracy arena. As of mid-2002, the Senate and the House are considering bills that would extend the Digital Millennium Copyright Act to hardware devices. The new requirement would ensure that all future digital devices sold in the United States include built-in anti-piracy technology. Sen. Fritz Hollings (D-SC) and Rep. Adam Schiff (D-CA) are the sponsors of the bills (Borland, 2002b).

After almost a decade of few digital audio developments in the 1990s, expect to see continued innovation, marketing, and debate in the decade of the 2000s. Which technologies and companies survive will largely depend on the evolving choices made by consumers, the courts, and the continued growth and experimentation with the Internet and digital technology.

BIBLIOGRAPHY

A2b. (2000, April). *A2b music technology*. Retrieved April 5, 2002 from http://www.a2bmusic.com/ faq.asp#two.

Borland, J. (2002a, February 22). *Napster court win puts labels in spotlight*. Retrieved April 5, 2002 from http://news.com.com/2100-1023-843521.html.

Borland, J. (2002b, March 28). *Net anti-piracy debate heads for House*. Retrieved April 5, 2002 from http://news.com.com/2100-1023-870954.html.

Cha, A. E. (2000, May 18). E-power to the people. *Washington Post*. Retrieved April 5, 2002 from http://www.washingtonpost. com/ac2/wpdyn?pagename=article&node=&contentId=A21559-2000May17.

Consumer Electronics Association. (2002, April). *Audio overview*. Retrieved April 5, 2002 from http://ce.org/ digitalamerica/audio/default.asp.

Disctronics. (2002a, April). *CD-Audio introduction*. Retrieved April 5, 2002 from http://www.disctronics.co.uk/ technology/cdaudio/cdaud_intro.htm.

Disctronics. (2002b, March 30). *Audio coding*. Retrieved April 5, 2002 from http://www.disctronics.co.uk/ technology/dvdaudio/dvdaud_audio.htm.

Dolby Labs. (2000, April). *Dolby Digital*. Retrieved April 3, 2000 from http://www.dolby.com/digital/ diggenl.html.

DTS. (2000, April). *DTS technology*. Retrieved April 3, 2000 from http://www.dtsonline.com/consumer/ index.html.

DVD-Audio or Super Audio CD? (2002, April). *CNET*. Retrieved April 5, 2002 from http://electronics.cnet.com/electronics/0-3219400-8-5812389-3.html?tag=st.ce.3219400-8-5812389-1.txt.3219400-8-5812389-3.

DVD Entertainment Group. (2002, March). *Consumers spend a record $16.8 billion buying and renting video*. Retrieved April 5, 2002 from http://www.dvdinformation.com/news/index.html.

DVD Forum. (2002, March). *Latest DVD-Multi news*. Retrieved April 5, 2002 from http://www.dvdforum.org/tech-dvdmulti.htm.

eBrain Consulting Services. (2002, March 19). *Audio: 2001 year-end numbers down, but 2002 future looks bright*. Retrieved April 5, 2002 from http://www.ebrain.org/services/serv_head.asp?headID= {4C839F27-2ADB-11D6-A20D-00508B44E4E6}.

Eckhouse, J. (2001, November 13). *CNET Review*. Retrieved April 5, 2002 from http://www.cnet.com/ software/0-3227898-1205-7861400.html?tag=dir-rev.

The future. (2002, April). *MiniDisc.org*. Retrieved April 5, 2002 from http://www.minidisc.org/ part_future.html.

Hiatt, B. (2001, July 27). With Napster weakened, RIAA hopes to settle landmark lawsuit. *MTV News*. Retrieved April 5, 2002 from http://www.mtv.com/news/articles/1445466/20010727/ index.jhtml?headlines=true.

Home Recording Rights Coalition (HRRC). (2000, April). *HRRC's summary of the Audio Home Recording Act.* Retrieved April 3, 2002 from http://www.hrrc.org/ahrasum.html.

Jacobi, J. (2001, July 5). *Sony CRX200E-A1 Double Density CD-R/RW.* Retrieved April 5, 2002 from http://computers.cnet.com/hardware/0-1095-405-5614619.html?tag=rev-rev.

MD FAQ. (2002, April). *MiniDisc.org.* Retrieved April 5, 2002 from http://www.minidisc.org/faq_sec_4.html.

MPEG. (1999, December). *MPEG audio FAQ.* Retrieved April 3, 2000 from http://tnt.uni-hanover.de/project/mpeg/audio/faq/#a.

NetMD info! (2002, April). *Minidisco.com.* Retrieved April 5, 2002 from http://www.minidisco.com/minipages/netmdinfo.html.

QSound Labs. (2000, April). *Technology.* Retrieved April 3, 2000 from http://www.qsound.com/tech/.

Recording Industry Association of America. (2002, April). *Digital music laws.* Retrieved April 5, 2002 from http://www.riaa.org/Copyright-Laws-4.cfm#1.

Recording Industry Association of America. (2001, February 12). *RIAA summary of the Ninth Circuit Decision in the Napster Case.* Retrieved April 5, 2002 from http://www.riaa.com/PR_story.cfm?id=371.

Recording Industry Association of America. (2000, April). *FAQ about RIAA's lawsuit against MP3.com.* Retrieved April 5, 2002 from http://www.riaa.com/tech/tech_pr.htm.

Reuters. (2002, March 28). Bertelsmann puts Napster buy on hold. *CNet News.* Retrieved April 5, 2002 from http://news.com.com/2100-1023-870723.html.

Schonfeld, E. (2002, March 12). Goodbye Napster, hello Morpheus (and Audiogalaxy and Kazaa and Grokster...). *Business 2.0.* Retrieved April 5, 2002 from http://www.business2.com/articles/web/0,1653,38874,00.html?ref=cnet.

Secure Digital Music Initiative. (2002, April). *Frequently asked questions about DMAT® and the Secure Digital Music Initiative.* Retrieved April 5, 2002 from http://www.sdmi.org.

Sony Electronics. (1999, February 19). *Expanded AV/IT applications create a world of new possibilities for "Memory Stick" IC recording media.* Retrieved April 3, 2002 from http://www.sony.com/SCA/press/feb_19_99.html.

Sony Electronics. (2000, April). *MiniDisc.* Retrieved April 3, 2002 from http://www.sel.sony.com/SEL/consumer/md/.

Sony Electronics. (2002, April). *SACD FAQS.* Retrieved April 5, 2002 from http://www.sel.sony.com/SEL/consumer/sacd/static/faqs.html.

Super Audio CD. (2000, April). *What is Super Audio CD?* Retrieved April 3, 2002 from http://www.superaudio-cd.com.

Swanson, S. (2002, February 25). Ups—and downs—in online music arena. *Information Week.* Retrieved April 5, 2002 from http://www.informationweek.com/story/IWK20020225S0016.

TargetTech. (2000, April). *DAT (digital audiotape).* Retrieved April 3, 2000 from http://www.whatis.com/dat.htm.

TDK. (2002, March 13). *TDK demonstrates functional 2 GB ML-enhanced CD-RW burner at CEBIT.* Retrieved April 5, 2002 from http://www.tdk.com/tecpress/cebit.html.

TDK introduces world's first 2 GB multi-mode CD/ML burner. (2002, January 8). *Business Wire.* Retrieved April 5, 2002 from http://www.cdmediaworld.com/hardware/cdrom/news/0201/ tdk_ml_drive.shtml.

U.S. Copyright Office. (2002a). *Copyright law of the United States of America and related laws contained in Title 17 of the United States Code, Circular 92.* Retrieved April 5, 2002 from http://www.loc.gov/copyright/title17/circ92.html.

U.S. Copyright Office. (2002b). *The Digital Millennium Copyright Act of 1998: U.S. Copyright Office Summary.* Retrieved April 4, 2002 from http://lcweb.loc.gov/copyright/legislation/dmca.pdf.

Watkinson, J. (1988). *The art of digital audio.* London: Focal Press.

Yoshida, T. (2000, April). *What are MiniDiscs?* Retrieved March 25, 2000 from http://www.minidisc.org/ieee_paper.html.

IV

TELEPHONY & SATELLITE TECHNOLOGIES

Local and long distance telephone revenues in the United States exceed those of all advertising media combined. Clearly, point-to-point transmission of voice, data, and video represents the single largest sector of the communications industry. The sheer size of this market has two effects: companies in other areas of the media want a piece of the market, and telephone companies want to grow by entering other media.

The Telecommunications Act of 1996 was designed to stimulate competition in the provision of these services, but, to date, comparatively little competition has emerged. Almost all local phone service in the United States is still provided by the traditional "local telephone company," referred to in Chapter 18 as the incumbent local exchange carrier (ILEC), and almost all of the competitive service providers that were going to provide competition to the ILECs in delivering broadband connectivity to the home have disappeared.

Ironically, the advanced technology that promised new markets and revenues for both incumbent and competitive service providers has been a major factor preventing a more competitive environment. The reason is the cost of the technology—those companies that made massive investments in new technologies have not yet received a significant return on their investments, leaving the telecommunications industry laden with debt and reeling from the aftermath of the technology bust.

The digital technology that generated this debt is, however, in place with the same potential to revolutionize tomorrow's communication as it had when it was conceived and purchased. These digital protocols have erased the distinctions in the transmission process for video, audio, text, and data. Because all of these types of signals are transmitted using the same binary code, any transmission medium can be used for almost any kind of signal (provided the needed bandwidth is available). Furthermore, the advance of digital compression technologies reduces the bandwidth needed to transmit a variety of signals, further blurring the lines dividing the transmission characteristics of communications media. Many of the organizational barriers remain, however, allowing division of these technologies and services into individual chapters.

The first chapter in this section discusses the basics of today's telephone network in the United States, with a focus on organizational factors that are playing a much more important role in transforming the telecommunications landscape than technological change. The following chapter focuses specifically on broadband networks. In addition to explaining how broadband technologies work and how much information they can transmit, this chapter discusses a variety of organizational, economic, and regulatory factors that will influence when and how each becomes part of the telephone network.

The application of these network technologies for home use is explored in Chapter 20. In exploring home network technologies, this chapter addresses the role played by the Internet in the diffusion of these technologies.

Satellites are a key component of almost every communication system. Chapter 21 explains the range of applications of satellite technology, including the history of the technology and the range of equipment needed (on the ground and in space) for satellite communication. Chapter 22 then explores one of the most important applications of early satellite technology—distance learning—which has since evolved to encompass virtually every communication medium.

The many, rapidly evolving forms of wireless telephony are reviewed in Chapter 23, along with explanations of the differences between traditional cellular telephony and newer incarnations such as PCS. The final chapter in this section discusses the range of teleconferencing technologies, from simple audioconferencing to videoconferencing and videophone systems that are primarily designed to facilitate face-to-face communication over distances. It also discusses the rapid evolution of group-based videoconferencing systems and the continued failure of one-to-one videophones.

In studying these chapters, you should pay attention to the compatibility of each technology with current telephone technologies. Technologies such as cellular telephone are fully compatible with the existing telephone network, so that a user can adopt the technology without worrying about how many other people are using the same technology. Other technologies, including the videophone and ISDN are not as compatible. Consumers considering purchase of a videophone or ISDN service have to consider how many of the people with whom they communicate regularly have the same technology available. (Consider: If someone gave you a videophone today, whom would you call?)

Markus (1987) refers to this problem as an issue of "critical mass." She indicates that adoption of interactive media, such as the telephone, fax, and videophone, is dependent upon the extent of adoption by others. As a result, interactive communication technologies that are not fully compatible with existing technologies are much more difficult to diffuse than other technologies. Markus indicates that early adoption is very slow, but once the number of adopters reaches a "critical mass" point, usage takes off, leading quickly to use by nearly every potential adopter. If a critical mass is not achieved, adoption of the technology will start to decline, and the technology will eventually die out.

One of the most important concepts to consider in reading this chapter (and the other chapters that include satellite technology) is the concept of "reinvention." This is the process by which users of a product or service develop a new application that was not originally intended by the creator of the product or service. Satellite technology is being reinvented almost daily as enterprising individuals devise new uses for these relay stations in the sky.

The final consideration in reading these chapters is the organizational infrastructure. Because of the potential risks and rewards, even the largest companies entering the market for new telephone services are hedging their bets with strategic partnerships and experimentation with multiple, competing technologies. In this manner the investment needed (and thus the risk) is spread over a number of technologies and partners, with the knowledge that just one successful effort could pay back all the time and money invested.

BIBLIOGRAPHY

Markus, M. L. (1987). Toward a "critical mass" theory of interactive media: Universal access, interdependence, and diffusion. *Communication research*, *14* (5), 491-511.

18

LOCAL AND LONG DISTANCE TELEPHONY

David Atkin, Ph.D. & Tuen-yu Lau, Ph.D.[*]

T he new millennium has ushered in a period of unprecedented turmoil in telecommunications, dampening expectations for a sector whose valuation swelled with the prospect of new technology during the 1990s. The U.S. Department of Commerce definition of telecommunications includes elements ranging from local exchange and long distance/international services to cellular telephony and paging (the latter elements are explored in other chapters of this book). As Bates, Jones, and Washington (2002) recount, the telephone medium—focused on the delivery of analog voice services over copper wire for its first century—is being transformed by digital transmission of voice, video, and data services via fiber optics networks.

Scholars note the telephone may help facilitate the rise of a global information superhighway, helping cultivate user skills in interactivity and scalability that are crucial in the operation of emerging information technologies (Neuendorf, et al., 1998; 2002). The technology convergence generating this inertia flows from the highly deregulatory Telecommunications Act of 1996 (P.L. 104-104, 1996), notable for its removal of entry barriers between local, long distance, and cable service providers. Yet, insofar as cross-sector competition in telecommunications is only beginning to take root, commentators (Chen & Wilke, 1999) suggest that it remains a "long distance" off.

The growth pains accompanying this media convergence underscore its rapid departure from the era of plain old telephone service (POTS), already one of the most ubiquitous and lucrative commu-

---

[*] David Atkin is Distinguished Faculty Research Professor and Assistant Chair, Department of Communication, Cleveland State University (Cleveland, Ohio). Tuen-yu Lau is the Director of the Digital Media Master's Program, School of Communications, University of Washington, (Seattle, Washington).

nications technologies in the world. In the United States, the $230.5 billion in gross revenues from local and long distance companies—30% of the world market—exceeds the combined revenues of all advertising media, even surpassing the gross national product (GNP) of most nations (Atkin, 1999). The profit potential of these two markets is the primary force behind the revolution in telephony, as other media companies seek to enter the lucrative telephone market, and telephone companies seek to enter other media. Using history as a guide, this chapter outlines the influence of changing regulations on the conduct of telephone companies, including implications for cross media competition.

Before addressing the history, however, it is important to examine the architecture of the telephone network. To understand the U.S. telephone system, you must first realize that the entire country is divided into 194 service areas known as LATAs (local access transport areas). Most phone calls made from one phone in a LATA to another phone in the same LATA are handled by the local phone company. Any call made between LATAs, however, must go through a "long-distance" connection. After the breakup of the Bell System in 1984, local phone companies were forbidden to provide interLATA service, although (as explained later in this chapter) in states where regulators deem there is sufficient local competition, they can offer interLATA and interstate long distance service.

The telephone network is often referred to as a "star" network because each individual telephone is connected to a central office with a dedicated circuit. As illustrated in Figure 18.1, the heart of the network is the central office that contains the switching equipment to allow any telephone to be connected to any other telephone served by the central office. To allow connections to other more distant telephones, central offices are, in turn, connected to each other by two networks. Local phone companies have one network that interconnects all central offices within a LATA, and long distance companies have their own, separate networks that connect central offices located in different LATAs.

Telephone switches are specialized, high-speed computing devices that route each call through the network to its destination. In the earliest days of the telephone network, a telephone operator made all of the connections by hand. Operators were then replaced by mechanical switches, which were faster and less expensive. The mechanical switches have since been replaced by electronic switches that add a host of functionality to the switching process, ranging from relaying the originating phone number to the destination (caller ID) to enabling a single phone line to connect with two or more other phone lines (three-way calling). These services are sometimes known as Class 5 services.

Today's telephone networks make extensive use of high-speed fiber optic cables to interconnect central offices. Fiber optics are also increasingly used in place of copper wire to allow more efficient connections between end users and the central office. Figure 18.2 illustrates a modern telephone network that uses fiber optics to aggregate signals from dozens of subscribers into a single connection to the central office.

Figure 18.1
**Traditional Telephone Local Loop Network
Star Architecture**

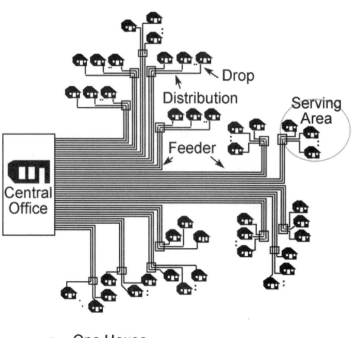

One House
≅ 500 homes
Serving Area Interface
——— Copper Pair (One Pair for Each Telephone)

Source: Technology Futures, Inc.

Figure 18.2
**Fiber-to-the-Feeder Local Telephone
Network**

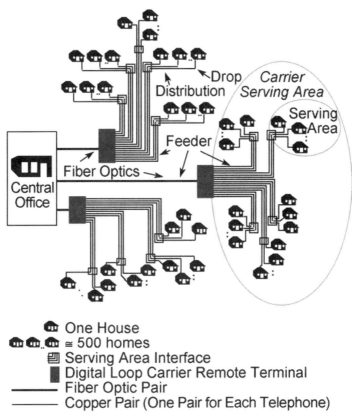

☗ One House
☗☗.☗ ≅ 500 homes
☖ Serving Area Interface
█ Digital Loop Carrier Remote Terminal
—————— Fiber Optic Pair
—————— Copper Pair (One Pair for Each Telephone)

Source: Technology Futures, Inc.

BACKGROUND

The Bell System dominated telephony in the century after Alexander Graham Bell won his patent for the telephone in 1876. After the original phone patent lapsed during the 1890s, the company came to be known as American Telephone & Telegraph (AT&T). This change ushered in a period of extensive competition as over 6,000 independent phone companies entered the fray, providing phone service and selling equipment (Atkin, 1996; Weinhaus & Oettinger, 1988). However, concerns over gaps in service standardization and interconnection prompted government oversight of telephony in 1910 (Mann-Elkins Act, 1910).

Meanwhile, AT&T's acquisition of independents intensified after 1910, forming the building blocks for what later became known as the regional Bell operating companies (RBOCs). In response, the Justice Department threatened its first antitrust suit against AT&T in 1913 (*U.S. v. Western Electric, Defendant's Statement*). But, by the 1920s, Congress was actually in favor of a single monopoly phone system (Willis-Graham Act, 1921). In the meantime, Bell worked to accommodate remaining independents, allowing interconnection with 4.5 million independent telephones in 1922 (Weinhaus & Oettinger, 1988). This industry rapprochement enabled Bell to focus energy on new ventures, such as "toll broadcasting" on radio stations (e.g., WEAF) (Brooks, 1976; Briggs, 1977).

Fearing telco domination of the nascent broadcast industry, Congress formalized a ban on telco-broadcast cross-ownership in the Radio Act of 1927 (and the succeeding Communication Act of 1934). The 1934 Act also granted AT&T immunity from antitrust actions, in return for a promise to provide universal phone service; 31% of U.S. homes had a phone at that time (Dizard, 1999).

After investigating complaints concerning AT&T's market dominance in the 1930s, the Federal Communications Commission (FCC) endorsed the industry's structure, characterizing it as a "natural monopoly" (FCC, 1939). By the late 1940s, however, the Justice Department began to feel uneasy about the sheer magnitude of AT&T's empire. It initiated antitrust proceedings against the phone giant in 1948, which culminated in a 1956 Consent Decree. Under the decree, the government agreed to drop its lawsuit in return for an AT&T pledge to stay out of the nascent computing industry.

The following decades were marked by deregulation aimed at ending AT&T's notorious exclusive dealing practice, euphemistically known as the "foreign attachment" restriction. Under the guise of protecting its network from problems of incompatibility and unreliability, AT&T alienated several non-monopoly companies seeking to attach consumer owned equipment to the network. In 1968, the FCC ruled against AT&T's ban of an acoustic coupler connecting radiotelephones. When allowing connection of this "Carterfone," the FCC and courts signaled that equipment not made by AT&T's manufacturing subsidiary, Western Electric, could be used in its network (*Carterfone*, 1968). During the following year, MCI was given permission to operate a long distance line, despite AT&T's objections (*Microwave Communications, Inc.*, 1969). Dissatisfied with this industry conduct, the Justice Department in 1974 initiated proceedings to dismember AT&T, reminiscent of its 1948 action. In particular, the complaint against AT&T alleged that they had:

(1)   Denied interconnection of non-Bell equipment to the AT&T network.

(2)   Denied interconnection of specialized common carriers with the Bell network.

(3)   Foreclosed the equipment market with a bias toward its Western Electric subsidiary.

(4)   Engaged in predatory pricing, particularly in the intercity service area (*U.S. v. AT&T*, 1982; Gallagher, 1992).

This time, with the aid of several interested non-monopoly firms, the Justice Department was in a much stronger position.

Thus, as demand for telecommunications services grew, the Bell System's regressive monopoly structure proved an impediment to growth and innovation (Jussawalla, 1993). After initially enjoying government protection during the first part of this century, the Bell monopoly gradually fell out of favor with regulators for inefficient and anticompetitive market conduct.

In fairness, AT&T achieved the burdensome goal of providing universal service with the highest reliability in the world. The company employed one million workers in 1982, claiming that it lost about seven dollars a month on its average telephone customer (Dizard, 1999). While that plea may be debatable, such costs necessitated the practice of cross-subsidization that kept local phone service prices low by charging high rates for long distance service. Known as the behemoth that worked, Bell was the largest company in the world, subsuming 2% of U.S. GNP. The company carried over one billion calls per day at the time of divestiture (*U.S. v. Western Electric*, 1982).

Whether or not AT&T was well compensated for its trouble, by 1982, even the company viewed its regulated monopoly as an impediment to progress. AT&T's willingness to consent to divestiture was, arguably, as much a function of self-interest as exogenous government pressure. Phone company executives recognized that it was in their own interest to discontinue the "voice-only" monopoly in which AT&T had become encased. So they negotiated a divestiture settlement, or consent decree, known as the Modified Final Judgment (MFJ) in 1982 (*U.S. v. AT&T*, D.D.C. 1982). Concern over the impact of the MFJ prompted the presiding judge, Harold Greene, to maintain oversight over all aspects of the MFJ, including the divestiture, for the next decade (Dizard, 1999).

The MFJ became effective in 1984, resulting in AT&T's local telephone service being spun off into seven new companies, known as the regional Bell operating companies, sometimes referred to as the "Baby Bells." The divestiture provided AT&T, which kept its long distance service and manufacturing arm (Western Electric), a convenient vehicle to exchange pedestrian local telephony for entry into the lucrative computer market. Perhaps more important, it enabled AT&T to cut most of its labor overhead without fear of union unrest or "bad press."

With recent deregulation allowing for the recombination of local and long distance services by a single provider, it is useful to explore whether AT&T's controversial conduct was an anomaly of that particular monopoly, or characteristic of vertically integrated capital-intensive phone utilities in general. To gain a better understanding of that question, we now examine the post-divestiture conduct of the telephone industry, and the implications of its entry into allied fields.

RECENT DEVELOPMENTS

After a series of FCC and court rulings relaxed restrictions on telco entry into cable (Telephone Company, 1992; *Chesapeake*, 1992), the RBOCs requested that the MFJ be vacated in 1994 (Atkin, 1996). In 1996, Congress and the President removed the ban as part of the Telecommunications Act. Since the law also removes the ban on telco purchases of cable companies in communities of over 35,000, we're likely to see new alliances between these industries. Although U S WEST (now Qwest) was one of the most aggressive entrants into cable, they recently moved to split their phone and cable businesses into two companies.

AT&T's bold move to deliver local service and become the nation's largest cable operator, bolstered by its acquisition of Tele-Communications, Inc. (TCI) and MediaOne, was dealt a setback when a deal to offer local phone service to Time-Warner's 12 million subscribers was tabled. As of mid-2002, the company was prepared to end its foray into cable by selling AT&T Broadband to Comcast Cable. It seems, then, that the strategy of simultaneously pursuing an in-region and out-of-region service in the two industries hasn't succeeded (Atkin, 1999). Telcos can also provide video programming in their own service area, but the new law prohibits joint telco-cable ventures in their home markets.

While this pending competition should lead to lower long-distance rates, local rates may not decline until AT&T, MCI, et al. obtain favorable terms for reselling Bell connections as their own services. New York City offered the first laboratory for local loop competition, as NYNEX competes against TCI and Cox's MFS Communications. By the mid-1990s, those cable-based competitors controlled over 50% of Manhattan's "special access" market—which involves the routing of long-distance calls to-and-from local lines—and faces further competition from MCI Communications (Landler, 1995). The Telecommunications Act of 1996 also maintains universal service mandates, but allows states and the FCC to decide how it should be funded.

With the new law resonating like the starting gun at a race, several companies issued threats about invading each other's markets. AT&T, for instance, issued a declaration of war on local phone companies everywhere, pledging to enter the market in all 50 states and win over at least one-third of the $100 billion sector within a few years. Their plan, which proved to be prohibitively expensive, was to offer service via alternative access providers while using the company's cable lines as a platform for speedy Internet access.

Despite this competitive rhetoric, it seems that telcos are much more interested in pursuing markets in telephony than in cable. Before being allowed to enter long distance markets, RBOCs must prove that they've opened their local phone networks to new rivals, following a 14-point checklist contained in the 1996 Act. The FCC set rules for implementing RBOC entry into long distance in August 1996, promulgating 742 pages of guidelines. These rules were immediately challenged by several Baby Bells, and the U.S. District Court issued a ruling that the guidelines trampled on state's rights (Mehta, 1998). All such cases were consolidated in the Eighth Circuit Court of Appeals, which granted a stay of the FCC's order; the Supreme Court ultimately upheld the FCC's authority to review such applications on a state-by-state basis.

In December 1999, Bell Atlantic—which merged with GTE to become Verizon—became the first Baby Bell to win regulatory approval for its application to provide long distance service. Providing service to New York State, this arrangement represented the first time since AT&T's divestiture that consumers could receive both local and long-distance service from a former Bell operating company. Their application to provide local service in Massachusetts was recently approved, as were others from SBC (Texas, Oklahoma, and Kansas). Several other applications are pending.

Such activity prompted Judge Greene to wonder whether the new law can deter phone giants from essentially reconstructing the Bell monopoly (Keller, 1996). Counter to that concern, AT&T actually initiated a split into three different companies in 1995. They include:

> ➢ A communication services firm, which retained the name AT&T.

> ➢ A network services firm named Lucent Technologies.

> ➢ A computer unit, named Global Information Solutions.

If telco entry into other industries is any indicator, this move toward computing and video will not prove as revolutionary as some might expect. As Bates, et al. (2002) note, new services can be grouped into several categories:

> ➢ Adding new types of content/uses that can utilize existing telephone networks.

> ➢ Value-added services complementing the basic service.

> ➢ New mechanisms for delivering existing and emerging services.

> ➢ Those new uses which were not feasible under the technical limitations of the old networks, but which may be under new broadband networks.

Years after winning the right to provide information service, the Baby Bells had little to show for their efforts (Chen & Wilke, 1999). Ameritech recently offered ISDN services in Ohio, including high-speed data and video applications, along with enhanced facsimile service—complete with mailboxes for storing and sorting fax messages. Pacific Telesis has offered voice-mail subscribers customized reports, delivered on a daily basis.

Beyond that, the Baby Bells have engaged in a few other modest projects involving information delivery (see Chapter 19). Despite blue-sky growth projections for video and data revenue—prompting telcos to assume $650 billion in debt worldwide—over 97% of fiber optic capacity goes unused today (Zuckerman & Solomon, 2001).

In addition to wireless, video on demand, and pay-per-view ventures, phone companies such as AT&T and MCI WorldCom have engaged in DBS (direct broadcast satellite) experiments (the latter in conjunction with Rupert Murdoch). MCI WorldCom also recently entered an alliance with Microsoft for the delivery of online products and services. AT&T's plans to develop high-speed Internet service were dealt a setback when an Oregon judge ordered the company to open its high-speed network to rival services such as America Online (Stern, 1999).

ANCILLARY SERVICES

Local and long distance service continues to dominate U.S. telco activities, generating $112 and $69.2 billion, respectively, at the turn of the millennium. There has also been prolific growth in wireless ($40 billion revenue) and Internet/broadband ($10 billion revenue) services, along with a host of others, including call waiting, call forwarding, call blocking, prepaid phone cards, computer data links, and audiotext (or "dial-it") service. Long distance companies, in particular, bear little resemblance to the firms they were in the late 1990s, as their core business keeps shrinking.

Perhaps the most robust growth has been in an area involving mass-audience applications of the telephone—800 and 900 services (Atkin & LaRose, 1994; LaRose & Atkin, 1992). Virtually nonexistent prior to divestiture, these services generated 10 billion calls annually during the 1990s, with over 1.3 million 800 numbers earning an estimated $7 billion in revenues for the industry (Atkin, 1995). Aside from dial-a-porn, audiotext presents hundreds of information options, including daily TV listings, national and international news, celebrity information, recordings of dead (and undead) celebrities, sports scores, horoscopes, and updates on popular television soap operas (Atkin, 1993; 1995). The $1 billion audiotext industry may thus serve as a bridge technology to more advanced information services (Neuendorf, et al., 2002). Yet the Psychic Friends Hotline—which targeted low income minority households with extensive promotions on cable—went out of business in 1998; they apparently didn't see the end coming!

As telcos implement DSL (digital subscriber line) technology and/or move to install fiber optics to the "last mile" of line extending to the home, providers will be able to provide a broad range of information and entertainment channels.

In the meantime, the proliferation of new telephone services is having another effect: telephone companies are running out of available phone numbers. The traditional solution of splitting an area code into two and allocating a new area code is proving more difficult to implement as the geographical areas for these area codes becomes smaller and smaller. One solution being tested in many areas is an "overlay" of the new area code in the same area as the old one, requiring that 10 digits be dialed for every local call.

A related issue is "local number portability." As competition emerges in the local telephone market, the FCC has been concerned that businesses and residences be able to keep their phone numbers when they choose to switch providers, adding slightly to the cost of local telephone service.

The ability to transmit digital signals such as DSL on the same phone lines that carry ordinary POTS signals created another controversy, as competitive local exchange carriers (CLECs) argued that existing phone companies should share the copper wires from their switching offices to homes or businesses. The incumbent phone companies (ILECs, incumbent local exchange carriers) argued against sharing, fearing a loss of up to half of their market share, which stood at 91% in 2002. This position has delayed the introduction of digital services because of the lack of availability of enough existing telephone lines to allow additional service to everyone who wanted it. In early 2000, the FCC solved the problem by mandating line sharing among local phone companies. As a result, U.S. residents can now get multiple services (POTS and DSL) provided by different companies on the same phone line.

Presenting a textbook case of the regressivity of monopolists, DSL was developed by a Bell engineer in 1989 and languished for nearly a decade because the RBOCs didn't want to sell against their other lucrative high-speed business Internet services (Dreazen, et al., 2002). The strong stock and bond markets of the 1990s helped fund CLECs in their efforts to increase DSL offerings, presenting a strong counterforce to the consolidated RBOCs. The number of local phone lines nearly doubed to 16.4 million (8.5% share) between 1999 and 2000 (Dreazen, 2001). Yet, where there were 330 CLECs competing against the Bells at the end of 2000, only 150 remained a year later. After $50 billion in high-yield telecom bonds were issued in 1998 and 1999, investors balked at the CLEC upstarts in 2000, with ICG's CEO noting "We've gone from full spigot to a situation where every capital source has shut down at the same time" (Dreazen, et al., 2001, p. A10).

CONSOLIDATION

Media merger and acquisition activity reached record levels in the wake of government deregulation during the 1990s, contributing to unprecedented levels of within-industry concentration. The impact of these mergers has been most dramatic in the area of local telephone service where, for instance, SBC is the local monopolist from Texas to Michigan.

SBC and Verizon account for 60% of phone lines in the United States, while Comcast Cable's pending (as of May 2002) merger with AT&T Broadband will leave the top three cable exhibitors in control of 65% of the nation's cable market. This leaves the audience's essential communication lines under the control of powerful oligopolies. Telco ambitions to sell voice, video, and data services are thus on hold, as the cable companies added 3.5 million high-speed modem subscribers in 2001, compared to only 2.5 million for DSL providers (Waters, 2001).

The forces of consolidation in telecommunications seem irresistible, given the large fixed costs, high barriers to entry and low marginal costs of serving each additional subscriber. As Wolfe (2002) notes, "[c]ritics of media concentration will now wonder how much more wheeling and dealing can go on before there are but one or two juggernauts controlling every image, syllable, and sound of information and entertainment" (p. A18). Observers thus maintain that the Telecommunications Act of 1996 contributes to a bigness complex that threatens rate competition, service quality, public access to the media, freedom of speech, and democracy itself (Atkin, 1999).

CURRENT STATUS

Declining stock valuations notwithstanding, the telephone companies remain in a formidable position to dominate video and information services, as they control more than half of U.S. telecommunications assets (Dizard, 1999). The industry handled 620 billion phone calls annually in the 1990s and has a long history of providing mass entertainment and information services (Atkin, 1995). The value of AT&T and its former segments has tripled since divestiture, as the telecommunications sector will soon subsume $1 trillion worldwide, one-sixth of the U.S. economy (Pelton, in press). Taken together, telephony and information services account for roughly one of 10 dollars spent in the United States—over $2,000/household annually—and their proportion of gross domestic product (GDP) is predicted to reach 17% by 2004.

While the Baby Bells are anxious to compete in long distance, the latter express concern that local companies still command 98% of local revenues over their regions. RBOCs retain control over 90% of the 155 million phone lines in the United States (Dreazen, 2001a), possessing "bottleneck control of the local loop and they'll be able to squash competition" (Berniker, 1994, p. 38). Given that MCI WorldCom recently paid $0.46 of every dollar it earns to the RBOCs in access charges, they have an interest in entering the $25 billion access charge business. This local exchange bottleneck helped motivate GTE to enter into its $70 billion merger with Bell Atlantic to create Verizon (Mehta, 2000).

At the time of divestiture, the Baby Bells represented roughly 75% of AT&T's assets. Between 1940 and 1980, the number of independent companies decreased by 77%, owing chiefly to consolidation among these companies, although some were acquired by AT&T (Weinhaus & Oettinger, 1988). By 1982, there were 1,459 independent telephone companies, producing 15.9% of the industry's revenues. Merger activity has accelerated since that time, but the United States still has hundreds of telephone providers.

Although the Telecommunications Act of 1996 returns telephone regulation to the federal government, the Baby Bells were previously governed by the judiciary, through Judge Harold Greene's review of the MFJ. The MFJ imposed several line-of-business restrictions on the RBOCs, including (1) the manufacture of telecommunications products, (2) provision of cable or other information services, and (3) provision of long distance services (*U.S. v. A.T.& T.*, D.D.C. 1982).

The FCC reified telco/cable cross-ownership restrictions out of fear that telcos would (1) engage in the predatory pricing characteristic of capital-intensive industries, and (2) use their natural monopoly over utility poles and conduit space to hinder competition with independent cable operators. Under these rules, former RBOCs were banned from providing cable service outside of their local access transport area (Telephone Company, 47 C.F.R. S. 63.54-.58, 1992). Although these terms of conduct have been subsequently deregulated, the "home court" advantage that RBOCs enjoy—in delivering a range of services—remains a concern. The federal appeals court in Washington, D.C. recently overruled the FCC, which had approved SBC's application to sell long distance service in Oklahoma and Kansas, because the company overcharges rivals who use their network (Dreazen, 2001b).

FACTORS TO WATCH

As the recent travails of AT&T suggest, the long distance sector has been open to competition for years, and may even be doomed as Internet telephony and other players push prices ever closer to zero. Aside from this convergence, telephone service will also be characterized by greater bandwidth capability, mobility, and globalization. As these trends unfold, we'll see an intensification of the debate over whether Internet telephony should be subject to the same taxes and regulation as POTS (as advocated by the telcos). Yet telephone companies are less anxious to see regulation applied to their television transmissions over the public switched telephone network (PSTN), claiming that they should be exempt from taxes levied in the form of cable franchise fees even when these competing wires deliver the same services.

Meanwhile, the industry's other basic service market—local exchange service—is in transition from a "natural regulated monopoly" to an "open competitive marketplace" (Bates, et al., 2002, p. 92). The catalyst, the Telecommunications Act of 1996, was passed amidst heady optimism concerning the benefits of deregulation for competition, consumer prices, the construction of an information superhighway, and resulting job creation in the coming information age. This radical deregulation was designed to encourage unprecedented media cross-ownership and competition by enabling (1) cable companies to enter local telephone markets and (2) local and long distance companies to enter each other's markets.

Shortly after its passage, however, the Act encountered heavy criticism in the popular press (Schiller, 1998), as even prominent Congressional supporters now seek hearings to investigate problems with industry conduct that the measure was designed to remedy (e.g., service, pricing) (Atkin, 1999). Preliminary data suggest that the Act has not lived up to its promise. Although rates in the competitive long distance sector have declined 13.1% since 1996, cable rates have risen 31.9% and local phone rates—still regulated—have risen 12.1% (Solomon & Frank, 2001).

Some five years after Act's passage, these telecommunications heavyweights are retreating to their neutral corners, as the proportion of customers who receive phone service over cable TV lines remains under 1%. Gene Kimmelman, co-director of Consumers Union, concludes: "It's an abysmal failure so far. The much-ballyhooed opening of markets to competition was a vast exaggeration" (Schiller, 1998, p. A1).

So even without faulty regulation, consumers are thus far paying more for telecommunications services in the new era of deregulation ushered in by the Act. While former FCC Chair Reed Hundt expressed dismay that competition was delayed by legal wrangling and industry "détente," he suggested that higher rates may be a necessary first step toward competition (Mills & Farhi, 1997), with longer-term trends favoring convergence, competition, and lower prices. One competitive bright-spot involves a group of smaller companies, such as Teligent, that are building their own advanced, high-speed communications facilities that will give them direct access to their customers, bypassing remnants of the old Bell system.

The ongoing litigation over other pending RBOC local service applications will have strong ramifications for the structure of the phone industry. Sprint's recent suspension of its local service initiative illustrates the difficulties long distance companies encounter in competing with the RBOCs, given the latter's advantages in capitalization and switching equipment control. AT&T's Michael Armstrong, concerned that carriers such as SBC/Ameritech and Verizon have thwarted competition, has threatened to terminate his company's costly local service initiatives; he seeks greater regulation in this area, complaining that the prices Bell charges for access to their networks are "exorbitant" (Dreazen & Solomon, 2001). In an effort to cut costs, AT&T announced in late 2001 that it plans to undergo yet another break-up into separate units for broadband, wireless, business, and consumer long distance.

Contrary to AT&T's hopes for greater local loop regulation, Congress moved in February 2002 to further deregulate high-speed Internet access by allowing the Bells to offer it without first demonstrating that their local phone networks are open to competition. Passing in February 2002, the bill—cosponsored by W.J. Tauzin (R-LA) and John Dingell (D-MI)—was tied up for two years in the House of Representatives; long distance opposition is likely to slow the measure's progress in the Senate (Wolfe, 2002). A related issue involves the ongoing redefinition of "universal service" mandates to include some form of lifeline access to video and information services (Auferheide, 1999).

While politicians might debate the need to maintain regulatory oversight until competition takes root, few would dispute that competition (and lower prices) will eventually be in the offing for telephony. The decade following divestiture saw competitive market forces dramatically reshape the telecommunications landscape. Consumer long distance rates dropped from an average of 40 cents a minute in 1985 to just 14 cents a minute by 1993 (Wynne, 1994). New long distance carriers helped spawn thousands of new jobs, with more likely to accompany a doubling of domestic spending on telecommunications projected by the decade ending in 2004.

The recent merger of SBC and Ameritech, in conjunction with those involving NYNEX and Bell-Atlantic as well as Pacific Telesis and SBC, raise the issue of whether the remaining Baby Bells (also Qwest—formerly U S WEST—and BellSouth) can go it alone. Although their territories range in profitability from the fast-growing south to the depopulated west, analysts expect that there remains plenty of opportunity for all RBOCs in the local phone industry. In fact, RBOCs enjoyed profit margins nearly double the 20% or so that their counterparts registered through the mid-1990s (Cauley, 1997), although they, too, were hurt by the 2001 meltdown in the telecommunications market.

Another recent merger proposal involving MCI WorldCom's $115 buyout of Sprint would have represented the largest takeover in history (Chen & Wilke, 1999); that merger proposal was later abandoned. AT&T, effectively blunted from access into local service, still commands the lion's share of the long distance market, followed by MCI and Sprint. There are over 100 other, smaller long-distance players (e.g., Metromedia), but their numbers have been thinned by the recent industry downturn. Table 18.1 reviews the top 10 worldwide telecom merger and acquisition deals, the five biggest of which rank among the 10 largest business deals in history.

SBC's growth and aggressive approach to local loop competition, in particular, seems to be reawakening fears about the dysfunctional conduct of local telephone monopolists. Following its acquisition of Ameritech, the company was besieged with complaints of poor service—racking up $188 million in penalties since 1999—along with charges of anticompetitive behavior (Waters, 2001). This new consolidation, combined with obvious telco cross-subsidization concerns, presents a basis for continued vigilance by regulators.

Table 18.1
Top Telecom Mergers through 2001

Target	Acquirer	Value (in Billions)	Announced
Ameritech	SBC Communication	$72.4	5/11/98
GTE	Bell Atlantic	$71.3	7/28/98
AirTouch Communications	Vodafone Group	$65.9	1/18/99
U S WEST	Qwest Communications	$48.5	6/14/99
AT&T Broadband*	Comcast	$44.0	12/21/01
MCI Communications	WorldCom	$43.4	10/1/97
Orange	Mannesmann	$35.3	10/20/99
Telecom Italia	Olivetti	$34.8	2/20/99
NYNEX	Bell Atlantic	$30.8	4/22/96
TCI Cable	AT&T	$30.0	6/24/98
Pacific Telesis	SBC Communications	$22.4	4/1/96

*Pending as of mid-2002

Source: Federal Communications Commission

AT&T's proposed sale of video assets to Comcast, meanwhile, suggests that telco entry into the video marketplace will not create the much-anticipated "revolution" in the broadcast and cable industries. This industry détente and consolidation is also bad news for various independent programmers, who supply specialty entertainment and information services. Speculation now centers on the possibility that the conservative FCC—whose Chair William Powell opposes ownership caps—may allow a Bell company to acquire a long distance firm such as Sprint (Dreazen, et al., 2002).

In sum, the industry-wide devaluation, consolidation, and retreat from competition presents regulators with the challenge of preventing undue concentration of our essential wired communication capabilities into ever fewer hands. Even so, the definition of telephony continues to evolve beyond a fixed line to include such methods as prepaid phone cards, which can also be used as a promotional and marketing tool. As Bates, et al. (2002,) note, "[W]hat we think of today as the telephone is increasingly just one piece of telecommunications goods and services being offered by an ever-increasing variety of suppliers" (e.g., Internet-flavored structures) (p. 199). If we're to maximize innovation, quality, and diversity in tomorrow's information grid, regulators must help channel telephony's vast capital and human resources into allied fields in a way that augments (rather than depletes) the existing cast of players.

BIBLIOGRAPHY

Atkin, D. (1993). Indecency regulation in the wake of Sable: Implications for telecommunication media. *1992 Free Speech Yearbook, 31,* 101-113.

Atkin, D. (1995). Audio information services and the electronic media environment. *The Information Society, 11,* 75-83.

Atkin, D. (1996). Governmental ambivalence towards telephone regulation. *Communications Law Journal, 1,* 1-11.

Atkin, D. (1999). Video dialtone reconsidered: Prospects for competition in the wake of the Telecommunications Act of 1996. *Communication Law and Policy Journal, 4,* 35-58.

Atkin, D. & LaRose, R. (1994). Profiling call-in poll users. *Journal of Broadcasting & Electronic Media, 38* (2), 211-233.

Auferheide, P. (1999). *Communications policy and the public interest: The telecommunications Act of 1996.* New York: Guildford.

Bates, B., Jones, K. A., & Washington, K. D. (2002). Not your plain old telephone: New services and new impacts. In C. Lin and D. Atkin (Eds.). *Communication technology and society: Audience adoption and uses.* Cresskill, NJ: Hampton.

Berniker, M. (1994, December 19). Telcos push for long-distance entry. *Broadcasting & Cable,* 38.

Briggs, A. (1977). The pleasure telephone: A chapter in the prehistory of the media. In Ithiel de Sola Pool (Ed.). *The Social Impact of the Telephone.* Cambridge, MA: MIT Press.

Brooks, J. (1976). *Telephone: The first hundred years.* New York: Harper and Row

Carterfone. In the matter of use of the Carterfone device in message toll telephone service. (1968). FCC Docket Nos. 16942, 17073; Decision and order, 13 FCC 2d 240.

Cauley, L. (1997, December 10). Genuine competition in local phone service is a long distance off. *Wall Street Journal,* A1, 10.

Chen, K. J., & Wilke, J. R. (1999, October 6). Kennard signals telecom deal will get tough FCC scrutiny. *Wall Street Journal,* A10.

Chesapeake and Potomac Telephone Co. v. U.S. (1992), Civ. 92-17512-A (E.D.-Va).

Dizard, W. (1999). *Old media, new media,* 3rd edition. New York: Longman.

Dreazen, Y. (2001a, May 22). Bells' rivals double local market share. *Wall Street Journal,* B6.

Dreazen, Y. (2001b, December 31). Court says SBC overcharges competitors, tells FCC to review long-distance decision. *Wall Street Journal,* A2.

Dreazen, Y., Ip, G., & Kulish, N. (2002, February 25). Why the sudden rise in the urge to merge and form oligopolies? *Wall Street Journal,* A10.

Dreazen, Y., & Solomon, D. (2001, February 8). ATT Chief says baby Bells may price company out of local service markets. *Wall Street Journal,* A8.

Federal Communications Commission. (1939). Investigation of telephone industry. *Report of the FCC on the investigation of telephone industry in the United States,* H.R. Doc. No. 340, 76th Cong., 1st Sess. 602.

Federal Communications Commission. (1992). Telephone company-cable television cross-ownership rules. *Second Report and Order, Recommendation to Congress, and Second Further Notice of Proposed Rulemaking,* 47 C.F.R. 63.54-63.58, 7 FCC Rcd. 5781

Gallagher, D. (1992). Was AT&T guilty? *Telecommunications Policy, 16,* 317-326.

Jussawalla, M. (1993). *Global telecommunications policies: The challenge of change.* Westport, CT: Greenwood Press.

Keller, J. (1996, February 12). AT&T and MCI explore local alliances. *Wall Street Journal*, A3.

Landler, M. (1995, April 3). The man who would (try to) save New York for NYNEX. *New York Times*, C 1, C6.

LaRose, R., & Atkin, D. (1992). Audiotext and the re-invention of the telephone as a mass medium. *Journalism Quarterly, 69*, 413-421.

Mann-Elkin's Act. (1910). Mann-Elkins Act, Pub. L. No. 218, 36 Stat. 539.

Mehta, S. (1998, January 5). Baby Bells cautious on quick entry to long-distance market after ruling. *Wall Street Journal*, B8.

Mehta, S. (2000, January 20). Bell Atlantic and GTE file with FCC to split off GTE's Internet backbone. *Wall Street Journal*, A4.

Microwave Communications, Inc. (MCI). (1969). FCC Docket No. 16509. Decision, 18 FCC 2d 953.

Mills, M., & Farhi, P. (1997, January 27). A year later, still lots of silence: Dial up the Telecommunications Act and get bigger bill and not much else. *Washington Post*, 18.

Neuendorf, K., Atkin, D., & Jeffres, L. (1998). Understanding adopters of audio information services. *Journal of Broadcasting & Electronic Media, 42*, 80-95.

Neuendorf, K., Atkin, D., & Jeffres, L. (2002). Adoption of audio information services in the United States. In C. Lin and D. Atkin (Eds.). *Communication technology and society: Audience adoption and uses.* Cresskill, NJ: Hampton.

Pelton, J. (in press). International telecommunications. In K. Anowkwa, C. Lin, & M. Salwen. *International Communication: Theory and cases.* New York: Wadsworth.

Schiller, Z. (1998, February 9). Local phone competition is still just a promise. *Cleveland Plain Dealer*, A1, 6.

Solomon, D., & Frank, R. (2001, December 21). Comcast deal cements rise of an oligopoly in the cable business. *Wall Street Journal*, 1.

Stern, C. (1999, June 21). Cabler's angst rises as courts fret 'net. *Variety*, 23-24.

Telecommunications Act of 1996, 104 Pub. L. 104, 110 Stat. 56, 111 (1996) (codified as amended in 47 C.F.R. S. 73.3555).

Telephone Company/Cable Television Cross Ownership Rules. (1992). 47 C.F.R. 63.54-63.58. *Second report and order, recommendations to Congress and second further notice of proposed rulemaking.* 7 FCC Rcd. 5781.

U.S. v. A.T.& T., 552 F. Supp. 131, 195 (D.D.C. 1982), *aff'd sub nom. Maryland v. U.S.*, 460 U.S. 1001 (1983).

U.S. v. Western Electric, Defendant's Statement, U.S. v. Western Electric Co. (1980). Civil Action No. 74-1698, pp. 169-170.

U.S. v. Western Electric Co. and AT&T. (1956). 1956 Consent Decree. Civil action no. 17-49, 13 RR 2143; 161 USPQ (BNA) 705; 1956 trade case. (CCH) Section 68246, at p. 71134 (D.C. N.J.).

Waters, R. (2001, July 18). Back to the future. *Investor's Business Daily*, 15.

Weinhaus, C. L., & Oettinger, A. G. (1988). *Behind the telephone debates.* Norwood, NJ: Ablex.

Willis-Graham Act. (1921). Pub. L. No. 15, 42 Stat. 27.

Wolfe, M. (2002, Feb. 21). Here comes another wave of mergers. *Wall Street Journal*, A18

Wynne, T. (1994, November 16). An earshocking proposition. *Cleveland Plain Dealer*, 11-B.

Zuckerman, G., & Solomon, D. (2001, May 11). Telecom debt debacle could lead to losses of historic proportions. *Wall Street Journal*, A1.

19

BROADBAND NETWORKS

Lon Berquist, M.A.[*]

Broadband technology allows for the digital delivery of voice, video, and data over a variety of transmission media. The terms "high-speed," "broadband," "advanced telecommunications," and "advanced services" are often used interchangeably to describe networks with considerable transmission capability. However, broadband networks typically are characterized by significant bandwidth, interactivity, and packet-based switching (Kirstein, et al., 2001).

BACKGROUND

The Federal Communications Commission (FCC) originally defined advanced telecommunications based on Section 706 of the Telecommunications Act of 1996. Section 706 defined "advanced telecommunications capability" as "high-speed, switched, broadband telecommunications capability that enables the users to originate and receive high-quality voice, data, graphics, and video telecommunications using any technology" (FCC, 2002). Subsequently, the FCC has used "advanced telecommunications capability," "advanced services," and "broadband" to describe services with an upstream (customer-to-provider) and downstream (provider-to-customer) transmission speed of more than 200 kilobits per second (Kb/s). The FCC distinguishes broadband from "high speed," which describes services with over 200 Kb/s capability in at least one direction. Supposedly, the FCC chose 200 Kb/s as the minimum transmission speed for broadband because it is the speed at which an Internet user can change Web pages as fast as one can flip through the pages of a book (OECD, 2001).

Transmissions speeds of 200 Kb/s are approximately four times faster than the speed of a standard phone line using a dial-up modem at 56 Kb/s. Although faster than dial-up, bandwidth of 200

---

[*] Telecommunication and Information Policy Institute, University of Texas at Austin (Austin, Texas).

Kb/s is hardly extraordinary. In fact, other technical standards-setting bodies define broadband with much greater bandwidth or transmission speeds. The International Telecommunications Union (ITU), for example, defines broadband transmission capacity as 1.5 million to 2 million bits per second (Mb/s) (OECD, 2001). Despite the varying definitions of broadband regarding transmission capability, all definitions include interactivity (two-way transmission) and packet-based switching.

Unlike telephone transmission which is circuit-switched (requiring a dedicated phone line to each sender/recipient), packet-switched networks break individual messages into separate packets of data that can travel together or separately over whatever lines are free. The receiving switch reassembles the data back into the original message for the recipient (Copeland, 2000). Packet switching is important for Internet protocol (IP) networks because it provides a sophisticated means for a data message to find its way to its intended destination, even if portions of the network are cut or damaged. See Figure 19.1 for an illustration of the difference between circuit-switched and packet-switched networks.

Figure 19.1
Packet & Circuit Switching

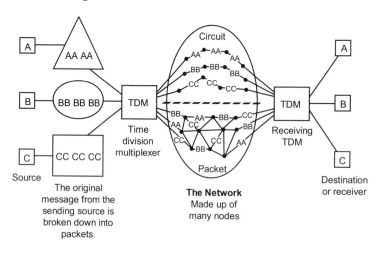

Circuit switching—In a circuit switched network, the source establishes a path to the sender and then transmits all the packets along the same path. Circuit switching is used to carry voice and video to ensure the order is maintained when received by the destination site.

Packet switching—In a packet switched network, individual packets must pass through the nodes on the network to be passed to the receiver. Throughout its route to the receiver, packets traverse any path and do not have to travel together. Packet number three of message one can be received before packet number one. If this does occur, the receiver will wait until all the packets arrive and then assemble them into the correct order.

Source: J. Hassay (1996)

The most common types of broadband technologies utilized by U.S. consumers are cable modems, digital subscriber line (DSL), terrestrial wireless, and direct-to-home satellite transmission. The primary advantage of broadband technology is to provide faster Internet connectivity, particularly when using the World Wide Web. Despite the advantages of broadband technology, the vast majority of Internet users (80% of online homes) still connect via standard telephone lines and modem with connection speeds no greater than 56 Kb/s (NTIA, 2001). Table 19.1 highlights the fact that cable modems (12.9% of online homes) and DSL (6.6% of online homes) are the primary alternative to dial-up Internet access.

Table 19.1
U.S. Home Internet Connection Type, 2001

Technology	Percentage
Dial-Up	80.0
Cable Modem	12.9
DSL	6.6
Other	0.5

Source: National Telecommunications and Information Administration

Although broadband connections make up a small fraction of the Internet universe, Nielsen/Net ratings has found that the number of broadband users almost doubled from 2000 to 2001, while the total number of Internet users grew at a less significant rate of 11% (Table 19.2). If this trend continues, broadband will become the primary means of connecting to the Internet, which will have a profound impact on how people use the Internet and how content is developed for the Internet.

Table 19.2
U.S. Broadband and Internet Growth

	November 2000	November 2001	Growth
Total Broadband Users	11.2 Million	21.3 Million	90%
Active Internet Universe	95.4 Million	105.5 Million	11%
% of Active Internet Universe	11.8%	20.2%	

Source: Nielsen/NetRatings

RECENT DEVELOPMENTS

CABLE MODEMS

As of mid-2002, deployment of cable modem service leads the race among high-speed data services in the United States. This is due, in part, to the fact that cable operators rapidly upgraded their passive one-way network to two-way hybrid fiber/coax (HFC) networks during the preceding decade. Once a cable system has been upgraded to allow two-way service, every subscriber is capable of receiving cable modem service. The telephone system's DSL service, cable's competitor, is distance sensitive and not available to phone customers too far from the central office (CO).

In the United States, cable modem service is available in 70 million homes, but estimates indicate there were just over seven million subscribers to cable's high-speed service as of the end of 2001 (NCTA, 2001). Installation of cable modem service runs from $50 to $200, with a monthly service fee of $30 to $50. A typical cable modem will provide high-speed Internet connections of up to 1.5 Mb/s downstream and 128 Kb/s upstream; however, the downstream bandwidth is usually shared with other cable subscribers on the same distribution node, which can impact download speeds among all the users on the node. Time Warner's Road Runner service leads the market with 1.9 million subscribers, followed by AT&T Broadband with 1.5 million and Comcast with 948,000. Should the proposed merger between AT&T and Comcast be approved, their newly formed company will become the cable modem leader among cable multiple system operators (MSOs) (see Table 19.3).

Table 19.3
U.S. Cable Modem Market, 2001

Cable MSO	Cable Modems
Time Warner	1,917,000
AT&T	1,512,000
Comcast	948,100
Cox	883,562
Charter	607,700
Cablevision	507,000
Adelphia	380,000
RCN	119,955
Mediacom	112,300
Insight	108,000
Other	75,000
Total	7,170,617

Source: Kinetic Strategies, Inc.

In March 2002, the FCC issued a Declaratory Order finding that cable modem service, rather than being a typical cable service or a regulated telecommunications service, is an unregulated "information service." This decision was in response to the desire among many cities that cable operators provide open access to competing Internet service providers (ISPs) (It's cable, 2002). Based on this decision, cable operators will not be required to allow network access to companies wanting to provide competing cable modem service within the cable operator's market.

Cable modems operate under the Data Over Cable Services Interface Specifications (DOCSIS) 1.0, but expectations are building for deployment of DOCSIS 1.1 certified modems that will allow cable systems to not only provide high-speed Internet connection, but also data telephony via voice over Internet protocol (VoIP) and multimedia services (Iler, 2001). As of mid-2002, cable provides phone service to 1.3 million cable subscribers (Coffield, 2001).

DSL

Digital subscriber line technology is utilized by telephone systems in order to transmit data over phone lines. DSL's advantage is it offers a dedicated line from the central office to the residence, providing both voice service and data with typical rates of 1.5 Mb/s downstream (though offering slower rates upstream). DSL service is similar to cable modems in that transmission is asymmetrical with a greater downstream capacity, adequate for most Web browsing and Internet use.

Installation costs for DSL can range from $50 to $200 dollars with a monthly service fee of $40 to $50. Increasingly, both cable systems and DSL providers are offering self-installation kits that allow subscribers to connect their own modems (whether a cable or DSL modem) and avoid an installation charge.

As of December 2001, approximately 3.5 million U.S. households subscribed to DSL, about half the number subscribing to cable modem service. The major incumbent local exchange carriers (ILECs) dominate the U.S. DSL market with a few competing data local exchange carriers (DLECs) offering service (Table 19.4). SBC, parent company of Southwestern Bell, Pacific Bell, and Ameritech, has almost 1.2 million DSL subscribers followed by Verizon's 975,000 and BellSouth's 463,000. Many DLECs, competitors offering exclusively DSL service, have suffered financially after tremendous investments failed to provide expected DSL subscriber growth. Northpoint Communications, Rhythms NetConnections, and Covad Communications all became Chapter 11 bankruptcy victims (Coffield & Smetannikov, 2001). Covad, however, has been revived and currently provides service to more than 300,000 subscribers (Brown, 2002).

Unlike cable systems, ILECs, under provisions highlighted in the Telecommunications Act of 1996, must provide network access to competitive local exchange carriers (CLECs) and competing DLECs. After heavy lobbying by incumbent carriers, the House of Representatives passed H.R. 1542 in February 2002 (Eilperin & Noguchi, 2002). The Internet Freedom and Broadband Deployment Act would free ILECs from having to share their lines with CLECs (Broadband legislative, 2002); however, it is unlikely the Senate will vote in favor of such a bill despite continued lobbying by the incumbent carriers.

Table 19.4
U.S. DSL Subscribers, September 2001

Operator	DSL Subscribers
SBC	1,187,000
Verizon	975,000
BellSouth	463,000
Qwest	405,000
Covad	346,000
Broadwing	55,000

Source: Converge Digest

SATELLITE

Satellite-based, broadband Internet service providers have not become a significant competitor among delivery technologies. Data-over-satellite is offered primarily by two competing services: Hughes Network System's DirecWay (formerly DirecPC) offered via DirecTV and StarBand service from EchoStar Communication Corporation's partnership with Gilat Satellite Network. As of mid-2002, DirecWay is received by 120,000 subscribers, and StarBand has 40,000 customers. Both use the newest two-way satellite dish uplink system that offers download speeds from 400 Kb/s with upload speeds of 128 Kb/s. Installation costs for satellite broadband can reach $500 with a monthly fee of $70. Competing satellite services such as WildBlue Communications, Inc. and Astrolink International LLC have joined the long list of telecommunications business failures (Arlen, 2002).

FIBER OPTICS

Fiber optics is utilized for long-haul communications transport facilities that transmit massive amounts of data for voice, video, and data networking. It is also used for connecting the long-haul facilities and the last mile. The last mile is the link between the middle mile and the last 100 feet to the subscriber (FCC, 2002). Both DSL and cable modem service rely on fiber optic cables for the bulk of the backbone of their networks (the middle mile), but the last mile transmission media consist of coaxial cable for cable modem service and twisted pair copper wire for the telephone system's DSL service. Even as fiber optics is getting closer to the home, DSL and cable modems dominate broadband delivery because of the universal availability of telephone networks and the fact that cable service is available to 70 million homes.

Because of the tremendous transmission capability of fiber optics, networks consisting exclusively of fiber will offer significant broadband capacity. Currently, however, there are few operational networks offering fiber-to-the-home (FTTH). Although the typical fiber optic network is built in high-tech communities such as Silicon Valley's Palo Alto (California) or large cities such as Chicago (Fuller, 2002), there have been recent fiber-to-the-home networks planned for smaller towns such as Kutztown (Pennsylvania) (Stump, 2002) and Osborne (Kansas) (Optical solutions, 2002). The FCC estimates there are 460,000 direct-to-subscriber fiber links throughout the United States, but only 0.6 % are connected to residences (FCC, 2002).

Costs for the tremendous bandwidth available via FTTH range from $45 per month for up to 45 Mb/s service or $100 per month for 100 Mb/s. Installation can cost over $2,000 (Metcalfe, 2000).

TERRESTRIAL FIXED WIRELESS

Fixed wireless technologies such as multipoint distribution service (MDS), multichannel, multipoint distribution service (MMDS), and instructional television fixed service (ITFS) in the 2.1-2.7 GHz wireless spectrum were initially meant to be used for distributing video as wireless cable television systems. The introduction of direct broadcast satellite as a more efficient means of video distribution dampened wireless cable's opportunities, but the prospects for data delivery via wireless spectrum was endorsed by a number of telecom giants such as Sprint, AT&T, and MCI WorldCom, which spent over $1 billion to purchase and build wireless broadband systems in over 40 cities (Gerwig, 2000). The 35-mile range of wireless cable, and the ability to provide thousands of residences or business users with downstream data rates similar to cable modem service (up to 1.5 Mb/s), appeared to be a good business model. Indeed, wireless cable utilizes the same cable modem as cable television systems, with the addition of an antenna and down-converter that adds to the expense.

In addition to the 2.1-2.7 GHz frequencies, the 27.5-30 GHz frequencies of local multipoint distribution service (LMDS) have been used for broadband wireless. LMDS, however, has a much more limited range of three to five miles, and widespread service is not yet available as of mid-2002.

Despite the comparable service and cost ($40 per month) to cable modems, wireless broadband has not been successful. The FCC estimated there were perhaps 20,000 subscribers to MMDS or LMDS service at the end of 2000. Because of the slow growth of fixed wireless broadband, AT&T Wireless announced it was closing its wireless operations, and Sprint Broadband Direct has stopped adding subscribers (Gohring, 2001).

NEW WIRELESS BROADBAND

Wireless phone providers are eagerly anticipating deployment of third-generation (3G) wireless services that utilize the ITU's worldwide standard that provides up to 2 Mb/s of voice, data, and multimedia communication well beyond the second generation's (2G) 14.4 Kb/s and the current 2.5G rate of 114 Kb/s that is limited to text messaging in addition to digital voice transmission.

As computer laptops gain in popularity, so will the development of wireless local area networks (WLANs) and personal area networks (PANs) in order to sustain mobile high-speed Internet connectivity. In the United States, the Institute of Electrical and Electronic Engineers (IEEE) has established various standards for WLAN networks, many of which have broadband capability. Wi-Fi (for wireless fidelity) is the name trademarked by the Wireless Ethernet Compliance Alliance (WECA) for wireless networks operating with the IEEE 802.11b standard offering throughput of up to 11 Mb/s over a short range of 10 to 500 meters. Wi-Fi is rapidly being deployed in airports, coffee shops, and other locations where laptop computer users congregate and a Wi-Fi antenna is nearby. Because Wi-Fi uses unlicensed spectrum (2.4 GHz), it is prone to interference and has weak security, but, nevertheless, use has increased to 15 million throughout the United States (Dornan, 2002). Many laptops now come with a Wi-Fi LAN network interface card, or one can be purchased for under $100. Soon, new iterations of IEEE 802.11 standards will be introduced that will increase bandwidth significantly.

Other WLAN technologies such as ultra wideband (UWB) promise even greater bandwidth and greater security due to its spread spectrum technology, but at a shorter distance (10 to 200 meters). Extremely short distance wireless technologies, such as Bluetooth with a broadband capability of 700 Kb/s, are termed wireless personal area networks (WPANs) and can be used primarily to connect to peripherals such as printers without the need for wires. (For more on wireless networking technologies, see Chapter 20.)

FACTORS TO WATCH

On the technology front, standards and bandwidth capability will continue to evolve. CableLabs, the cable industry's technical standards research and development consortium, has begun planning

DOCSIS 2.0, which touts a symmetrical data rate of up to 30 Mb/s. Because DOCSIS 2.0 modems have greater computer-control capability, they should bring the costs of cable modems down to $50, from the current average of $130 (Brown & Stump, 2002).

On the satellite horizon, EchoStar and Hughes Electronics are set to merge (see Chapter 6). If the merger is successful, the United States would have one dominant DBS and data-over-satellite service provider (Arlen, 2002).

Fiber optics, already the most dynamic transmission medium available, looks to be gaining even more capacity as researchers from Bell Labs have discovered it is possible to transmit 100 terabits per second (Tb/s) on a single fiber optic strand, or approximately the equivalent of one billion simultaneous phone conversations or 20 billion short e-mail messages (Thompson, 2001).

The FCC has opened up new Ku-band satellite frequencies for licensing (10.7 GHz-14.5 GHz) with expectations that they will be used for satellite Internet service. The FCC will also begin auctioning spectrum in the 700 MHz range that was originally allocated for UHF TV channels 60 through 69 (Steinke, 2000).

Economist Robert Crandall completed an economic analysis of the benefits of broadband technology. His analysis suggests broadband, because of its direct benefits in the technology sector and indirect benefits in online consumer use (Table 19.5), could contribute as much as $500 billion to the economy (Crandall & Jackson, 2001). Surprisingly, recent international rankings show the United States has fallen behind other nations in broadband deployment (Table 19.6).

Because of concerns about the economic lull in the U.S. telecommunications market, the potential economic impact of broadband, and the role of U.S. capabilities in a broadband world, broadband deployment has become a significant political issue.

TechNet, a lobbying group made up of industry firms such as Cisco, Intel, 3Com, IBM, and Microsoft, has suggested the United States should set a goal of having 100 Mb/s available to 100 million homes by 2010. In the interim, they insist a goal of 6 Mb/s should be achieved in half of all U.S. households by 2005 (Thibodeau, 2002). The Telecommunications Industry Association (TIA) has urged President Bush to name a cabinet level "Broadband Czar" to help implement broadband policy (Hearn, 2001).

In addition, a coalition of lobbyists, including the TIA, the Information Technology Industry Council, with the Semiconductor Industry Association, the National Association of Manufacturers, the Business Software Alliance, and the Consumer Electronics Association, have formed the High Tech Broadband Coalition to lobby legislators and the FCC.

Table 19.5
Summary of Annual Consumer Benefits from Universal Broadband Deployment ($ Billions per Year)

Source	Estimate
Direct Estimates	
Broadband Subscriptions	$427
Household Computer & Network Equipment	33
Alternative Estimates	
Shopping	257
Entertainment	142
Commuting	30
Telephone Services	51
Telemedicine	40
Total Benefits	*$520*

Source: Crandall & Jackson (2001)

Table 19.6
International Broadband Penetration, June 2001

Country	Broadband Penetration (%)
Korea	13.91
Canada	6.22
Sweden	4.52
United States	3.24
Netherlands	2.74
Austria	2.36
Denmark	2.33
Belgium	2.27
Iceland	1.99
Luxembourg	1.60

Source: OECD (2001)

Those not satisfied with the pace of broadband growth argue that better quality content is needed to attract consumers to broadband networks. At the same time, content providers complain that broadband is not developing fast enough for them to produce content specifically to take advantage of broadband's speed. The legal copyright issues regarding Napster, MP3, and other streaming audio software and services have been reported widely in the media. Copyright concerns about video streaming are now becoming more apparent. According to MPAA executive Jack Valenti, over 350,000 feature films are being downloaded illegally every day (Valenti, 2002). As broadband sub-scribership increases, there will likely be greater battles concerning distribution of copyrighted material as the availability of so much bandwidth will allow copyrighted materials such as movies to be more widely distributed, and with greater ease.

BIBLIOGRAPHY

Arlen, G. (2002). Satellite's high-flying expectations. *Multichannel News, 23* (12), 36.

Broadband legislative forecast for 2002. (2002). *Broadband Networking News, 12* (3), 1.

Brown, K. (2002). Covad closes chapter 11, opens new plan. *Multichannel News, 23* (5), 38.

Brown, K., & Stump, M. (2002). CableLabs moves ahead on DOCSIS, cable home. *Multichannel News, 23* (12), 33, 37.

Coffield, D. (2001). Weaving broadband: MSOs integrate voice and data. *Interactive Week, 8* (36), 35-36.

Coffield, D., & Smetannikov, M. (2001). DSL failures illustrate Telecom Act problems. *Interactive Week, 8* (31), 13.

Copeland, L. (2000). Packet-switched vs. circuit-switched networks. *Computerworld, 34* (12) 74.

Crandall, R. W., & Jackson, C. L. (2001, July). The $500 billion opportunity: The potential economic benefit of widespread diffusion of broadband Internet access. *Criterion Economics.*

Dornan, A. (2002, February 6). Emerging technology: Wireless LAN standards. *Network Magazine.* Retrieved April 23, 2002 from http://www.networkmagazine.com/article/NMG20020206S0006/1.

Eilperin, J., & Noguchi, Y. (2002, February 28). House passes Internet access legislation: Senate opponents pledge to block measure that helps "baby Bell" companies. *Washington Post*, A10.

Federal Communications Commission. (2002). *Inquiry concerning the deployment of advanced telecommunications capability to all Americans in a reasonable and timely fashion, and possible steps to accelerate such deployment pursuant to Section 706 of the Telecommunications Act of 1996.* CC Docket 98-146.

Fuller, M. (2002, January). Chicago's CivicNet to bring fiber to every neighborhood. *Lightwave.*

Gerwig, K. (2000). Wireless fixations: Service providers focus on point-to-multipoint rollouts. *Tele.com, 5* (6), 27.

Gohring, N. (2001). MMDS shifts gears: Operators hope high-speed second generation will yield profit. *Interactive Week, 8* (38), 46.

Hassay, J. (1996). Broadband network technologies. In A. E. Grant (Ed.). *Communication technology update,* 5th edition. Boston: Focal Press.

Hearn, T. (2001, October 7). Bush urged to name broadband czar. *Multichannel News.* Retrieved April 15, 2002 from http://www.tvinsite.com/index.asp?layout=print_page&doc_id=&articleID=CA177456.

Iler, D. (2002). Road to PacketCable passes DOCSOS 1.1. *Multichannel News, 22* (48), 110.

It's cable service. (2002). *Warren's Cable Regulation Monitor, 10* (7), 1.

Kirstein, M., Burney, K., Paxton, M., & Bergstrom, E. (2001). Moving toward broadband ubiquity in U.S. business markets. *Cahners In-Stat Group.* Report No.: BB0101UB.

Metcalfe, B. (2000). Faster than DSL or CTM, fiber optics to the home: Build it and they will come. *Infoworld, 22* (5), 116.

National Cable & Telecommunications Association. (2001). *Industry statistics.* Retrieved April 15, 2002 from http://www.ncta.com/industry_overview/indStat.cfm?indOverviewID=2.

National Telecommunications and Information Association. (2002). *A nation online: How Americans are expanding their use of the Internet.* Retrieved February 5, 2002 from http://www.ntia.doc.gov /ntiahome/dn/index.html..

Organisation for Economic Cooperation & Development. (2001). *The development of broadband access in OECD countries*. Report of the Directorate for Science, Technology and Industry, Committee for Information, Computer and Communications Policy.

Optical solutions wins FTTH contract. (2002). *Communications Today, 8* (66). Retrieved April 4, 2002 from http://www.telecomweb.com.

Steinke, S. (2000). Internet on thin air. *Network Magazine, 15* (3), 22.

Stump, M. (2002). Kutztown, Pa., muni has lots of fiber. *Multichannel News, 23* (12), 36, 37.

Thibodeau, P. (2002, January 16). TechNet pushes for broadband boost: Industry group says the government must make high-speed Internet access a national priority. *Computerworld*. Retrieved April 1, 2002 from http://www.pcworld.com/news/article/0,aid,79863,00.asp.

Thompson, J. (2001). Fiber gets a boost: Fiber-optic technology evolves. *Boardwatch, 15* (12), 10.

Valenti, J. (2002, February, 25). Movies get framed. *Washington Post*, A23.

20

HOME NETWORKS

Jennifer H. Meadows, Ph.D. & August E. Grant, Ph.D.[*]

The idea of setting up a computer network at home seemed completely unusual just a few years ago. However, with the increased use of computers in the home and the increasing number of homes with more than one computer, the idea of a home network is not only possible, in many cases, it is desirable.

Most homes used to only have one personal computer with one Internet connection. With the rapid growth of personal computer ownership came the rapid growth of multiple computer households. There are over 65 million homes in the United States with a PC, and 25 million of those homes have more than one PC (Teger & Waks, 2002). With multiple computers in a household comes the problem of sharing the Internet connection, as well as the various computer peripherals in the home such as printers and scanners. A home network solves that problem by allowing multiple computers to share an Internet connection as well as those peripherals. As discussed below, this sharing is only the beginning of what a home network can do.

One of the primary drivers behind home networking is the increase in broadband access to the home. As discussed in Chapter 19, more than 10% of U.S. households are connected to the Internet using digital subscriber line (DSL), cable modems, or broadband satellite service. Ironically, the key attribute for home networking of these broadband services is not the speed, but the fact that the connection is always on. Once a broadband user becomes accustomed to instant access, without spending any time to dial up or connect, the utility of having this access available to all computers in a home

---

[*] Jennifer Meadows is an Associate Professor in the Department of Communication Design, California State University, Chico (Chico, California). Augie Grant is Senior Consultant for Focus 25 Research & Consulting and a Visiting Associate Professor in the College of Mass Communications and Information Studies at the University of South Carolina (Columbia, South Carolina).

268

becomes even more salient. The value of this connection was underscored by a 2002 survey of computer users indicating that the always-on connection was more important than speed for people considering upgrading from dial-up to broadband Internet access in their homes (Demand for, 2002).

The key device in most home networks is a residential gateway, sometimes known as a cable/DSL router. These are devices that interconnect all of the computers and other devices that use IP data streams to create the home network, in turn connecting the home network to the outside broadband connection, thereby allowing different streams of information to be routed intelligently throughout the home. The capability of routing any type of data stream to set-top boxes, telephones, and other devices will eventually allow audio, video, and telephone signals to be distributed throughout the home in the same manner as computer data streams are routed.

Looking into the future, here are some possible applications of a home network and residential gateway:

> You order a boxing match that is sent to the downstairs television set, while Internet radio is sent to one receiver upstairs and another household member researches travel destinations on the Web—all at the same time.

> Telephone calls to your teenager are routed automatically to her bedroom, while business calls go to the home office.

> The dishwasher breaks and uses the home network to contact the repair facility before you know something is wrong.

> On your way home from a trip, you hear of a hard freeze coming to your area. You contact your home network from your wireless Web phone and turn the heat on so the house is comfortable by the time you get home and the pipes don't freeze.

This chapter will briefly review the development of home networks and residential gateways, discuss the types and uses of these technologies, and examine the current status and future developments of these exciting technologies.

BACKGROUND

Networking was once thought to be within the domain of the office or institution, not the home. Those few homes that did have networks typically had a traditional Ethernet network that required an expensive type of telephone wiring called Category 5 (Cat 5) wiring. Such a network needed a server, hub, and router. The network would need to be administered, requiring one household member to have computer network expertise. The installation and administration of these early home networks required a major allocation of time, money, and effort on the part of the user. Clearly, they were not for the average home computer user because the user would need extensive computer and networking expertise as well as a means of financing the network.

Several factors changed the environment to allow home networks to take off:

> *The Internet.* As discussed in Chapter 12, Internet use and growth has been exponential. More and more people want access, and competition for Internet access within the home is quickly becoming a common occurrence

> *Computer and peripheral sales.* The number of multiple computer homes is rapidly growing, facilitating a need for shared peripherals and Internet access. Who wants to buy several printers and have multiple Internet service provider (ISP) accounts when these resources can be shared by all the computers in a home?

> *Broadband.* The availability of broadband connections such as DSL and cable modems to the home is growing. These connections carry much more information than a traditional phone line, allowing for faster and enhanced Web browsing as well as the delivery of telephony, video, and audio services.

> *New consumer electronic devices.* There is a plethora of new consumer electronic devices that work in concert with the Internet such as portable Web pads, Internet audio receivers, MP3 players, and even Internet enhanced appliances. Technologies such as direct broadcast satellites and digital video recorders also can be used to great benefit on home networks. (See Chapters 6 and 16.)

The five major uses for home networks include resource sharing, communication, home control, scheduling, and entertainment (Enikia, 1999). Sharing one broadband connection and computer peripherals within the home is an example of resource sharing. Communication is enhanced when one computer user can send files to another computer in the house for review, or perhaps send reminders. A family can keep a master household schedule that can be updated by each member from his or her computer. Home control includes being able to remotely monitor security systems, lighting, heating and cooling systems, etc. Being able to route digital video and audio to different players within the home are examples of entertainment. The residential gateway allows the home network to have these multiple functions.

The next section will discuss the recent growth in home networking and residential gateways, including the types of networks and how they work.

RECENT DEVELOPMENTS

Recent changes in technology have made affordable and easy-to-use home networking and residential gateways available.

HOME NETWORKING

There are four basic types of home networks:

> Traditional.

> Phone line.

> Wireless.

> Power line.

When discussing each type of home network, it is important to consider the transmission rate, or speed, of the network. High-speed Internet connections and digital audio and video require a faster network. Regular file sharing and low-bandwidth applications such as home control may require a speed of 1 Mb/s (Megabits per second) or less. The MPEG-2 digital video and audio from DBS services requires a speed of 3 Mb/s, DVD requires between 3 Mb/s and 8 Mb/s, and compressed high-definition television (HDTV) requires around 20 Mb/s. Many of the current home networking technologies operate at about 10 Mb/s, an important limit considering future generations of digital media. Speed concerns may not matter once the next generation of home networking technologies is introduced, as new technologies are being developed with top data speeds of 50 Mb/s to 100 Mb/s.

Traditional networks use Ethernet, which has a data transmission rate of 10 Mb/s to 100 Mb/s. This is the kind of networking commonly found in offices and universities. As discussed earlier, traditional Ethernet has not been popular for home networking because it is expensive and difficult to use. To direct the data, the network must have a server, hub, and router. Each device on the network must be connected, and many computers and devices require add-on devices to enable them to work with Ethernet. Thus, despite the speed of this kind of network, its expense and complicated nature make it somewhat unpopular in the home networking market, except among those who build and maintain these networks at the office.

Some new housing developments come "Internet ready" and include Cat 5 wiring with routers and hubs. For example, the properties at the Irvine Ranch in Irvine (California) include homes, condos, and apartments that come with home networks (The Irvine, n.d.). Homes under construction are more likely to have networking wiring built in. The extra wiring is typically offered as a premium enhancement to the home at prices ranging from a few hundred dollars (for simple wiring and no hardware) to tens of thousands of dollars (for complex systems that route any type of data, audio, video, or telephone signals).

New homes represent a small fraction of the potential market for home networking services and equipment, so manufacturers have turned their attention to solutions for existing homes. These solutions almost always are based upon "no new wires" networking solutions that use existing phone lines or power lines, or they are wireless.

Phone lines are ideal for home networking. This technology uses the existing random tree wiring typically found in homes and runs over regular telephone wire—there is no need for Cat 5 wiring. The technology uses frequency division multiplexing (FDM) to allow data to travel through the phone line without interfering with regular telephone calls or DSL service. There is no interference because each service is assigned a different frequency. The Home Phoneline Networking Alliance (HomePNA) has presented two open standards for phone line networking. HomePNA 1.0 provided data transmission rates up to 1 Mb/s and was replaced by HomePNA (HPNA) 2.0, which allows data transmission rates up to 10 Mb/s and is backward compatible with HPNA 1.0 (Thomasson, 2001).

In addition to the presence of a router or residential gateway that has HPNA, each device on an HPNA network must have an adaptor that connects the device to a phone jack. The most common

271

adaptors are internal cards that plug into expansion slots of computers and external adaptors that connect between the phone jack and a computer's USB port. As of mid-2002, these adaptors cost between $30 and $60 each, and a separate one is required for each device on the network. (Although the Home Phoneline Alliance advocates that manufacturers build HPNA capability into computers and Internet appliances, few devices were built with HPNA as of mid-2002.)

HPNA 2.0's 10 Mb/s speed is adequate for most data applications, including Web surfing and file sharing, but some consider it too slow for entertainment applications such as HDTV. Speed will not be a problem in the next generation of the standard. The Home Phoneline Networking Alliance is working with its member companies to develop HPNA 3.0, which will offer speeds up to 100 Mb/s throughout the home network, and will be backward compatible with HPNA 1.0 and HPNA 2.0. The standard is expected to be released by the end of 2002, with HPNA 3.0 products appearing in the latter half of 2003 (HomePNA unveils, 2001).

One of the most talked about types of home network is wireless. There are four major types of wireless home networking technologies as of the beginning of 2002: Wi-Fi (otherwise known as IEEE802.11b High-Rate), 802.11a, HomeRF, and Bluetooth.

Wi-Fi, HomeRF, and Bluetooth are based on the same premise: Low-frequency radio signals from the instrumentation, science and medical (ISM) bands of spectrum are used to transmit and receive data. The ISM bands, around 2.4 GHz, are not licensed by the FCC and are used mostly for microwave ovens and cordless telephones. (802.11a, on the other hand, operates at 5 GHz.)

Wireless networks are configured with a receiver that is connected to the wired network or gateway at a fixed location. Transmitters are either within or attached to electronic devices. Much like cellular telephones, wireless networks use microcells to extend the connectivity range by overlapping to allow the user to roam without losing the connection (WECA, n.d.).

Wi-Fi is the consumer friendly label attached to IEEE 802.11b High Rate, the specification for high rate wireless Ethernet. Wi-Fi can transfer data up to 11 Mb/s and is supported by the Wireless Ethernet Compatibility Alliance, founded in 1991 by a group of companies, including 3Com, Cisco, and Nokia, interested in wireless networking (WECA, n.d.). Wi-Fi operates at 2.4 GHz and utilizes direct sequence spread spectrum technology. Wi-Fi has been widely adopted worldwide for both enterprise and home markets. It is predicted that it will be the dominant wireless networking technology through 2002 (Handley, 2002c). The inhibiting factors to adoption of Wi-Fi include limited bandwidth, radio interference, and security. The maximum bandwidth of Wi-Fi is only 11 Mb/s. Radio interference affects the network because the 2.4 GHz band used by Wi-Fi is home to a myriad of technologies including cordless phones, microwave ovens, and the competing but incompatible Bluetooth standard. Finally, Wi-Fi uses WEP (wireless encryption protocol) for security, but WEP has been broken, calling the security of Wi-Fi networks into question (Brown, 2001; Geir, n.d.).

At the same time as the 802.11b specification was released, the 802.11a specification was released as well. 802.11a operates within a higher frequency band, 5 GHz, using OFDM (orthogonal frequency division multiplexing). With data throughput up to 54 Mb/s, 802.11a clearly has a speed advantage over 802.11b's 11 Mb/s and does not have the RF interference problems that come with operating at 2.4 GHz. Because it operates at a higher frequency, 802.11a only works at shorter distances than 802.11b. It has been suggested that 802.11a will soon be the standard for wireless networking. However, the big concern is interoperability with existing 802.11b networks. Bridging technologies have and are being developed to allow this to happen (Brown, 2001, Geir, n.d.).

HomeRF, like Wi-Fi, was developed primarily for the home networking market. HomeRF is an open specification using shared wireless access protocol (SWAP). The technology is described as a blend of TDMA/DECT cordless phone for voice, 802.11, and open air for data (About HomeRF, n.d.). HomeRF 2.0 can transmit data up to 10 Mb/s up to 100 feet. Developers claim that HomeRF handles voice better than Wi-Fi because it was designed for voice and data including cordless phone use, while 802.11 was designed for data only (About HomeRF, n.d.).

HomeRF also touts that the technology is not subject to interference like Wi-Fi and Bluetooth. Wireless technologies using the unlicensced 2.4 GHz (Wi-Fi, Bluetooth, and HomeRF) are subject to interference from other devices using the spectrum such as microwave ovens and cordless phones, and it has been shown that this interference can and does cause problems for wireless networks. HomeRF claims to alleviate this problem by using "intelligent hopping" algorithms that detect interference and allow the packets to hop to interference-free frequencies (About HomeRF, n.d.)

While HomeRF and Wi-Fi can transmit data up to 11 Mb/s for up to 150 feet, Bluetooth was developed for short-range communication at a data rate of 1 Mb/s. Bluetooth technology is built into devices, and Bluetooth-enhanced devices can communicate with each other creating an ad hoc network. The technology works with and enhances other networking technologies. For example, students with a Bluetooth enabled laptop computer and Bluetooth enabled cellular phone in their backpack can surf the Web while sitting in the park because the computer links to a phone that can connect to an Internet service provider (Frequently asked, n.d.-a).

Wireless networks require both a central access point costing $200 to $1,000 and a separate wireless card for each computer or other device on the network. As of mid-2002, wireless PCMCIA cards for laptops and internal cards for computers ranged in price from $50 to more than $200, allowing a complete wireless network with multiple nodes for less than $500.

The developers of power line networking realized that all of the devices that a consumer might want to network are already plugged into the home's electrical sockets. In addition, there are almost always power outlets in the rooms where a networked device would be placed. Why not network over these power lines? Inari and Enikia are two companies that pioneered power line home networking, developing separate standards for power line networking. Because the power line networking industry was a few years behind its phone line and wireless competitors, many of the companies in this industry banded together to form the Homeplug Powerline Alliance in 2001 (Handley, 2002b). The primary purpose of the organization was to create an open specification for power line networking, allowing the industry to more effectively compete with other networking systems (Gardner, 2001).

Homeplug introduced their v1.0 specification in 2001 (Gardner, et al., 2002). Homeplug's specification uses OFDM for transmission (Gardner, et al, 2002). In the past, power line networking technologies have been hampered by a lack of standards in addition to immature technologies and regulatory issues. Homeplug v1.0 is intended to overcome these problems.

With power line networking, household electrical wires are used for data transmission, with data rates up to 12 Mb/s as of mid-2002. With these products, a gateway is plugged into an electrical outlet and a modem. Network interface adapters are then connected between electrical outlets and the USB, Ethernet, or parallel ports of a networked device such as a computer.

The data speed available over power line networking has been limited by the hostile nature of electrical lines. Data rates are affected by change and flux created by power surges, lightning, and brown outs. However newer technologies including those by Enikia, Inari, and Linksys are addressing the problem, promising speeds comparable to those achieved by other home networking technologies (Inari, n.d.; Alliance, n.d., PowerLine Networking, n.d..).

Usually, a home network will involve not just one of the technologies discussed above, but several. It is not unusual for a home network to be configured for HPNA, Wi-Fi, and even traditional Ethernet. Table 20.1 compares each of the home networking technologies discussed in this section

Table 20.1
Comparison of Home Networking Technologies

	How it Works	Specifications & Standards	Speeds	Reliability	Cost	Privacy
Conventional Ethernet	Uses Cat 5 wiring with a server and hub to direct traffic	IEEE 802.3 IEEE 802.5	10 Mb/s to 100 Mb/s	High	High	Secure
HomePNA	Uses existing phone lines and OFDM	Home Phoneline Networking Assn. HomePNA 1.0 HomePNA 2.0 IEEE and WTU	1.0: Up to 1 Mb/s 2.0 Up to 10 Mb/s	High	Low	Secure
IEEE 802.11b Wi-Fi	Wireless. Uses electro-magnetic radio signals to transmit between access point and users. Direct sequence spread spectrum	IEEE 802.11b Wi-Fi WECA 2.4 GHz	Up to 11 Mb/s	High to Moderate	High to Low	Secure
IEEE 802.11a	Wireless. Uses electro-magnetic radio signals to transmit between access point and users. OFDM	IEEE 802.11a 5 GHz	Up to	High	Moderate	Secure
HomeRF	Wireless. Uses electro-magnetic radio signals to transmit between access point and users. TDMA and 802.11x	HomeRF Working Group 2.4 GHz	10 Mb/s	High to Moderate	Moderate	Secure
Bluetooth	Wireless. Uses radio frequency at 4.2 GHz	Bluetooth Special Interest Group	1 Mb/s	High to Moderate	Moderate	Secure
Powerline	Uses existing power lines in home.	Proprietary and HomePlug HomePlug v1.0	1 Mb/s to 10 Mb/s	Moderate	Low	Secure

Source: Meadows & Grant

RESIDENTIAL GATEWAYS

The residential gateway, also known as the cable/DSL router, is what makes the home network infinitely more useful. This is the device that allows users on a home network to share access to their broadband connection. As broadband connections become more common, the one "pipe" coming into the home will most probably carry numerous services such as the Internet, phone, and entertainment. A residential gateway seamlessly connects the home network to a broadband network so all network devices in the home can be used at the same time.

Figure 20.1
HomePortal Residential Gateway

Source: 2Wire, Inc.

The current definition of a residential gateway has its beginnings in a white paper developed by the RG Group, a consortium of companies and research groups interested in the residential gateway concept. The RG Group determined that the residential gateway is "a single, intelligent, standardized, and flexible network interface unit that receives communication signals from various external networks and delivers the signals to specific consumer devices through in-home networks" (Li, 1998).

Since the RG Group's white paper, there has been significant development of the residential gateway concept, and three groups are presently working on standards, which they see as enabling increased adoption and development of this technology: The Telecommunications Industry Association's TR-41.5 committee, the International Organization for Standardization and the International Electrotechnical Commission (ISO/IEC), and the Open Services Gateway Initiative (OSGi) (Li, 1998).

The reason behind all these groups working on specifications is linked to the function of the residential gateway. The definition taken by the groups discussed above envisions a residential gateway as a single centralized device, thus the need for standardization. Another concept of the residential gateway envisions multiple gateways in and out of the home attached to specific devices such as a set-top box or a modem.

OSGi's specification notes that a service gateway should:

> ➢ Be open and independent of platform.

> ➢ Be application independent.

> ➢ Support multiple services.

> ➢ Support multiple local network technologies.

> ➢ Support multiple device access technology.

> ➢ Coexist with other standards.

> ➢ Be secure (FAQs, n.d.).

Residential gateways can be categorized as complete, home network only, and simple.

A *complete residential gateway* operates independent of a personal computer and contains a modem and networking software. This gateway can intelligently route incoming signals from the broadband connections to specific devices on the home network. Set-top box and broadband-centric are two categories of complete residential gateways. A broadband-centric residential gateway incorporates an independent digital modem such as a DSL modem with Internet protocol (IP) management and integrated HomePNA ports. Set-top box residential gateways use integrated IP management and routing with the processing power of the box (The emerging, 1999). Complete residential gateways also include software to protect the home network, including a firewall, diagnostics, and security log.

Home network only residential gateways interface with existing DSL or cable modems in the home. These route incoming signals to specific devices on the home network, and typically contain the same types of software to protect the home network found in complete residential gateways.

Simple residential gateways are limited to routing and connectivity between properly configured devices. Also known as "dumb" residential gateways, these have limited processing power and applications, and only limited security for the home network.

Within the industry, manufacturers faced the dual task of defining the term "residential gateways" in general so that users would get to know the product category while, at the same time, differentiating themselves from their competition. For example, 2Wire, a company that develops residential gateways, has specified that residential gateways should:

> ➢ Be user installable—It should be an easy procedure that anyone can do.

> Be remotely configurable—With technical support from the manufacturer, the home user should be able to configure and update the RG through the Internet.

> Support all types of home networking.

> Have firewall and virtual private network capability—There should be privacy and security using a firewall to hide computers within the home and a VPN (virtual private network) to have a secure private network through the public network.

> Support both versions of DSL (2Wire develops RGs for use with a DSL connection).

> Be controllable through a browser—All controls should run through a browser so the user who needs to change something can access the controls using a browser window on a home PC.

> Should support multiple services such as Internet, voice, data, and entertainment (The emerging, 1999).

In 2000, more than 18 companies, most of them start-ups, were designing or marketing complete residential gateways. The market for these full-featured residential gateways, which cost between $300 and $1,000, proved to be limited, as most users who wanted to set up a home network chose the less expensive ($100 to $200) simple residential gateways, most commonly marketed as cable/DSL routers. The market for residential gateways was further eroded by the collapse of the competitive DSL market (see Chapter 19), which eliminated an important distribution channel that was expected to subsidize part of the expense of the residential gateway. Most of the companies selling complete residential gateways left the market, and the remainder were forced to lower prices to compete with the simpler, lower-priced offerings. In the meantime, the number of vendors offering any type of residential gateway has grown as a new set of players has entered the market. Parks Associates, for example, reviewed a total of 81 companies in the residential gateway market in its 2002 report on home networking (Residential gateway, 2002).

WORKING TOGETHER: THE HOME NETWORK AND RESIDENTIAL GATEWAY

A home network controlled by a residential gateway or central router allows multiple users to access a broadband connection at the same time. Household members do not have to compete for access to the Internet, printers, scanner, and even the telephone. The home network allows for shared access to printers and peripherals. Using appropriate software, household members can keep a common schedule, e-mail reminders to each other, share files, etc. The residential gateway or router allows multiple computers to access the Internet at the same time by giving each computer a "virtual" IP address, with the household only needing one external IP address. The residential gateway routes different signals to appropriate devices in the home. For example, Web pages are sent to the specific computer requesting them at the same time that entertainment signals in the form of radio, video, and games can be routed to the stereo, television, or digital video recorder attached to the network (such as a ReplayTV unit).

Home networks and residential gateways are key to what industry pundits are calling the "smart home." Although the refrigerator that tells us we are out of milk may seem a bit over the top, utility management, security, and enhanced telephone services are just a few of the potential applications of this technology. Before these applications can be implemented, however, two developments are necessary. First, appropriate devices for each application (appliance controls, security cameras, telephones, etc.) have to be configured to connect to one or more of the different home networking topologies (wireless, HPNA, or power line). Next, software, including user interfaces, control modules, etc. needs to be created and installed. It is easy to conceive of being able to go to a Web page for your home to adjust the air conditioner, turn on the lights, or monitor the security system, but these types of services will not be widely available until consumers have proven that they are willing to pay for the service.

Of the three types of information residential gateways were predicted to distribute throughout the home—data, entertainment, and voice (telephone)—only data has proven its value to consumers as of mid-2002. Consumers are definitely interested in sharing Internet access across multiple computers, but the market for other networkable devices such as Internet radios, set-top boxes, and IP telephones has yet to materialize.

CURRENT STATUS

The market for home networking and residential gateways is being primarily driven by low-cost computers, increases in the number of multiple PC homes, increases in online households, and increases in the number of homes with broadband connections. Estimates of the penetration of home networks vary widely, with the most liberal estimates—30% of U.S. households—including any type of connection between computers (including printer sharing). The most conservative estimates—less than 10% of U.S. households—include only households that have "complete" home networks with a central server or router and a wired or wireless network topology.

The rapid growth in each of the drivers listed above, especially broadband connections, will cause comparable increases in the number of home networks. Wireless networking is growing fast, with the majority stake being held by Wi-Fi. HomeRF dropped in its share of node shipments from 45% of the market in 2000 to 30% in 2001 (Handley, 2002d).

Home networking companies are using two distinct strategies to market to consumers. The first is a traditional retail strategy, through which consumers buy the needed equipment from retail stores and online retailers. The second strategy is to treat home networking as an enhanced service that is offered by companies that also offer DSL or cable modems. Instead of purchasing the necessary hardware, the consumer pays a monthly fee to the broadband service provider, which typically also provides technical support for the home network.

Consumers who are reluctant to deal with the complexities of home networks are more likely to take advantage of the second option, as it minimizes the upfront investment and assures them of technical support as they get their network up and running. The companies that provide home networking as a service enjoy the promise of a continuing revenue stream that, in the long run, far exceeds the hardware cost, gambling that the amount of technical support they must provide will decrease dramatically over time. SBC, for example, offers this service to its DSL subscribers through Pacific Bell, Southwestern Bell, and other subsidiaries.

Home networking technologies are available from a variety of traditional computer brands such as Apple and its Airport (Wi-Fi), Intel, and Gateway. In addition, there are a number of companies specific to networking and residential gateways that are major players in the market such as 2Wire, Linksys, and D Link.

As of mid-2002, most residential gateways include an Ethernet port for connection to a broadband modem, a USB connection (for a nearby computer), and support for one type of network (HPNA, wireless, or Ethernet). A few manufacturers, such as 2Wire, support almost every network types in a single gateway, including Ethernet, USB, and HPNA, with Wi-Fi also available as a fourth network type.

The only home networking protocol that has not made inroads in the marketplace as of mid-2002 is power line networking. The late start for power line networking systems delayed creation of standards until long after HPNA and wireless standards were approved, and power line network data speeds continue to lag more mature topologies. Over time, however, the ubiquity of electrical sockets and the relative security of power line networks compared to wireless networks should allow for a substantial market niche. In fact, Cahners In-Stat/MRD predicts that the power line home networking market will go from $18 million in 2001 to $190 million in 2002 and $706 million in 2006 (Handley, 2002b).

FACTORS TO WATCH

Home networks are one of the fastest-growing communication technologies, with a double-digit growth rate expected through at least 2007. In tracking this market, the following factors should be considered:

BROADBAND ACCESS

The number of homes with a broadband connection has grown from less than one million in 1998 to more than 20% of U.S. homes in 2002, with continued strong growth expected through 2005. Broadband penetration is an important factor because broadband access is the primary driver behind the growth of home networking. Home networks allow homes to use the broadband network efficiently and get the most out of it. As discussed in Chapter 19, the two major players in broadband access are cable and DSL. Which one will emerge the dominant player remains to be seen. The demise of competitive service providers slowed the growth of DSL in 2001, but local phone companies have set their sights on catching up and overtaking cable modems. One major tool that phone companies will need to make DSL more competitive is the ability to stretch the reach of DSL beyond three miles from the telephone company switching office. These DSL "loop extension" technologies range from adding amplifiers to a line to stretch the reach of DSL to moving the telephone switch (and related DSL equipment known as "mini-DSLAMs") to equipment pedestals closer to the homes served.

Cable modem service, on the other hand, is available to any home passed by the coaxial cable. One problem with cable and cable modems is that users all share the same cable, thus if many people are on at the same time in the neighborhood, the speed of the network can be significantly slowed. In addition, a cable modem requires that all upstream traffic share a comparatively small amount of bandwidth, so even within the home, there can be problems if multiple users want to upload a document at the same time. As of March 2002, cable is leading the competition for customers, but that may soon change.

One battle that has not been joined in the broadband war is over price. Rather than dropping, prices for broadband service remained stable through mid-2002, with some DSL companies even raising prices slightly. History suggests that it will only be a matter of time before a price war breaks out between DSL, cable modem, and other broadband service providers. This will certainly fuel diffusion of broadband access, which, in turn, should be expected to fuel the demand for home networking products and services.

Figure 20.2
Growth in Broadband Subscribership

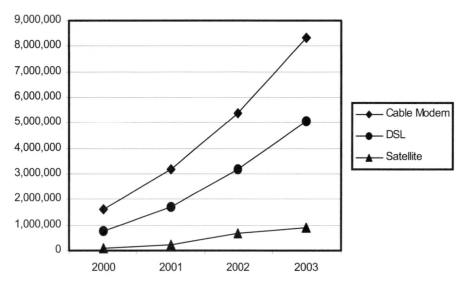

Source: Meadows & Grant

NEW CAPABILITIES FOR RESIDENTIAL GATEWAYS

The capabilities of residential gateways will certainly expand if the product achieves widespread use. For example, the same HPNA technology that transports large amounts of data over ordinary telephone lines can also transmit additional telephone voice lines throughout a home without the need for additional wiring. This technology can be connected through a residential gateway that has PBX-type functionality (telephone switching) to a broadband connection such as a DSL line that can also support numerous virtual phone lines to create home telephone systems as sophisticated as in any office—without adding any additional wiring to the home.

One key to the success of any home networking technology is the number of companies that build connections for that type of network into their products. It is not practical to have Internet radios,

Webpads, and smart appliances that must each have its own dedicated modem to connect to the Internet. Widespread adoption of any home network technology by manufacturers of peripherals will be a good predictor of success of the technology.

Broadband networks and home networks do present some security threats. An increasing concern of cyber-crime in this new century makes protecting the home network all the more important in many users' minds. The residential gateway provides one means of easy-to-use security—many have a firewall that hides the home network from potential intruders. Look for these security features to become more advanced.

MORE NETWORKED DEVICES IN HOMES

The term "smart home" is not likely to fade away in the near future (for better or worse). Entertainment devices, appliances, Webpads, phones, and security systems are just a few of the networked devices on the drawing board as of mid-2002. Look for these devices to hit the market in 2003 and 2004 as home networks proliferate. The key here is that before users are going to want networked devices, they are going to need a home network. This fact points to the importance of standards. You wouldn't buy a PAL television set for the United States would you? The wireless and phone line networking industries both had organizations working on standards since the 1990s, while power line networking companies did not begin working on a common standard until mid-2000 (Alliance overview, n.d.). This fact alone could slow the adoption of power line networking technology.

FASTER AND MORE ROBUST NETWORKING TECHNOLOGIES

While the current home networking technologies have speeds up to 11 Mb/s (not counting traditional Ethernet), look for faster and stronger networking technologies in the future. These will be cheaper and easier to use than anything available now and will be fast enough to accommodate whatever new networked devices are yet to be developed.

The next generation of home networking standards will increase home network speed from the 10 Mb/s range to between 50 Mb/s and 100 Mb/s. This increase in speed will not have a great effect on services coming into the home over cable modems and DSL links that offer average speeds closer to 1 Mb/s, but it will enable new opportunities in sharing entertainment content throughout the home, especially HDTV and high-quality DVD video. The entry of power line networking into the fray will also increase competition among the competing network topologies, raising consumer awareness of home networking, but possibly adding further confusion to the market.

Another interesting development in wireless networking is 802.11g. This standard has fast speeds of up to 54 Mb/s with mandatory compatibility with 802.11b. 802.11g would then overcome one of the barriers of 802.11a—the fact that it requires extra bridging technology to work with existing 802.11b networks (Brown, 2001).

While it is easy to be optimistic about the future of home networking and residential gateways, several factors must be considered when pondering their future success. First, the standards issue must be addressed. The industry may just let the market play out to see who the winner will be. Second, the industry must remember to keep the user in mind. It's wonderful to read about smart washing machines and refrigerators, but if the electronic devices are not easy-to-use and useful in the home, then what's the point? Finally, before any of this can happen, broadband connections have to be readily available at a reasonable price, and the services provided over these networks must be things that people want and cannot get elsewhere for less or with more ease.

BIBLIOGRAPHY

About HomePNA. (n.d.). *HomePNA.* Retrieved March 26, 2002 from http://www.homepna.org/about/faq.html.

About HomeRF. (n.d.). *HomeRF.* Retrieved from http://www.homerf.org/learning_center/faq.html.

Alliance overview. (n.d.). *HomePlug Powerline Alliance.* Retrieved March 26, 2002 from http://www.homeplug.com/about.

Brown, B. (2001). *802.11a—Fast wireless networking.* Retrieved April 21, 2002 from http://www.extremetech.com/article/0,3396,s%253D1034%2526a%253D19380,00.asp.

Demand for broadband growing slowly, survey shows. (April 23, 2002). *Telecom Direct.* Retrieved April 23, 2002 from http://www.telecomdirect.pwcglobal.com/telecom/direct:TIH/Telecom_Buzz/Internet/ InternetArt::/Article/reuters042302c.

The emerging residential gateway market. (1999). *2Wire, Inc.* Retrieved on April 22, 2000 from http://www.2wire.com/company/articles/article_resgatemarket.html.

Enikia, Incorporated. (1999). High-speed home networking over AC power lines! *Company brochure.* Piscataway, NJ: Enikia Incorporated.

FAQs. (n.d.). *Open Services Gateway Initiative.* Retrieved March 26, 2002 from http://www.osgi.org/about/ faqs.asp.

Frequently asked questions. (n.d.-a). *Bluetooth.* Retrieved April 24, 2002 from http://www.bluetooth.com/ util/faq1.asp.

Frequently asked questions. (n.d.-b). *Wireless Ethernet Compatibility Alliance.* Retrieved March 26, 2002 at http://wirelessethernet.com/benefitsfaq.asp.

Gardner, S. (2001, October). *Home networking 101.* Presented at the Broadband Home Conference. San Jose, California.

Gardner, S, Markwalter, B., & Yonge, L. (2002). HomePlug standard brings networking to the home. *CommsDesign.* Retrieved March 26, 2002 from http://www.commsdesign.com/main/200/12/ 0012feat5.htm.

Geir, J. (n.d.). The big question 802.11a or 802.11b. Retrieved April 21, 2002 at http://www.802.11-planet.com/columns/article/0m4000,1781_961181m00.html.

Handley, L. (2002a). Consumers adopt home networks to keep up with the Joneses. *In-Stat MDR.* Retrieved April 10, 2002 from http://www.instat.com

Handley, L. (2002b). Powerline home networking to get new life with arrival of HomePlug Standard. *In-Stat MDR.* Retrieved January 30, 2002 from http://www.instat.com.

Handley, L. (2002c). 802.11b WLAN market ends 2001 with a bang. *In-Stat MDR.* Retrieved February 27, 2002 from http://www.instat.com.

Handley, L. (2002d). Wireless home networking becomes more desirable: HomeRF continues to lose share to 802.11x. *In-Stat MDR.* Retrieved January 7, 2002 from http://www.instat.com.

Home Networking 101 (n.d.). *HomePNA.* Retrieved March 26, 2002 from http://www.homepna.org/ homenetworking101/breakdown.html.

HomePNA unveils market requirements for 3.0 specifications. (2001, November 12). *HomePNA.* Retrieved April 20, 2002 from http://www.homepna.org/news/presssr.asp?ReleaseId=13.

Inari. (n.d.). Retrieved on April 22, 2002 at http://www.inari.com.

The Irvine Company. (n.d.) Irvine Ranch. Retrieved on April 25, 2002 from http://www.irvineranch.com.

Li, H. (1998). Evolution of the residential-gateway concept and standards. *Parks Associates.* Retrieved April 22, 2000 from http://www.parksassociates.com/media/jhcable.htm.

PowerLine Networking. (n.d.). *Linksys*. Retrieved April 23, 2002 from http://www.linksys.com/edu/part6.asp.

Residential gateway report: 2002. (2002). *Parks Associates*. Retrieved April 20,2002 from http://www.parksassociates.com/reports&services/reports/tocs/RGreport2002.htm.

Teger, S., & Waks, D. (2002, February 25). Sandy and Dave's report on the broadband home. *BBH Central*. Retrieved February 26, 2002 from http://www.broadbandhomecentral.com/report.

Thomasson, D. (2001, October). *HomePNA: Home networking with no new wires*. Presented at the Broadband Home Conference. San Jose, California.

WECA overview. (n.d.). *Wireless Ethernet Compatibility Alliance*. Retrieved March 26, 2002 from http://www.wi-fi.org

21

SATELLITE COMMUNICATIONS

Carolyn A. Lin, Ph.D.[*]

rtificial satellites have been serving the world's population in nearly every aspect of its cultural, economic, political, scientific, and military life during the past three decades. Since the successful launch of the very first communication satellite, Intelsat I in 1965, the satellite was envisioned as the ultimate vehicle for linking the world together into a global village. As we begin the 21st century, that vision is being augmented as companion communication technologies in Internet, digital voice, data, and video transmission, as well as networking systems, are evolving rapidly.

In a multimedia, multichannel communication environment, where wired and wireless communication technologies compete against each other for a finite pool of consumers, satellite technology has been able to maintain its favorable position in the marketplace as a broadband and global coverage communication delivery system. This is because satellite technology is uniquely suited for delivering a wide variety of communication signals to accomplish a great number of different communication tasks. There are three basic categories of non-military satellite services: fixed satellite service, mobile satellite systems, and scientific research satellites (commercial and noncommercial).

> ➤ Fixed satellite services handle hundreds of millions of voice, data, and video transmission tasks across all continents between fixed points on the earth's surface.

> ➤ Mobile satellite systems help connect remote regions, vehicles, ships, and aircraft to other parts of the world and/or other mobile or stationary communication units, in addition to serving as navigation systems.

---

[*] Professor & Coordinator, Multimedia Advertising Certificate Program and 2002 Distinguished Faculty Research Award Winner, Department of Communication, Cleveland State University (Cleveland, Ohio).

> ➤ Scientific research satellites provide us with meteorological information, land survey data (e.g., remote sensing), and other different scientific research applications such as earth science, marine science, and atmospheric research.

The satellite's functional versatility is imbedded within its technical components and its operational characteristics. Looking at the "anatomy" of a satellite, one discovers two modules (Miller, et al., 1993). First is the spacecraft bus or service module, which consists of five subsystems:

(1)   The *structural subsystem* provides the mechanical base structure, shields the satellite from extreme temperature changes and micro-meteorite damage, and controls the satellite's spin function.

(2)   The *telemetry subsystem* monitors the on-board equipment operations, transmits equipment operation data to the earth control station, and receives the earth control station's commands to perform equipment operation adjustments.

(3)   The *power subsystem* is comprised of solar panels and backup batteries that generate power when the satellite passes into the earth's shadow.

(4)   The *thermal control subsystem* helps protect electronic equipment from extreme temperatures due to intense sunlight or the lack of sun exposure on different sides of the satellite's body.

(5)   The *altitude and orbit control subsystem* is comprised of small rocket thrusters that keep the satellite in the correct orbital position and keep antennas pointing in the right directions.

The second major module is the communications payload, which is made up of transponders. A transponder is capable of:

> ➤ Receiving uplinked radio signals from earth satellite transmission stations (antennas).

> ➤ Amplifying received radio signals.

> ➤ Sorting the input signals and directing the output signals through input/output signal multiplexers to the proper downlink antennas for retransmission to earth satellite receiving stations (antennas).

The satellite's operational characteristics are literally "out of this world" and reach deep into outer space. Satellites are launched into orbit via a space shuttle (a reusable launch vehicle) or a rocket (a non-reusable launch vehicle). There are two basic types of orbits: geostationary and non-geostationary. The geostationary (or geosynchronous) orbit refers to a circular or elliptical orbit incline approximately 22,300 miles (36,000 km) above the earth's equator (Jansky & Jeruchim, 1983). Satellites that are launched into this type of orbit at that altitude typically travel around the earth at the same rate that the earth rotates on its axis. Hence, the satellite appears to be "stationary" in its orbital position and in the "line-of-sight" of an earth station, with varying orbital shapes, within a 24-hour period. Therefore, geostationary orbits are most useful for those communication needs that demand no interruption around the clock, such as telephone calls and television signals (see Figure 21.1).

Figure 21.1
Satellite Orbits

Source: Technology Futures, Inc.

Non-geostationary (or non-geosynchronous) orbits are typically located either above or below the typical altitude of a geostationary orbit. Satellites that are launched into a higher orbit travel at slower speeds than the earth's rotation rate; thus, they can move past the earth's horizon to appear for a limited time in the line-of-sight from an earth station. Those satellites that are launched into lower orbits (known as LEO, for low earth orbit) travel at higher speeds than the earth's rotation rate, so they race past the earth to appear more than once every 24 hours in the line-of-sight from an earth station. Non-geostationary orbits are utilized to serve those communication needs that do not require 24-hour input and output, such as scientific land survey data.

BACKGROUND

During the early years of satellite technology development, the most notable event was the identification of the geostationary orbit by the English engineer Arthur C. Clarke in 1945 and the successful launch of the first artificial satellite, Sputnik, by the Soviet Union on October 4, 1957. On January 31, 1958, the first U.S. satellite, Explorer 1, was launched. In July 1958, Congress passed the National Aeronautics and Space Act, which established the National Aeronautics and Space Administration (NASA) as the civilian arm for U.S. space research and development for peaceful purposes. In August 1960, NASA launched its first communication satellite, Echo I—an inflatable metallic space balloon, or "passive satellite," designed only to "reflect" radio signals (like a mirror).

The next period of satellite technology development was marked by an effort to launch "active satellites" that could receive an uplink signal, amplify it, and retransmit it as a downlink signal to an earth station. In July 1962, AT&T successfully launched the first U.S. active satellite, Telstar I. In February 1963, Hughes Aircraft launched the first U.S. geostationary satellite, Syncom, under a contract with NASA (Divine, 1993).

In tandem with these experimental developments came the creation of the Communication Satellite Corporation (Comsat), a privately-owned enterprise overseen by the U.S. government, via the passage of the Communications Satellite Act of 1962. Meanwhile, an international satellite consortium, the International Telecommunications Satellite Organization (Intelsat), was formed on August 20, 1964. Before undergoing privatization in fall 2001, Intelsat was a nonprofit cooperative organization. Intelsat continues to coordinate and market satellite services, while member states are both investors and profit sharers of the revenues. The 19 founding member nations—including the United States, 15 western European nations, Australia, Canada, and Japan—agreed to designate Comsat to manage the consortium (Hudson, 1990). The era of global communication satellite services began with the launch of Intelsat-I into geostationary orbit in 1965. Intelsat-I was capable of transmitting 240 simultaneous telephone calls.

Other important issues during this period involved satellite transmission frequency and orbital deployment allocation policy. Both issues were thrust into the management of the World Administrative Radio Conference (WARC) in 1963. WARC is a technical arm of the International Telecommunications Union (ITU), an international technical organization founded in 1865 to regulate radio spectrum use and allocation. Given that a geostationary satellite's "footprint" can cover more than one-third of the earth's surface, the ITU divides the world into three regions:

> Region 1 covers Europe, Africa, and the former Soviet Union.

> Region 2 encompasses the Americas.

> Region 3 spans Australia and Asia, including China and Japan.

The ITU also designated radio spectrum into different frequency bands, each of which is utilized for certain voice, data, and/or video communication services. Major categories of frequency bands include:

> L-band (0.5 to 1.7 GHz) is used for digital audio broadcast, personal communication services, global positioning systems, and non-geostationary and business communication services.

> C-band (4 to 6 GHz) is used for telephone signals, broadcast and cable TV signals, and business communication services.

> Ku-band (11 to 12/14 GHz) is used for direct broadcast TV, telephone signals, and business communication services.

> Ka-band (17 to 31 GHz) is used for direct broadcast satellite TV and business communication services.

With the passage of time, the satellite's status as the most efficient technology to connect the world seems firmly established. However, an array of other challenges remains concerning issues related to national sovereignty and the orbital resource as a shared and limited international commodity. At the 1979 WARC, intense confrontations on the issue of "efficient use" versus "equitable access" to the limited geostationary orbital positions by all ITU member nations took place, pitting western nations that owned and operated satellite services against the developing nations that did not. For the developed nations, equitable access to orbital positions suggests a waste of resources, as many developing nations lack the economic means, technical know-how, and practical needs to own and operate their own individual satellite services. The developed nations consider an "efficient use" (first-come, first-served) system to be more technically and economically feasible. By contrast, developing nations consider equitable access a must if they intend to protect their rights to launch satellites into orbital positions in the future, before developed nations exhaust the orbital slots. This dispute would carry over to the new millennium before it was fully resolved.

International competition for the share of satellite communication markets was not limited to the division between developed and developing nations. In response to U.S. domination of international satellite services, western European nations, for instance, established their own transnational version of NASA in 1964. It was renamed the European Space Agency in 1973, and was responsible for research and development on the Ariane satellites and launch programs. Modeled after the Intelsat system, Eutelsat was formed in 1977 to serve its western European member nations with the Ariane communication satellites launched by the European Space Agency. A parallel development was the establishment of the Intersputnik satellite consortium by the former Soviet Union, which served all of the eastern European Communist bloc, the former Soviet Union, Mongolia, and Cuba. Other developing nations owned and operated their own national satellite services during this early period as well, including Indonesia's Palap, which was built and launched by the United States in 1976.

The success of Intelsat, initially as an international nonprofit organization providing satellite communication services, was repeated in another international satellite consortium—the international maritime satellite (Inmarsat) system—founded in 1979. Inmarsat now provides communication support for more than 5,000 ships and offshore drilling platforms using C-band and L-band frequencies. Inmarsat is also used to conduct land mobile communications for emergency relief work for such natural disasters as earthquakes and floods.

RECENT DEVELOPMENTS

During the 1980s, the satellite industry experienced steady growth in users, types of services offered, and launches of national and regional satellite systems. By the mid-1990s, satellite technology had matured to become an integral part of a global communication network in both technical and commercial respects. The following discussion will highlight a few of these important landmark events, which set the stage for the future of satellite communication in 2000 and beyond.

Due to a new allocation of frequencies in the Ku-band and Ka-band, direct-to-home (DTH) satellite broadcasts or direct broadcast satellite (DBS) services became more economically and technically viable as higher-power satellites can broadcast from these frequencies to smaller-sized earth receiving stations (or dishes). This development helped stimulate the television-receive-only (TVRO) dish industry and ignite an era of DBS services around the world (see Chapter 6). The best success story of this particular type of satellite service can be found in Europe and Japan. For the Europeans, direct-to-home satellite services are deemed an economical means to both transport and share television programs among the various nations that are covered by the footprint of a single satellite, as the pace of cable television development has been slow due to lack of privatization in national television systems. The Japanese utilized this service to overcome poor television reception owing to mountainous terrain throughout the island nation.

As the economic benefits of satellite communication became apparent to those nations that owned and operated national satellite services during the 1970s, many countries followed suit in the 1980s to become active players as well. In particular, developing countries saw satellite services as a relatively cost-efficient means to achieve their indigenous economic, educational, and social developmental goals in the vastly-underdeveloped regions outside of their selected urban centers. Following the example of Indonesia (Palap in 1976), India (Insat in 1983), Brazil (Brasilsat in 1985), and Mexico (Morelos in 1985) were among the countries that became national satellite system owners.

A parallel development also occurred with regional satellite consortia. For instance, Arab nations also formed their own satellite communication system, Arabsat (in 1985), to serve their domestic and regional needs. Hispasat (in 1989) provides Spanish-language content to the United States, Spain, Portugal, and all the Latin American countries. Asiasat (in 1989), a private operation jointly owned by a Chinese and a Luxembourg corporation, serves both China and Southeast Asia by carrying Chinese language television programs. Two other satellite systems are also competing to capitalize on the vast Chinese and Indochina markets. Chinasat, the Chinese state owned-and-operated entity, went into service in May 1997. Not to be outdone, Sinosat, a private joint venture between German banks and Chinese enterprises, started its service in July 1998.

This growth in the number of satellite services available in the marketplace also provided the impetus for free market competition in both domestic and international satellite communications markets. Intelsat had been a near-monopoly for most of its existence. This monopoly status was successfully challenged with Federal Communications Commission (FCC) approval of five private satellite systems (including PamAmSat and Orion) to enter the international satellite communication service market in 1985. As a result, international competition for global satellite services began, when both PamAmSat and Orion launched their satellite services in 1988 (Reese, 1990).

Subsequently, the FCC adopted two separate policies that permit all U.S. satellite systems to offer both domestic and international satellite services and allow private satellite networks authorized by the World Trade Organization (WTO) to provide their services to the U.S. market, respectively, in 1996 and 1997. Later, a slew of deregulatory activities aimed at disengaging national government stakes took place. Inmarsat was the first to become privatized on April 15, 1999, followed by Comsat in August 2000. Eutelsat became fully privatized in July 2001, and Intelsat followed suit two weeks later. These privatization developments were encouraged by the "Orbit Act" (the Open-market Reorganization for the Betterment of International Telecommunication Act) of 2000, passed by the U.S. Congress, which embraced full and open competition in the satellite communication marketplace and ordered the full privatization of intergovernmental satellite organizations (NTIA, 2000).

Another aspect of this unstoppable growth trend in the satellite industry involves the expansion of satellite business services. The first satellite system launched for business services was satellite business systems (SBS) in 1980. Its purpose was to provide corporations with high-speed transmission of conventional voice and data communication as well as videoconferences to bypass the public switched telephone network. As earth uplink and downlink station equipment became more affordable, a number of corporate satellite communication networks were launched (e.g., Sears, Wal-Mart, K-Mart, and Ford). These firms leased satellite transponder time from SBS providers to perform such tasks as videoconferencing, data relay, inventory updates, and credit information verification (at the point-of-sale) using very small aperture terminals (VSATs)—earth stations (or receiving antennas) that range from 3-feet to 12-feet in diameter. In many areas, large and small satellite earth stations were clustered in "teleports," allowing a concentration of satellite services and a place to interconnect satellite uplinks and downlinks with terrestrial networks. Today, more than 500,000 VSAT terminals can be found in more than 120 countries providing such services as rural telephony, distance learning, telemedicine, disaster recovery, and offshore networks, in addition to corporate and government applications (VSAT Global Forum, 2002).

As the market grows, digital transmission techniques have become the norm because they allow increased transmission speed and channel capacity. For instance, voice and video signals can be digitized and transmitted as compressed signals to maximize bandwidth efficiency. Internet via satellite is another fast-growing service area (including streaming media services) targeting corporate users by all major satellite services in the United States and Europe. Other digital transmission methods, such as time division multiple access (TDMA), demand assigned multiple access (DAMA), and code division multiple access (CDMA), can be utilized to achieve similar efficiency objectives. TDMA is a transmission method that "assigns" each individual earth station a specific "time slot" to uplink and downlink its signal (for example, two seconds). These individual time slots are arranged in sequential order for all earth stations involved (e.g., earth stations 1 through 50). Since such time allotment and sequential order is repeated over time, all stations will be able to complete their signal transmission in these repeated sequential time segments using the same frequency.

DAMA is an even more efficient or "intelligent" method. In addition to the ability to designate time slots for individual earth stations to transmit their signals in a sequential order, this method has the capability to make such assignments based on demand instead of an a priori arrangement. CDMA utilizes the spread spectrum transmission method to provide better signal security, as it alternates the frequency at which a signal is transmitted and hence allows for multiple signals to be transmitted at different frequencies at different times.

Yet another important technological advance during this period involves launching satellites that have greatly-expanded payload capability and multiband antennas (e.g., C-band, Ku-band, and/or Ka-band receiving and transmitting antennas). These multiband antennas enable more versatile services for earth stations transmitting signals in different frequency bands for different communication purposes, as well as more precise polarized spot beam coverage by the satellite. A spot beam focuses a signal from a particular transponder on a small area of the satellite's footprint, increasing the strength of the signal in that part of the footprint and allowing the same frequency to be used for a different signal in another part of the footprint. Such refined precision in spot beam coverage allows for targeted satellite signals to be received by designated earth receiving stations, which can be smaller and more economical due to the strength of the more focused spot beam signal. This, in turn, allows the satellite to better serve greater numbers of individuals and facilitate growth in the business service sector.

CURRENT STATUS

Even though national governments have opened their domestic markets to international competition in accordance with the initiatives of the World Trade Organization in the past few years, the larger picture of the satellite and radio communication market still depends on the standards and regulations set by the ITU. The ITU currently has 189 member nations. It successfully resolved the single-most politically contentious issue ever faced by the organization: balancing equitable access to and efficient use of geostationary orbital positions for all member nations. The World Administrative Radio Conference of 1995 presented regulations that ensured the rights of those member nations. These regulations enable those who have actual usage needs to obtain a designated volume of orbit/ spectrum resources on an efficient "first-come, first-served" basis through international coordination. Additionally, these regulations also guarantee all nations a predetermined orbital position associated with free and equitable use of a certain amount of frequency spectrum for the future. Subsequently, during WARC '97, decisions involving frequency allocation for non-geostationary satellite services were resolved, as shown in Table 21.1.

Table 21.1
Frequency Allocation for Non-Geostationary Satellite Services

Service	Frequency Band	Frequency Allocation
Fixed Satellite Service	Ka-band	18.9/19.3 GHz and 28.7/29.1 GHz
Mobile Satellite Systems	Ka-band	19.3/19.7 GHz and 29.1/29.5 GHz
Low-Speed Mobile Satellite Systems	L-band	454/455 MHz

Source: C. A. Lin

Non-geostationary satellites are launched into low-earth orbits (LEOs) or medium-earth orbits (MEOs) to provide two-way business satellite services. An example of this type of service is Teledesic, which proposes to provide high-speed global voice, data, and video communication, as well as "Internet-in-the sky" services. A similar, voice-only service is offered by Iridium (see Figure 21.3). Recent market emergence in cellular radio, personal communication services (PCS), and global positioning systems (GPS) represents another reason non-geostationary mobile satellite services are on the rise. Cellular radio and PCS both provide wireless mobile telephone services. When radio signals are being relayed via a satellite utilizing digital transmission technologies, these signals can reach a wide area network to accommodate global mobile communication needs (Mirabito, 1997). For instance, Globalstar's LEO satellites can transmit a phone signal originated from a wireless phone unit or a fixed phone unit to a terrestrial gateway, which then retransmits the signal to existing fixed and cellular telephone or PCS networks around the world. GPS is a mobile satellite communication service that interfaces with geographic data stored on CD-ROMs, originally developed for military use. Today, GPS is also used as a navigational tool to pinpoint locations in remote regions and disaster areas, and on vehicles, ships, and aircraft for civilian applications. (For more on the use of satellites for personal communications, see Chapter 23.)

Figure 21.3
Iridium System Overview

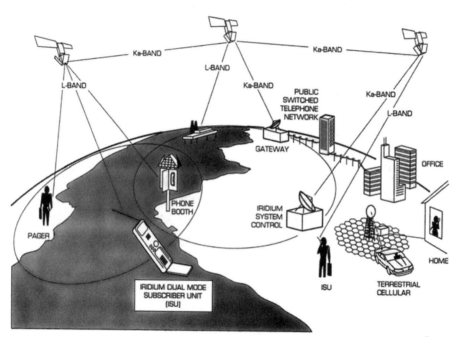

Source: Iridium, Inc.

The increasing interest in launching business satellite services in non-geostationary orbits is an outgrowth of strong market demand and increased congestion in the existing services provided through GEO satellites. The industry has seen an almost exponential market growth around the world toward the end of the 20th century. Due to the rapid expansion of the constellations of satellites launched by competing systems, several major players went through financial instabilities and were eventually reinvented under new management. For instance, ICO (Intermediate Circular Orbit) went into bankruptcy protection in August 1999 (Glasner, 1999) and was later integrated into Teledesic in May 2000 to keep its 12 MEO satellite service afloat (New life, 2000). Another service that emerged from bankruptcy in 2000 was the LEO satellite service provider Orbcomm, which had its first launch in December 1997 (Lloyd, 2001). Similarly, Iridium first launched its system in May 1997, declared bankruptcy in May 1999 (Iridium falls, 1999), and emerged from bankruptcy in December 2000 to run its 66 LEO satellite fleet (Iridium, 2001). Most recently, even with an ongoing financial restructuring due to filing bankruptcy in February 2002, Globalstar—which launched its system in February 1998—is still maintaining services to its clients with its 48-satellite LEO operation (Globalstar, 2002). The up-and-coming Teledesic is positioned to begin a 30-satellite LEO service in 2005 and to eventually increase it constellation to 288 (Teledesic, 2002).

To help accommodate the rapid development of mobile telecommunications, the recent WRC-2000 (World Radiocommunication Conference of 2000) reached the following major decisions (ITU, 2000):

(1)   Three additional frequency bands have been designated for international terrestrial mobile telecommunications: 806-960 MHz, 1710-1885 MHz, and 2500-2690 MHz.

(2)   To prevent frequency interference between geosynchronous and non-geosynchronous systems in order to provide high-quality communications services including telephony, television, and Internet applications, power limits were set for each type of satellite network services.

(3)   The rules for Ku-band (10-18 GHz) sharing between the two were defined.

(4)   Additional spectrum allocations were made for a new generation of radio-navigation-satellite services such as Europe's Galileo and Russia's GLONASS (global navigation satellite system) and the U.S. GPS.

FACTORS TO WATCH

As the satellite communication industry continues to grow at an accelerated pace, the most promising developments involve competition between mobile satellite systems operated in non-geostationary orbits (i.e., low- or medium-earth orbits) and fixed satellite services operated in geostationary orbits. As the former represents a more nimble application of services ranging from telephony and Internet to telemedicine and distance education, the latter remains the staple for large-scale voice, data, and video communication solutions.

Experts (e.g., Pelton, 1998) have predicted that satellites launched into LEOs and MEOs will provide an efficient global telecommunications network. In conjunction with other mobile communication services, such as the experimental high-altitude long-endurance (HALE) aircraft, these systems are valuable because they "eliminate the 'last mile' problem faced by land-based carriers" (Reagan, 2002, p. 67) to provide wireless services that can reach even the most remote regions of the world.

This growth phenomenon is largely sustained by two main factors. First and foremost is the stable supply of satellite launch services. For instance, the Chinese Long March satellite launch program emerged as a major player in the commercial satellite launch industry in 2000, on par with the European Space Agency's Ariane launch program and the Russia-based launch services (including joint-ventures with Germany, France, and Ukraine). The Chinese government has planned to launch manned space flights in 2003 with the successful reentry of its Shenzhou I-III spacecrafts, following in the footsteps of Russia and the United States (China's Shenzhou, 2002). Domestic commercial rocket-launch systems in the United States (i.e., Boeing's Delta series, Orbital's Pegasus, and Lockheed-Martin's Athena, Titan, and Atlas series) have also replaced use of NASA's space shuttle for commercial satellite launches, which were discontinued after the 1986 space shuttle Challenger incident.

The second growth factor concerns the continuing trends toward deregulation and privatization for both national and international satellite services. As the role of the WTO grew, it helped accelerate the deregulatory policies in national telecommunications industries. The arrival of the European Union also hastened those privatization actions, which included market mergers as well as shakeups. On the international front, the United States, Russia, China, European ministries of Posts, Telephone, and Telegraphs (PTTs), and their Japanese counterparts, for instance, have all sped up their efforts to approve international joint ventures, deregulate their domestic telecommunications industries, and privatize government-owned-and-operated entities when necessary.

These market developments were instrumental in helping reformulate the organizational goals of the ITU in the 21st century. The ITU, cognizant of the rising trend of global trade liberalization, also began to accommodate the rapid growth of global commerce by making its technical regulatory role a facilitating force instead of an intergovernmental bureaucratic drag. Such an ideological shift is evidenced by the recent Americas Telecom 2000 Conference (held in April 2000), where a demonstration that delivered digital video and Internet traffic was conducted jointly by Alcatel Espace (France) and ATC Teleports (United States) (Intelsat, 2000). Internet communication, in many ways, symbolizes the technological freedom of information flow between people and societies. To further pursue these social and economic developmental agendas, the ITU had two major goals. The first goal addressed the issues of the digital divide in "information and communication technologies" between developed and underdeveloped nations (Zeitoun, 2002). The second goal concerned the recognition of universal access and competition as the guiding principles for national telecommunications model building (Perrone, 2002).

These two goals were also the objectives for the World Telecommunications Development Conference (WTDC-02) held in March 2002. WTDC-02 was charged with the mission of providing a blueprint for the future of worldwide telecommunications. The conference generated a six-strategy action plan that addresses the digital divide issue in technical, societal, and economic terms, in addition to the information and communication technologies measures that can be used to foster socio-economic development of the world's populations, especially the most impoverished. Further work on advancing this humanistic objective via telecommunications development has been slated to take place in two future summits in 2003 (in Geneva) and 2005 (in Tunis).

The international space station orbiting at 240 miles above the earth appears as "a steady white pinpoint of light moving slowly across the sky" from ground sighting (NASA, 2002), and is illustrative of a new era in space exploration and telecommunications. What this new era symbolizes is the formulation of new ideas that utilize telecommunications technology solutions, especially those that are satellite- and space-communication based, to bring about social and economic developments that benefit the different peoples of our vast global village.

BIBLIOGRAPHY

China's Shenzhou 3 Spacecraft returns to Earth safely. (2002, April 1). *Space Daily*. Retrieved April 3, 2002 from http://www.spacedaily.com/news/china-02zg.html.
Divine, R. A. (1993). *The Sputnik challenge*. New York: Oxford Press.
Glasner, J. (1999, November 2). Craig McCaw's master plan. *Wired News*. Retrieved March 8, 2002 from http://www.wired.com/news/business/0,1367,32247,00.html.
Globalstar. (2002, February 15). Globalstar, creditors finalize agreement on debt restructuring and new business model. *Press release*. Retrieved March 11, 2002 from http://www.globalstar.com/EdiwebNews/300.html.
Hudson, H. E. (1990). *Communication satellites: Their development and impact*. New York: The Free Press.

International Telecommunications Union. (2000, June 2). World Radiocommunication conference concludes on series of far-reaching agreements. *New release, press, and publication information.*

Intelsat. (2000, April 9). Intelsat showcases satellite solutions for interactive multimedia applications, rural telephony, and Internet connectivity at Americas Telecom. *Press release.* Rio de Janeiro, Brazil, 1-2.

Iridium falls out of orbit. (1999, August 13). *Wired News Report.* Retrieved March 8, 2002 from http://www.wired.com/news/business/0,1367,21267,00.html.

Iridium Satellite LLC. (2001). The world's first global handset. *About Iridium.* Retrieved March 11, 2002 from http://www.iridium.com/corp/iri_corp_story.asp?storyid=4.

Jansky, D. M., & Jeruchim, M. C. (1983). *Communication satellites in the geostationary orbit.* Dedham, MA: Artech House.

Lloyd, W. (2001, May 1). Orbcomm FCC ECFS submissions. *Lloyd's Satellite Constellations.* Retrieved March 11, 2002 from http://www.ee.surrey.ac.uk/Personal/L.Wood/constellations/orbcomm.html.

Mirabito, M. M. A. (1997). Wireless technology and mobile communication. *The new communications technologies.* Boston: Focal Press.

Miller, M. J., Vucetic, B., & Berry, L. (1993). *Satellite communications: Mobile and fixed services.* Norwell, MA: Kluwer Academic Publishers.

NASA Human Space Flight. (2002, April 1). *Sighting opportunities: Viewing them from the ground.* Retrieved April 3, 2002 from http://spaceflight.nasa.gov/realdata/sightings/help.html.

National Telecommunications and Information Administration. (2000). *Open-market Reorganization for the Betterment of International Telecommunications (ORBIT) Act.* Pub. L. No. 106-180, U.S. Department of Commerce.

New life. (2000, May 17). *Wired News Report.* Retrieved March 8, 2002 from http://www.wired.com/news/business/0,1367,36408,00.html.

Pelton, J. N. (1998). Telecommunications for the 21st century. *Scientific American, 278,* 80-85.

Perrone, L. F. (2002, March 26). Harmonizing idealism with solvency: The basic challenge of the modern telecommunication regulator. *ITU News Magazine.* Retrieved April 2, 2002 from http://www.itu.int/itunews/issue/2002/02/harmonizing.html.

Reagan, J. B. (2002). The difficult world of predicting telecommunication innovations: Factors affecting adoption. In C. A. Lin & D. J. Atkin (Eds.). *Communication technology and society: Audience adoption and uses.* Cresskill, NJ: Hampton Press.

Reese, D. W. E. (1990). *Satellite communications: The first quarter century of service.* New York: Wiley Interscience Publications.

Teledesic. (2002). *About Teledesic: Fast facts.* Retrieved March 11, 2002 from http://www.teledesic.com/about/about.htm.

VSAT Global Forum. (2002). *Technology Trends.* Retrieved March 12, 2002 from http://www.gvf.org/vsat_industry.

Zeitoun, T. (2002, March 22). Transforming the digital divide into digital opportunities: ITU–D's challenge over the next four-year period. *ITU News Magazine.* Retrieved April 2, 2002 from http://www.itu.int/itunews/issue/2002/02/transforming.html.

22

DISTANCE LEARNING TECHNOLOGIES

John F. Long, Ph.D.[*]

Most forms of communication technology have undergone or are in the process of the conversion from analog to digital platforms. The distance learning community has been a direct beneficiary of this metamorphosis. The fundamental rationales for engaging distance learning—retraining, a non-traditional student clientele, and geographic isolation—remain stable. However, the infrastructure, software, and curricular designs have become increasingly sophisticated and adaptive.

Distance learning systems typically employ an amalgam of communication technologies. The actual system architecture is chosen based upon a number of factors. Costs, geographical location of participants, visual quality requirements, sites supported, interactivity, and pedagogical goals jointly define what parameters must be placed on the system design and which technologies should be utilized for optimal outcomes. Technically, conferencing technologies are systems that provide a multi-user interface, communication and coordination within the group, shared information space, and the support of a heterogeneous, open environment that integrates existing single-user applications. These systems can be categorized according to a time/location matrix using the distinction between same-time (synchronous) and different times (asynchronous), and between same place (face-to-face) and different places (distributed).

In concert with the established social needs for distance education, technological developments have provided a means to accomplish and extend directives articulated by education and business.

---

[*] Professor and Chair, Department of Communication Design, California State University, Chico (Chico, California).

The rapid rise of digital telecommunications and transformation have provided a plethora of delivery systems, collaboration tools, and databases that act as a backbone for contemporary distance education efforts. These resources and technologies are no longer restricted to an elite community and, as a result, are changing the nature of education and training forever (Capell, 1995). Digitization of resources empowers new models of education and an enhanced instructional experience for the traditional classroom. The varied forms of technology in classroom applications have risen dramatically since 1995.

The component processes that comprise distance education include learning, teaching, communication, and instructional character. Characterization of distance education from the systems perspective is critical. Anything that happens in one part of the system affects other parts of the system (Moore & Kearsley, 1996). Decisions to employ a new technology in a learning system must be grounded in the pedagogical soundness of the overall system. It remains imperative that the learning environment incorporate evolving technologies, as opposed to allowing emerging technologies to drive the learning system.

BACKGROUND

Distance education has gone through several phases of development. Over a century ago, correspondence study, using mail as the enabling technology, allowed distance learners to obtain degrees and train for professional endeavors. Many of these home study schools continue to operate today. For example, Education Direct (formerly ICS) has provided services since 1890, and the American Association for Collegiate Independent Study (AACIS) offers courses for nearly four million students annually.

The second development phase could be characterized as an integrated approach to distance education. These systems used multiple media such as correspondence, radio, television, audiotapes, and telephone conferences. The primary examples of these 1960s to 1970s projects are the University of Wisconsin's AIM Project and Britain's Open University (White 1996). The Articulated Instructional Media (AIM) Project hypothesized that learners could acquire skills from broadcast media and, at the same time, derive an "interactive" experience through correspondence or telephone conferencing. The original conference for AIM became the basis for the British Open University. The OU model incorporated the public broadcasting resources of the BBC with correspondence materials, offering degree education to any adult. Since 1971, the OU has served over one million students (Witherspoon, 1997).

The emergence of broadcast media and, later, teleconferencing, represents another phase of distance education. By the end of 1922, 74 colleges offered classes on the radio (Barnouw, 1966). Spurred by the Ford Foundation, educational TV broadcasts from universities began in the 1930s and were omnipresent by the 1950s. Collegiate consortia developed in 1961 through implementation of the Midwest Program on Airborne Television Instruction (MPATI). This configuration broadcast educational material to six states from transmitters on DC-6 airplanes (Smith, 1961).

Distance learning took a considerable leap forward in the mid-1960s with the launching of Intelsat 1. This modest beginning in satellite technology, offering one television channel or 240 telephone circuits, provided the hope for distance insensitive relays of educational material. During the next decade, several universities experimented with the Applications Technology Satellite (ATS). Because of diverse territories and extreme population dispersion, the Universities of Hawaii and Alaska were two of the first to utilize ATS for education and health training (Rossman, 1992).

As satellite technology improved and increased, its viability in the educational marketplace surged forward. Business, the military, industry, and institutions of higher learning realized the potential that satellites provided. The U.S. military has developed numerous distance education networks using satellites. Using two-way audio and one-way video, the Air Force Technology Network (ATN) was able to reach 18,000 students in 69 sites (Moore & Kearsley, 1996). To complement its videoconferencing infrastructure, the military moved to link 16 sites for fully-interactive exchange of video, voice, and data.

Business and industry have made substantial use of satellites as well. Nearly half of Fortune 500 companies engaged in corporate training were using this technology in the early 1990s (Irwin, 1992). IBM's Interactive Satellite Educational Network (ISEN) broadcasts courses to over 50 locations. The system was configured as one-way video/two-way audio with digital keypads for testing responses. Wang Laboratories uses a similar system for international training, as did Ford Motor Company and Aetna Life & Casualty. Currently, there are over 14,000 receive sites for corporate education (Dominick, et al., 2000).

Because of expertise in the academic fields and corresponding needs in private industry, a virtual university, similar in concept to the British Open University, was founded in 1985. The National Technological University (NTU) offers advanced degrees using digitally compressed video by satellite. Offering more than 400 courses, its clients are corporations whose employees require advanced training (such as AT&T, GE, and Kodak). Because of its digitally compressed signal, NTU was able to substantially reduce transmission costs and increase channel capacity (Witherspoon, 1997).

Other higher education consortia have developed to defray institutional costs and share expertise. The Adult Learning Satellite Service (ALSS), operated by PBS, offered courses to public and private entities also using satellite-delivered compressed video. Through generous funding from Congress, the Star Schools Program initiated regional satellite consortia designed to enhance educational opportunities in grades K-12. Most programs were advanced to educate audiences not served by traditional methods.

The next phase of distance learning technologies utilizes computer networking and multimedia. The Internet offers students the opportunity to take courses and complete degree programs directly through their computers. Unlike satellite conferencing, the student does not have to attend a remote site to receive instruction and interact with the instructor. Computer networking also provides asynchronous learning alternatives through the use of electronic mail, bulletin boards systems, and the like. Groups such as the Open University, University On-Line, and CalCampus transform regular university classes into digital format. These classes can then be replayed live or delayed through a Web browser. Graphics, digital images, and computer simulations can be added to the presentation (Open University, 2000). Most programs using the Internet have faced technological limitations due to modem speeds. The narrow bandwidth available through the public Internet cannot yet simulate virtual classrooms as in satellite teleconferencing. The emerging Internet 2 addresses these bandwidth challenges.

Other ambitious distance education arrangements came about in the late 1990s. The Western Governors University (WGU) brought together 17 western states to create a virtual university. An interactive "smart catalog" and technological nerve center of WGU identifies the media delivery system for each class. The principal means of delivery are satellites using compressed digital video and interactive Web-based designs (Blumenstyk, 1998).

A similar design emerged with the California Virtual University (CVU). In this model, over 300 California public and private higher education institutions would provide access to the expected 500,000 new students during the next decade. At the center of this cooperative was an interactive Internet catalog of technology-mediated distance learning offerings combined with a Web site and intranet tools available to all campus Web sites (Design team, 1997). By April 1999, the CVU lay in ruins due to faculty opposition and lack of funding. The WGU, although still operational, is a mere shell of its expectations (Downes, 1999).

One of the more significant projects of the 1990s that influenced the scope of distance learning was the creation of Internet 2 (I2). In October 1996, President Clinton adopted the central goals as the Next Generation Internet (NGI) Initiative and committed the administration to a system 1,000 times faster than the existing Internet. A very elaborate Internet 2 partnership was developed that includes industry, universities, and government resources. As of mid-2000, over 150 universities were involved in R&D on collaborative projects related to Internet 2 (Olsen, 2000). (For more on Internet 2, see Chapter 12.)

In order to facilitate development and use of I2 infrastructure, the National Science Foundation (NSF) and private industry has funded various phases for state or regional internetworking. System architectures required a gigaPOP (a network point of presence capable of transmitting speeds exceeding one gigabyte per second). In contemporary design and experimentation, geographically proximate institutions share the gigaPOPs. These provide for high-bandwidth, desktop-to-desktop communication between institutions. One prototype, VITALnet, interconnects Duke, UNC-Chapel Hill, North Carolina State, and an Internet service provider (ISP), MCNC, at 2.4 Gb/s. System advocates recognize its value in distance collaborative learning and telemedicine (NC Giganet, 1997). The California Research and Education Network-2 (CalREN-2) has initiated a venture involving research campuses in two major high-speed clusters or aggregation points (APs) forming distributed giga-POPs. The APs connect to each other using OC12 lines (high-speed lines capable of transmitting at 655 Mb/s), and OC3 or DS3 lines to regional universities. The OC12 lines then connect APs to the national I2 and vBNS for national collaborative efforts. (vBNS is a specialized nationwide IP network that supports high-performance, high-bandwidth applications. Originating in 1995 as the very high performance Backbone Network Service (vBNS), vBNS+ is the product of a five-year cooperative agreement between Worldcom and the National Science Foundation.)

With universal high-bandwidth networking capability a reality, the issue of instructional software that would serve as the basis of distributed instruction became an issue. Harnessing information in a comprehensible fashion must be addressed. In a series of experiments, Sun Microsystems developed the Distributed Tutored Video Instruction (DTVI) platform. This form of desktop videoconferencing (DVC) allows for text and object-oriented interaction among visually present learners. Sun's digital version of DTVI is called "Kansas." This virtual collaborative learning environment permits participants to move about in 2D space, and to make or break connections with others to collaborate and have independent discussions. The "Kansas" window establishes the instructional feasibility of a network-based virtual space (Sun Microsystems, 2000).

Figure 22.1
"Kansas" DTVI System

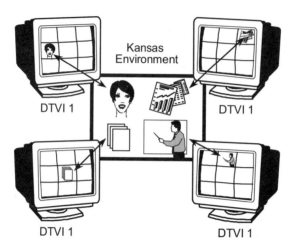

Source: J. F. Long

By the late 1990s, other technology firms such as Cisco, 3Com, and Microsoft formed partnerships with universities to provide integrated network solutions. A simultaneous buildout prompted by the NSF under its I2 directive provided an environment capable of more sophisticated and virtual environments for teaching, learning, and research. As a result of increased bandwidth, processor speeds, video compression, and implementation of standards such as H.323, Web-based learning began to assume its role as the information technology of choice.

As a result, various proprietary learningware emerged to address instructional needs in asynchronous, synchronous, and collaborative learning formats. At this time, about 35 of these products were developed for asynchronous teaching methods. The most standard of these have been "WeBCT," "Topclass," and "Courseinfo." Alternate versions of groupware were developed to support synchronous Web-based environments. These configurations allow students to participate in real time over a network, instantly receiving information that is text based, a video stream, or a combination. Microsoft's "NetMeeting" and White Pine's "Meetingpoint" are examples of this form of videoconferencing. Due to connectivity issues concerning video streaming, only small numbers of users were able to interact simultaneously.

RECENT DEVELOPMENTS

The use of distance technology continued to increase in 2001. Seventy percent of colleges now offer distance learning programs, and infrastructure spending is up 13% over 2000 (Olsen, 2001). Research and development spearheaded by the National Science Foundation have also received moderately improved federal support from the Bush administration. Funding for innovative distance learning programs at universities channeled through the Department of Education is also receiving support at the federal level (Carnevale, 2001).

Web-based strategies have emerged as the focal point of educational investment for institutions and corporations alike. Academic portals such as the Western Governors University, California Virtual University, and Canadian Virtual University consolidate Web-based offerings from a large field of member institutions. Commercial portals such as Peterson's Guides, Fathom, and Telecampus operate as aggregating services for member universities in an international scope (Jackson, 2002). IBM, Lucent, and Sun Microsystems offer Web-based training to upgrade employee skills. The U.S. Army in collaboration with PricewaterhouseCoopers has initiated eArmyU beginning with 12,000 soldiers. Its goal is to enroll 80,000 soldiers by 2005, and it is currently engaged with 23 colleges (Arnone, 2002). In addition, many colleges and corporations are continuing to collaborate on degree programs. For example, Babson College and the University of Texas at Austin have entered into arrangements offering Master degrees to Intel and IBM employees, respectively (Carnevale, 2002).

In an effort to effectively reach this ever-expanding and differentiated clientele, increasingly sophisticated virtual learning environments (VLEs) have been developed. The VLE is learning management software that brings together the functionality of computer-mediated communications (e-mail, newsgroups, objects) and the online ways of delivering course materials, such as the Internet. In asynchronous academic Web-based instruction, about 10 VLEs remain as viable alternatives. Of these, WeBCT and Blackboard now dominate this client segment. This trend is not surprising in that Blackboard has recently partnered with Microsoft, which promotes the product through its sales channels. WeBCT and Eduprise (a provider of e-learning services for education) have developed in concert a highly-integrated system. This software also interfaces with PeopleSoft systems, an enterprise software quickly moving into academic administration management.

The corporate sector brings different sets of learning management systems to Web-based instruction. These are moderately different compared with the academic software in terms of flexibility. Corporate asynchronous software emphasizes assessment metrics and compatibility with proprietary corporate databases. Docent, Inc. offers several integrated products in its continuing alliance with Global 2000 companies. It recently announced the release of Docent Peak Performance, a complete system enabling companies to assess employee competency and manage professional development of strategic objectives (Docent, 2002).

Another corporate-based software suite receiving considerable acclaim is Click2Learn. It, too, is deployed to disseminate corporate knowledge and improve productivity and performance. Click2Learn has recently been awarded contracts to enhance defense readiness programs for the National Guard in deploying its Aspen Enterprise Learning Platform. Driven in part by the events of September 11, 2001, the military selected this system because the National Guard and other military personnel can rapidly create content based upon reusable learning objects. This criterion is crucial in adjusting to defense training plans (Click2Learn, 2002).

In an important breakthrough, Click2Learn became the first vendor to meet the compliance requirements for the Sharable Content Object Reference Model (SCORM). The purpose of SCORM is to standardize technical specifications and to establish a unified content model. Specifically, it proposes to develop a means to allow for the interoperability of learning tools and content on an international scale. First released in 2000, the Advanced Digital Learning (ADL) Initiative has now developed several iterations in improving its adaptability. This collaborative effort by government, industry, and higher education is making significant progress in developing compatibility between competing learning management server (LMS) products. At the most recent ADL "Plugfest," representatives from 128 organizations worldwide gathered to test products and promote the acceleration of industry standards (ADL Plugfest, 2001). The developing concept of the ADL initiative is beginning to gain momentum. An academic ADL co-laboratory already exists at the University of Wisconsin. The Electronic Campus, providing access to 7,000 courses and 250 degree programs, will partner with the Wisconsin project. The Electronic Campus, the Distance Learning Policy Board, and the Educational Technology Cooperative (forming the Southern Regional Education Board) have aligned with ADL, making a very powerful statement concerning the viability of this initiative (ADL Plugfest, 2001).

Software manufacturers and network design specialists have made substantial progress in synchronous Web-based environments. These solutions have provisions for text-chat, live streaming audio, PowerPoint presentations, application sharing, and co-browsing features. "Centra Symposium" is the industry leader and offers the largest array of differentiated products. The features of "Horizon Live," such as "Office Hours Live," make it a distinctive option in interactive technology. The attributes of "Lotus Learningspace," such as its scalability, make it a favored choice for corporate training.

The H.323 Multi-user Control Unit (MCU) protocol remains a significant subset of synchronous delivery systems. Currently, Microsoft's NetMeeting is the leading software client (endpoint) for H.323 systems. The Intel Proshare System and CUSeeMe have made substantial inroads. Problematic with this protocol is the lack of functional multipoint gateways. Most industry experts concur that H.323 will be the universal standard by 2005, taking the place of the H.320 videoconferencing protocol. As universities and research consortia continue to join the I2 initiative, there should be even greater interest in this standard of high bandwidth available. Currently, bandwidth is a problem for Internet synchronous delivery, especially with high volumes of streaming video content (Jackson, 2002).

In order to advance collaborative projects, the Global Terabit Research Network (GTRN) has proposed an international partnership to link high-speed education networks. The consortium claims that there will be end-to-end operability and addresses research on a global scale in high-energy physics, astronomy, climatology, and earth science. The prevailing economics of optical cable is driving the projects. There is an oversupply of transatlantic cable capacity, which has caused a huge reduction in costs for connections. As a result, collaborations such as the Large Hadron Collider in Geneva requiring petabytes (1 million gigabytes) of connectivity and experimentation with the Hubbell Space Telescope can continue to proceed with large cadres of researchers on multiple projects in widely-dispersed locations (Olsen, 2002).

Designed originally for federal agencies and research universities, Internet 2 projects have broadened and filtered quickly into community colleges, as well as to secondary and elementary schools. One project now underway is the Virtual Harlem Project. This permits students to take 3D tours through early 20th century Harlem. Partnerships are beginning to develop, in part, to defray costs, but with a long-term goal of a national education network (Evelyn, 2001). In the near future, thousands of educational institutions plan to connect to the I2 backbone network.

In spite of the recent economic downturn in the technology sector, many corporations still engage in strong partnerships with universities on collaborative projects. For example, Cisco Systems has an alliance with Carnegie-Mellon on a 3D brain mapping initiative. AT&T is interfacing with Dartmouth on development of public key infrastructure labs to facilitate scalability. Lucent, IBMm and 3Com continue to support high-bandwidth experimentation in distributed virtual environments (Lanier, 2001).

The resources that continue to support Internet 2 are extensive. Based upon the direction of most research and collaboration, the primary outcome will be the continued development of tele-immersion, combining aspects of virtual reality architecture and bandwidth to provide the groundwork for a most unique form of human interaction.

CURRENT STATUS

Contemporary projects continue to focus on the ubiquitous access to online services and how to provide them. Business has made the most substantial inroads using wireless technology. Lucent Technologies has coupled with Carnegie-Mellon University to provide 400 wireless access points throughout campus. Wireless local area networks (WLANs), transmitting data via radio or infrared signals, can reach access points up to 10 miles away (Charp, 2002). At the federal level, the NSF-funded High Performance Wireless Research and Education Network has made significant inroads in establishing the usability of broadband wireless to American Indian communities. This organization has leveraged additional projects with Hewlett-Packard (Twist, 2001).

Other technological advances include wired network research. Grid computing is becoming the enabling technology for multi-institutional collaboration. This advancement facilitates the use of virtual laboratories for distributed-computing experiments by effectively standardizing portals. Complimentary research is examining bandwidth and how to optimize it. Demonstrations will soon be underway of bandwidths from 2.5 Gb/s to 10 Gb/s in creating a dedicated global research experimental network (What is, 2002). Many public and private organizations have created guidelines and standards for this emergent network. Because it is the Department of Defense's recognized research component, it is likely that the Defense Research and Engineering Network (DREN) will set the user standards for these unique high-speed networks (DREN, 2002).

The most comprehensive projects continue to be those under the directive of Internet 2. With high-bandwidth networks now established internationally, collaborative projects engaging tele-immersive visualization technology have begun to emerge. A recent venture with STAR TAP, a Chicago-based exchange point, provided a platform for 3D visualization of earthquake data between Chilean and U.S. researchers (Americas tele-collaborate, 2001). Recent developments on collaborative research have enabled the Advanced Research and Education Network Atlas (ARENA). This project functions to optimize the utility of network backbones such as vBNS, gigaPoPs, and federal agency networks (Internet 2, 2001).

Online learning continues to show tremendous growth potential. An increasing number of college campuses regard course management systems as critical in their institutional planning. A recent survey determined that three-quarters of institutions have identified a single standard for its management software, up 15% from 2000. This same study found that 20% of all college courses use management tools, up 6% from the previous year (Green, 2001). Another study predicts that, by 2004, 75% of U.S. college students will have taken at least one online class. Moreover, the demographics of these students is changing in that the population of students aged 25 and older is growing, possibly exceeding 18- and 19-year-old students within 10 years (Trends in, 2001). With this surge in the non-traditional student base, the number of universities offering Web-based instruction will have to embrace a similar growth curve.

Learning effectiveness from the standpoint of instructors and students lends support to continued online course development. Most research indicates that there is typically no significant difference in terms of online and traditional learning environments. However, some evidence suggests that students learn better online (California Distance Learning Project, 2000). The reasons for these findings are varied: a function of age, employment status, or motivation. Others posit that satisfaction and the ability to learn online is a function of learning style. Active learners may be more difficult to satisfy in a virtual environment when compared with intuitive learning styles (Mourtos & McMullin, 2001). More general data suggests that 30% of those taking online classes are less satisfied when compared with taking traditional classes. The least satisfied group was under 18 years old (35%), whereas those over 40 reported they were more satisfied with online courses (30%) (National Center for Education Statistics, 2001).

FACTORS TO WATCH

During the past few years, the use of the Internet for Web-based training has greatly expanded distance-learning opportunities. Virtual universities have developed, institutional and business use has risen exponentially, and standards have been adopted. Bandwidth size has increased, as have processor speeds. Interoperability and scalability have become huge issues and will continue to be addressed.

Organizations such as the Advanced Distributed Learning initiative will likely have a major impact as it continues to develop iterations of SCORM conformance. This XML (extensible markup language)-based packaging of content should do much in providing portability and interoperability between learning environments. Once established, conformity should perpetuate and accelerate the LMS market.

Another area that will receive greater attention as bandwidth swells is that of video streaming. The MPEG-4 standard and its various developments have done much for interactive learning environments. Philips' Webcine has done much to assure that this standard can be applied in broadband networks. Furthermore, MPEG-7, known as the Multimedia Content Description Interface, has been launched. This multimedia standard is designed for fast and exceptionally accurate media retrieval systems and content description. MPEG-21 standards will eventually be designed to provide the capability for multimedia content delivery and consumption in broadband network interactive environments.

BIBLIOGRAPHY

Academic Technologies for Learning. (1999). *What happened at California Virtual University*. Retrieved April 08, 2001 from http://www.atl.ualberta.ca/downes/threads/column041499.html.

ADL Plugfest 4 proves e-learning specifications work. (2001). *ADL News*. Retrieved March 12, 2002 from http://www.adlnet.org/news-events/news/full_story.CFM.html.

Americas tele-collaborate via STAR TAP. (2001). *STAR TAP News*. Retrieved March 08, 2002 from http://www.startap.net/grid2001.html.

Arnone, M. (2002). Army's huge distance-education effort wins many supporters in its first year. *Chronicle of Higher Education*. Retrieved March 12, 2002, from http://chronicle.com/weekly/v48/i22/22a03301.html.

Barnouw, E. (1966). *A tower in Babel: A history of broadcasting in the United States*. New York: Oxford Press.

Blumenstyk, G. (1998, February 6). Western Governors U. takes shape as a new model for higher education. *Chronicle of Higher Education, 44* (8), 21-24.

California Distance Learning Project. (2000). *College.com study shows positive results for online education*. Retrieved April 02, 2002 from http://www.cdlponline.org/dl2000/positiveresults.html.

Capell, P. (1995). *Report on distance learning technologies*. Pittsburgh: Carnegie-Mellon Software Engineering Institute.

Carnevale, D. (2001). Bush budget proposal seeks modest gains in technology-research spending. *Chronicle of Higher Education*. Retrieved March 02, 2002 from http://chronicle.com/free/2001/04/2001041001t.html.

Carnevale, D. (2002). Colleges tailor online degrees for individual companies. *Chronicle of Higher Education*. Retrieved March 10, 2002 from http://chronicle.com/free/2002/01/200212801u.html.

Charp, S. (2002). Instructional networks-some emerging tools. *T.H.E. Journal*. Retrieved March 12, 2002 from http://www.thejournal.com/magazine/vault/articleprintversion.CFM?aid=3815.html.

Click2Learn. (2002). *Click2Learn wins additional $ 2 million from the military for training services*. Retrieved March 10, 2002 from http://home.click2learn.com/en/news/pr.html.

Defense Research and Engineering Network. (2002). *DREN main page*. Retrieved March 08, 2002 from http://www.hpcmo.hpc.mil/htdocs/DREN/index.html.

Design team of the California Virtual University. (1997). *California Academic Plan*. California Virtual University, Draft 3.0.

Docent. (2002). *Docent announces Docent peak performance*. Retrieved March 02, 2002 from http://www.docent.com/products/product_CDS.html

Dominick, J., Sherman, B., & Messere, F. G. (2000). *Broadcasting, cable, the Internet, and beyond: An introduction to modern electronic media*, 4th edition. New York: McGraw-Hill.

Downes, S. (1999). *What happened at California Virtual University*. Retrieved from http://www.atl.ualberta.ca/downes/threads/column041499.html.

Evelyn, J. (2001). Internet 2 project will broaden access for community colleges and schools. *Chronicle of Higher Education*. Retrieved March 02, 2002 from http://chronicle.com/daily/2001/03/2001/2001030201t.html.

Green, K. (2001). eCommerce comes slowly to the campus. *The Campus Computing Project*. Retrieved April 02, 2002 from http://www.campuscomputing.net/summaries/2001/index.html.

Internet 2.(2001). *Overview of ARENA*. Retrieved March 19, 2002, from http://www.internet2.edu/arena.html.

Irwin, S. (1992). *The business television directory*. Washington, DC: Warren Publishing.

Jackson, R. (2002). *Virtual universities and learning portals*. Retrieved March 19, 2002 from http://www.outreach.utk.weblearning.courses.html.

Lanier, J. (2001). Virtually there. *Scientific American*. Retrieved April 06, 2001 from http://www.sciam.com/2001/0401/issue/0401/lanier.html.

Moore, M., & Kearsley, G. (1996). *Distance education: A systems view*. Belmont, CA: Wadsworth.

Mourtos, J., & McMullin, K. (2001). *A comparison of student learning and satisfaction in online and onground engineering courses*. College of Engineering, San Jose State University. Retrieved April 02, 2002 from http://www.engr.sjsu.edu/nikos/fidp/pdf/bangkok.html.

National Center for Education Statistics. (2001). *NEDRC table library*. Retrieved April 02, 2002 from http://nces.ed.gov/surveys/npsas/table_library/tables/npsas23.asp.

NC Giganet. (1997). *Background information*. Retrieved from http://www.ncgni.org/background.html.

Olsen, F. (2000). Internet 2 aims to build digital-video network for higher education. *Chronicle of Higher Education*. Retrieved April 04, 2001 from http://www.chronicle.com/free/v46/i33/33a04901.

Olsen, F. (2001). Survey finds another increase in campus spending on information technology. *Chronicle of Higher Education*. Retrieved April 04, 2001 from http://chronicle.com/daily/2001/04/2001040401t.html.

Olsen, F. (2002). Managers of fast networks see growing need for a new international backbone. *Chronicle of Higher Education*. Retrieved March 22, 2002 from http://chronicle.com/daily/2002/02/200202121t.html.

Open University. (1997). Retrieved April 20, 1998, from http://www.online.edu.

Rossman, P. (1992). *The emerging worldwide electronic university*. Westport, CT: Greenwood Press.

Smith, M. (1961). *Using television in the classroom*. New York: McGraw-Hill.

Sun Microsystems Laboratories. (2000). *The Kansas Project*. Retrieved May 04, 2001 from http://www.sunlabs.com/researcg/distancelearning/kansas.html.

Trends in online learning: Demographics and statistics. (2001). *TeleEducation New Brunswick*. Retrieved April 02, 2002 from http://teleeducation.nb.ca/content/lotw2001/trends.html

Twist, K. (2001). Native networking trends: Wireless broadband networks. *The Benton Foundation*. Retrieved March 02, 2002 from http://www.benton.org/DigitalBeat/db092001.html.

vBNS. (1998). *vBNS university case studies*. Retrieved March 04, 2000 from http://www.vbns.net/press/case.

What is I grid 2002? (2002). *STAR TAP News*. Retrieved March 08, 2002 from http://www.startap.net/grid2002.html

White, J. (1996). Britain's open road. *The Distance Educator, 2* (2), 8-11.

Witherspoon, J. (1997). *Distance education: A planner's casebook*. Boulder, CO: Western Interstate Commission for Higher Education.

23

WIRELESS TELEPHONY

Jennifer H. Meadows, Ph.D. & August E. Grant, Ph.D.[*]

Mobile telephones are so ubiquitous today that it is sometimes difficult to realize that they were available only to an exclusive few just 20 years ago. In that brief time span, they have become an indispensable tool for conducting business, a convenience for consumers, and one of the most visible indications of the pervasiveness of technology in human culture.

A walk down almost any street in all but the most remote corners of the world will include a view of someone talking into a mobile telephone. The sound of a mobile phone "ringing" has gone from being a curiosity to an annoyance to an accepted interruption in daily life. An extreme example of the role played by mobile phones is the set of conversations between passengers on the hijacked airliners and others on the ground during the September 11 terrorist attacks in the United States. Armed with information obtained through their mobile phones, the passengers on Flight 93 fought back, undoubtedly saving hundreds of lives in the process.

Countless lives are saved daily through use of this technology, as emergency personnel can be summoned to the scene of an accident or medical emergency more quickly than ever. Saving lives is just one of the many reasons that mobile telephones have become so popular. The same ability to provide a rapid response or communicate to and from almost anywhere has revolutionized the way business is conducted. You no longer have to be in a fixed location to be in touch. For many, time spent waiting in airports, driving, riding on trains, etc. has been transformed from a nuisance to a productive time.

---

[*] Jennifer Meadows is an Associate Professor in the Department of Communication Design, California State University, Chico (Chico, California). Augie Grant is Senior Consultant for Focus 25 Research & Consulting and a Visiting Associate Professor in the College of Mass Communications and Information Studies at the University of South Carolina (Columbia, South Carolina).

This is not to say that everyone is enjoying the revolution. The freedom to communicate offered by wireless telephony has brought with it new issues in social behavior, as a range of leaders from etiquette experts to lawmakers debate when and where we are allowed to use our telephones. The debate over negative impacts includes mostly unproven allegations, from use of the phones causing automobile accidents to radiation from the phones causing cancer in users.

Today's wireless telephony technologies include cellular phones, PCS, and satellite service. Although each uses a different part of the spectrum and has different technology, they share a number of features in common. The most important feature is the heart of the network, sometimes known as the mobile telephone switching office (MTSO). This intelligent switching system automatically assigns slices of the spectrum to individual telephone calls, switching frequencies and transmitting towers as a user moves from one part of a service area to another, and interconnecting the mobile phone call with the public switched telephone network (PSTN).

This chapter explores these and other factors regarding wireless telephony. In addition to the social issues, the history, technology, marketplace, and future of wireless telephony are examined.

BACKGROUND

The pioneers of radio technology at the turn of the last century were much more interested in developing a technology for interpersonal communication than one for broadcasting, and the pioneers of telephony, including Alexander Graham Bell, were more interested in distributing mass entertainment than providing a medium for interpersonal communication. It is doubly ironic that the telephone and radio have crossed paths twice, with the diffusion of the telephone providing interpersonal communication and radio technology providing mass communication.

A century later, each technology has "returned" to its roots, with cable television providing mass communication over wires and mobile telephones providing interpersonal communication through the airwaves. In the case of mobile telephony, the primary reason for the delay was the development of technology that would allow efficient use of radio waves for point-to-point communication.

The hobbyists who pioneered radio a century ago used it primarily to talk to each other over long distances. As the number of radio transmitters increased, however, the airwaves became crowded and the government was forced to exercise control over who used radio and for what purpose. The radio spectrum was divided into bands that were allocated for different purposes including radio broadcasting, amateur (ham) radio, law enforcement, and military communication. As technology developed to allow use of higher frequencies, additional bands were allocated for television broadcasting, satellite communication, etc.

Notably absent from the spectrum allocation was mobile telephony. One small slice of spectrum was allocated for mobile use beginning in 1947, but this band allowed a maximum of about 25 simultaneous telephone conversations. The 25-conversation limit applied to any geographic area, from the smallest community with radiotelephone service, to large cities such as New York City. Since the average person only used their telephone a fraction of the time, service could be offered to up to 700 subscribers in a market.

The enabling technology for today's mobile telephone system was cellular technology, developed in the 1970s by AT&T. Utilizing the principle of frequency hopping that was co-developed and patented during World War II by actress Hedy Lamar, the premise of cellular telephony was that an area could be divided into small cells ranging from about one to seven miles in diameter that were each served by a low-power transmitter and receiver (Nies, 2000). As illustrated in Figure 23.1, the use of low-power transmitters allowed the same frequencies to be used over and over in nearby (but not adjacent) cells. The components of a cellular telephone system are diagramed in Figure 23.2.

Figure 23.1
Diagrammatic Cellular Cluster Frequency Pattern

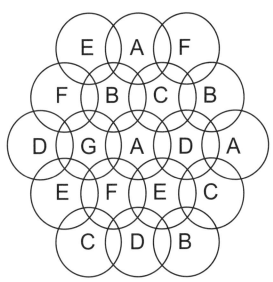

Source: Technology Futures, Inc.

Recognizing the need for spectrum for mobile telephony, the Federal Communications Commission (FCC) allocated a portion of the upper part of the UHF-TV band for mobile telephony in 1970. Over the next dozen years, AT&T's Bell Labs developed the supporting technologies, including the necessary switching technology and the mobile phones themselves. The system that developed, known as the U.S. Advanced Mobile Phone System (AMPS), used analog transmissions and digital switching. Trials of this cellular service occurred in Chicago in 1979, and the first regular service commenced, again in Chicago, in 1983 (Murray, 2001).

At the same time that AT&T was developing the technology, the FCC was experimenting with new regulatory paradigms. As a new service, cellular telephony was an obvious place to experiment. With the goal of establishing competition in the cellular telephone market, the FCC decreed that it would divide the cellular spectrum in two, awarding half to the incumbent local exchange carrier (ILEC—another term for the local phone company), and the other to a competitive company. The FCC's initial plan to award the competitive licenses to the most "worthy" applicants proved impractical. In an effort to provide "fairness," the FCC switched to use of a lottery to choose the competitive provider from all qualified applicants (Murray, 2001).

Figure 23.2
Cellular Telephone Network Architecture

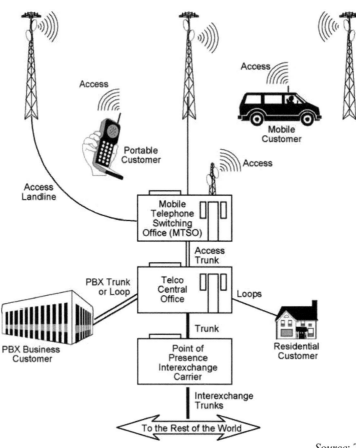

Source: Technology Futures, Inc.

The result of this experiment in regulation was a new "gold rush," with the prize of cellular spectrum being much more valuable than any precious metal. Although many winners went on to construct their own cellular telephone systems, the majority simply sold their licenses for a handsome profit to larger companies that had the capital to build and operate cellular telephone systems (Ellen, 1989).

Technology and regulation were not the only enabling factors in cellular telephony. The fact that virtually all mobile telephone systems were interconnected with the PSTN meant that the early adopters of the technology could immediately reach or be reached by virtually anyone with a phone. This compatibility with existing technologies eliminated the need for a "critical mass" of mobile telephone users. (Other wireless technologies that were not interconnected with the PSTN such as CB radio have not enjoyed the same advantage, with use being limited to a relatively small community.)

One of the best measures of the demand for cellular telephony is industry revenue, which surpassed $1 billion in 1988, $5 billion in 1991, $10 billion in 1994, and $20 billion in 1996 (CTIA, 2002). The rapid growth of the industry led to an unexpected problem. Theoretically, as usage increased, individual cells could be split into multiple cells using even less power, allowing even more calls to simultaneously use the same spectrum. As a practical matter, however, the cellular industry quickly discovered that there was a practical limit to how small a cell could be (about a mile diameter), thereby limiting the number of simultaneous users.

As cellular use exploded in the most crowded metropolitan areas, the FCC and the cellular industry responded with two means of serving additional customers. First, the FCC agreed to allocate spectrum for a new class of wireless telephony known as PCS (personal communication services). As opposed to its analog cellular cousin, PCS was designated as an all-digital technology that would provide new data services in addition to voice service, make more efficient use of the limited spectrum, and allow use of telephones that were smaller and used less power, thus increasing battery life.

The FCC ultimately allocated 120 megahertz of spectrum for broadband PCS service into six blocks of frequencies, holding auctions to allocate these channels on a regional and local basis. From 1994 to 2001, the FCC held seven rounds of auctions, collecting more than $35 billion from companies for the privilege of providing PCS service (including $17 billion in a contested auction, discussed below). One of the goals of the new system was increased competition, with existing cellular providers limited to bidding on only two of the six available frequency blocks in each service area (FCC, 2002). (The FCC also allocated 20 megahertz for unlicensed PCS applications and 3 MHz for "narrowband" PCS such as paging.)

At the same time that PCS was being developed, the incumbent cellular service providers found another solution to spectrum crowding by adopting digital technology to supplant their analog cellular service. Digital cellular service allowed a greater number of subscribers to use the same spectrum that was formerly used for analog cellular service, and it allowed providers to offer a host of new services to their subscribers, including paging and caller ID.

Two digital standards were proposed for use in the United States: TDMA and CDMA. TDMA (time division multiple access) allows more than one caller to use the same channel, with each phone being assigned a specific fraction of time for transmitting its digital signal, allowing three digital conversations to share a single channel. CDMA (code division multiple access), on the other hand, is a spread-spectrum technology that uses the entire spectrum assigned to the service to transmit digital signals rather than one channel at a time.

In the meantime, European cellular companies adopted a third digital standard, GSM (group standard mobile), a variant of TDMA. That standard soon spread through most of the world, allowing a cellular customer to roam from country to country with a single telephone and service provider.

As technology was developing and the industry was growing, the organizational infrastructure of the cellular industry was undergoing dramatic consolidation. Mergers among providers and buyouts of smaller companies by larger companies soon reduced the number of companies providing cellular phone service in the United States to a small fraction of the number that had been awarded licenses in the FCC's cellular telephone lotteries. Among the more notable acquisitions was the 1994 purchase of McCaw Communications (Cellular One), built by cellular pioneer Craig McCaw, by AT&T for $14.5

billion. Another example is Verizon Wireless, which was created out of a complex series of mergers and acquisitions involving dozens of separate cellular and PCS systems including Vodafone, GTE Wireless, Bell Atlantic Corporation, AirTouch Cellular, and Primeco Personal Communications.

Consumers who traveled benefited from consolidation as the larger cellular providers began offering rate plans that did not include "roaming" charges, allowing a person to use their phone virtually anywhere in the United States with no additional charge. The larger companies also had greater access to capital, allowing them to make needed investments in improved technology including digital service.

In the mid-1990s, satellite technology was poised to offer even more wireless competition through MSS (mobile satellite service), a service that was designed to use a network of low-earth orbit (LEO) satellites to replace terrestrial towers to receive and transmit phone calls. The advantage of MSS was that service could be available to and from almost any point on earth. The high cost of constructing MSS networks and the related high costs for service stalled MSS technology in the late 1990s. Iridium was the first operational MSS system that used a constellation of 66 satellites to begin offering mobile phone service via satellite in 1997; Iridium was bankrupt in 1999. (For more details on MSS and LEO satellites, see Chapter 21.)

RECENT DEVELOPMENTS

Wireless telephony is probably undergoing the greatest change of any technology discussed in this book. This section will discuss changes in four major areas: 2.5G wireless services, 3G wireless services, cell phone legislation, and wireless telephony market changes.

2.5G

Wireless telephony is often referred to in terms of its generations. First generation (1G) refers to analog phone service. 2G refers to digital cellular and PCS services with enhanced calling features such as caller ID and one-way, slow data transmission. 3G refers to third generation wireless service. 3G wireless services have data transmission rates up to 2 Mb/s, and this high speed allows broadband services. 2.5G is an intermediary stage between 2G and 3G (Introduction to, 2002). While most of the world is working toward the deployment of 3G, telecommunications companies are implementing 2.5G services to start the process. 2.5G services are digital like 2G, but they also employ packet switched data transmission to allow enhanced information services.

Japan's NTT DoCoMo instituted the world's first 2.5G service when it offered i-mode service in February 2000. With i-mode, users can talk, send e-mail, and surf i-mode Web pages with their i-mode enabled phones. Unlike 2G phones, i-mode phones are always connected, and users are charged for the amount of data they transfer rather than the time they are connected. When it began, i-mode phones used cHTML (compact HTML) for Web pages. At the same time, Web-enabled phones in the United States used WAP (wireless application protocol). The first generation of WAP was difficult to use and limited in its capabilities. Users were still charged per minute for use, and wireless Web use did not take off in the United States. cHTML, on the other hand, is a compact form of HTML, the basic building block of Web pages. The user experience is better and faster, and page development is easier. i-mode has been incredibly popular in Japan; in fact, it is the number one way that Japanese people connect to the Internet. As of May 2002, there were 62 million i-mode subscribers in Japan. That number is close of half the population (Changing to, n.d.). Later versions of WAP are now better able to handle faster and more detailed Internet browsing. In fact, NTT DoCoMo participated in the development of WAP 2.0, which should be released in late 2002 (Competitive advantages, n.d.).

NTT DoCoMo's i-mode service has been spreading throughout the world including the United States. NTT DoCoMo bought 16% of AT&T Wireless in 2001, and they rolled out their mMode service in the United States in April 2002 (Chidi, 2002). The service uses a GSM/GRPS network. (GRPS stands for General Packet Radio Service, and it allows for data transfer up to 144 Kb/s.) AT&T Wireless touts GSM/GRPS as a marker of "the beginning of a new standard in wireless" (Our technology, n.d.). Advantages include use of the GSM world standard, an always-on data connection, payment only for the amount of data used, easy-to-use devices, long battery life, and security (Our technology, n.d.).

Users sign up for mMode packages: Mini, Mega, or Max. Each package has a small monthly service charge ranging from $2.99 to $12.49. Users then pay a data per kilobyte charge of either 1 cent or 2 cents. Mega and Max plan holders receive 1 or 2 megabytes. mMode service is divided into three general areas: connect, manage, and entertainment. Examples of connect are e-mail and phone. Manage includes calendar, finding an ATM, etc. Entertain includes playing games, restaurant reviews, etc. (AT&T Wireless, 2002). As of May 2002, mMode service was available in 14 cities including Seattle, Dallas, Miami, and Kansas City. AT&T Wireless plans to cover 100% of the U.S. population by the end of 2002. Compatible phones include the Sony Ericsson T68, Nokia 8390, and the Motorola Timeport Model P73821. Prices range from $199 for the Nokia and Ericsson phones to $79 for the Motorola phone (AT&T Wireless, 2002).

The launch of 2.5G services in the United States and the rest of the world is just the beginning of what may be a revolution in wireless communication. Third generation wireless is the logical next step in this revolution.

3G

In 1999, the International Telecommunications Union (ITU) issued a worldwide standard for 3G wireless communications, IMT-2000 (International Mobile Telecommunications 2000 initiative). Table 23.1 reviews important capabilities of 3G systems including data transmission rates of 144 Kb/s for vehicular traffic, 384 Kb/s for pedestrian traffic, and 2 Mb/s for indoor traffic. The ITU also allocated frequencies for IMT-2000. These frequency bands include 1885-2025 MHz, 2110-2200 MHz, 806-960 MHz, 1710-1885 MHz, and 2500-2690 MHz (IMT 2000 Global, n.d.; Why 3G, n.d.).

In the United States, the FCC and the National Telecommunications Industry Association (NTIA) looked at the 2500-2690 MHz band and the 1755-1850 MHz band for possible 3G deployments. Reports issued by the FCC and NTIA indicated that using these bands would be a burden on existing licensees and users. 2500-2690 MHz is used for MMDS (multichannel multipoint distribution service), MDS (multipoint distribution service), and ITFS (Instructional Television Fixed Service) (Third generation, n.d.). 1755-1850 MHz is the federal government band and is used by the Department of Defense. The FCC is now examining the 700 MHz band for 3G. This band is used for UHF television stations 52 through 69. Broadcasters were given these licenses for free, but are required to give them up by 2007 in exchange for digital television spectrum. The spectrum was to be auctioned off on June 19, 2002; however, the wireless phone industry requested a delay because of fears that broadcasters would charge billions of dollars to wireless carriers who wanted the spectrum earlier than 2007 (Mark, 2002; NTIA, 2001; FCC, 2001). In early May 2002, two bills were introduced; one in the House and one in the Senate, that would delay the auction (Mark, 2002).

Table 23.1
3G System Capabilities

Capability to support circuit and packet data at high bit rates:

> 144 kilobits/second or higher in high mobility (vehicular) traffic
> 384 kilobits/second for pedestrian traffic
> 2 Megabits/second or higher for indoor traffic

Interoperability and roaming

Common billing/user profiles:

> Sharing of usage/rate information between service providers
> Standardized call detail recording
> Standardized user profiles

Capability to determine geographic position of mobiles and report it to both the network and the mobile terminal

Support of multimedia services/capabilities:

> Fixed and variable rate bit traffic
> Bandwidth on demand
> Asymmetric data rates in the forward and reverse links
> Multimedia mail store and forward
> Broadband access up to 2 Megabits/second

Source: FCC

Despite the spectrum issues in the United States, 3G rollouts are underway in Europe and Asia. Japan's NTT DoCoMo instituted the first 3G system, FOMA, in October, 2001 (Revolutionary 3G, 2001). FOMA has data transfer rates up to 384 Kb/s. FOMA supports full-motion video transferring including videoconferencing, music and game distribution, and other bandwidth heavy data transfers. FOMA supports or will support interactive (videoconferencing, telephone), point-to-point (Web, remote medical diagnosis), one-way (digital newspapers, karaoke), and multipoint services (mobile TV, advanced car navigation) (Revolutionary 3G, n.d.). 3G phones are designed to accomplish a variety of tasks.

FOMA uses W-CDMA (wideband CDMA), one of a group of operation technologies that all comply with the IMT-2000 standard. In 1999, the ITU listed the approved terrestrial radio interfaces. These were CDMA Multi-Carrier (CDMA 2000), Wideband CDMA (W-CDMA or CDMA-Direct Spread), CDMA Time Division Duplex (UTRA-TDD), TDMA-Single Carrier (UWC-136 or EDGE), and FDMA/TDMA (DECT) (LaForge, 2001). The three leading interfaces are WCDMA, CDMA 2000, and EDGE. Each one has it advantages relative to the regulatory and economic environment (What is, n.d.; 3G network, n.d.).

WCDMA is favored in Europe and Asia. It is the IMT-2000 interface for standards in the 2 GHz band. WCDMA supports transmission rates up to 2 Mb/s and can be built off existing GSM technology. In Europe, wireless carriers have new frequency spectrum to build out WCDMA networks. This is not the case in the United States, as opening new spectrum has been delayed until television stations migrate their analog broadcasts from the upper end of the UHF band to digital transmissions at lower frequencies. This delay has forced U.S. wireless carriers to use their existing frequencies for 3G.

In Europe and most of Asia, GSM was adopted as the wireless standard. In the United States, CDMA and TDMA (which is similar to GSM) are used. Because of the scarcity of spectrum, wireless carriers are looking to 3G solutions that allow them to use their existing spectrum and have backward-compatibility with existing technology. cdma2000 was developed by the CDMA Development Group, and is backward compatible with cdmaOne, the existing 2G CDMA standard (3G overview, n.d.). cdma2000 can transmit speeds up to 2 Mb/s and uses the existing frequency spectrum. CDMA 2000 is clearly the choice for U.S. carriers presently using CDMA. For carriers using GSM and TDMA, the choice is EDGE (Enhanced Data Rates for Global Evolution). This technology allows GSM and TDMA users to offer 3G services on existing frequencies. EDGE, though, can only reach speeds to 384 Kb/s, and some actually claim that EDGE is a 2.5G technology. The evolution of wireless service from TDMA is suggested to go to EDGE and then to WCDMA. EDGE is being touted in the United States as a means to start 3G services while Europe waits (All eyes, 2002; 3G and UMTS, n.d.; Carlson, n.d.). Figure 23.3 shows possible avenues for 3G migration.

Figure 23.3
3G Migration

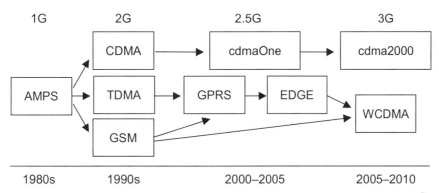

Source: J. Meadows

WIRELESS PHONE LEGISLATION

While it is easy to get caught up in the excitement of new 2.5G and 3G services, it is important to note significant changes in the use of wireless phones. Few people in the United States have not experienced the distracted driver talking on the phone. In fact, studies have found that driving while on the phone can be as hazardous as drunk driving (Flynn, 2002). Keep in mind that other distractions such as eating, using the car stereo, and putting on makeup have also been blamed for accidents. Several cities began the push to ban driving while talking on a handheld mobile phone. In November 2001, New York became the first state to ban the use of handheld phones by drivers. Infractions bring a $100 fine for the first offense up to $500 for the third or greater offense. Similar legislation is being considered in other states, and bans are in place in more than 23 countries including Japan, Italy, and Great Britain (New York cell, 2001). It is interesting to note that studies have found that hands-free mobile phones pose no less of a distraction to drivers than handheld phones.

While municipalities begin to implement bans for driving and talking, other institutions are beginning to reconsider mobile phone bans. The events of September 11, 2001 brought to light the importance of the mobile phone for immediate communication. Many schools, which had mobile phone bans, are now reconsidering those bans. Some schools have repealed the ban and are allowing students to carry a phone with parental permission (Some schools, 2001). The phone calls from passengers on Flight 93 on September 11 to loved ones highlighted the issue of in-flight mobile phone bans. While the aviation industry claims that wireless devices can interfere with airplane operations and air-to-ground communication, others in the wireless industry indicate that there is no evidence to support that claim. Although the rule has been questioned, the FAA has indicated that it is not going to change it (Stout, 2001).

Where and when to talk on a wireless phone will always be a source of controversy. Restaurants have signs requesting patrons turn off their phones, professors have notices in syllabi to turn off phones, even movie theatre openings, which used to ask patrons to remove crying babies, now have requests to turn off the phone (not that it seems to stop anyone from taking a call!).

Another long-standing controversy is health issues surrounding the use of wireless phones. Claims that cell phones cause brain cancer have been around since the first phones were put into use. What is problematic is that there is no conclusive evidence that radio waves from wireless phones cause health problems. On the other hand, the General Accounting Office released a report on mobile telephone health effects and concluded that there was not enough evidence to indicate there is not a risk (Cell phone health, 2001). In 2000, the British government issued a report suggesting that children be limited in their wireless telephone use because of possible long-term effects (FDA to study, 2000). What has been suggested is that users make sure to point the antenna away from the head or use an ear piece to keep the antenna as far away from the head as possible.

WIRELESS MARKET CHANGES

The wireless telephony market in the United States has been going through a series of changes including rebirth, consolidation, and growth. Rebirth would be an appropriate term for Iridium Satellite LLC, the satellite telephony provider. The initial company, Iridium LLC, went bankrupt in 1999 and could not find a buyer for its system of 66 satellites. Iridium even asked permission to allow its satellites to burn up in the atmosphere. At the last moment, in December 2000, Iridium LLC was bought by investors and became Iridium Satellite LLC. Globalstar is the other satellite telephony provider. Both Iridium and Globalstar have a disappointing number of subscribers, but the war on terrorism and September 11, 2001 both brought the technology to light. War correspondents in Afghanistan were able to bring video news reports to world television viewers with the help of satellite phones. This example highlights the limited uses for satellite telephones. While useful in places with limited or no landline and terrestrial wireless, the service is too expensive for the average consumer (About Globalstar, n.d.; Our story, n.d.).

In the United States, wireless providers are consolidating. There are a limited number of major wireless providers in the United States, along with a host of companies providing regional service. The major companies include Verizon Wireless, Nextel, Cingular, AT&T Wireless (which was spun off from AT&T in October 2001), Sprint PCS, and VoiceStream.

This consolidation is allowing wireless providers to offer attractive packages with no long distance or roaming charges because coverage areas have expanded tremendously. Verizon Wireless is a joint venture between Verizon Communications and Vodafone and uses CDMA. VoiceStream is partly owned by Deutsche Telekom and is the only U.S. service to use GSM exclusively. Rumors of preliminary merger talks were reported in Spring 2002 between Cingular and AT&T Wireless. This merger would create a company with 40% market share. Industry insiders expect Verizon Wireless may look to buy Sprint PCS should a Cingular/AT&T Wireless merger happen (Nowlin, 2002).

One other factor affecting competition is the fact that one block of spectrum allocated for PCS use in the United States remains unused due to legal wrangling. In 1996, NextWave Telecom was the winning bidder for a large slice of PCS spectrum, promising to pay $4.7 billion for the spectrum. The company went bankrupt before they paid for the spectrum, so the FCC, in effect, repossessed the spectrum and reauctioned it for almost $17 billion. NextWave argued in bankruptcy court that bankruptcy laws prevented the FCC from taking back the asset, with the Washington Court of Appeals agreeing that the FCC overstepped its boundaries in repossessing the spectrum (Glasner, 2002). As of May 2002, the case is before the U.S. Supreme Court, which will ultimately decide who gets the spectrum and the related economic windfall.

CURRENT STATUS

The Cellular Telecommunications & Internet Association (CTIA) has provided detailed statistics on the U.S. cellular telephone industry twice a year since the 1980s. Their latest statistics indicate that more than 62% of adult Americans have cellular telephones, with more than 135 million U.S. subscribers. After falling each year from 1988 to 1998, the average monthly bill for cellular service began rising in 1999 and was $45.56 in 2001. The growth in the number of subscribers has produced steady increases in revenues for the U.S. mobile telephone industry, which had gross revenues of $58.7 billion in 2001 (CTIA, 2002).

Since the first auctions of PCS spectrum, the U.S. government has collected more than $35 billion, with additional revenues pending as the auction of the 700 MHz spectrum commences (FCC, 2002). It should be noted that taxes on mobile phone bills net both the federal and state governments more than $1 billion per year as well.

Mobile telephone use is as popular in most parts of the world as it is in the United States, with more than a billion users. Western Europe has more cellular phone users than any region on earth, with more than 367 million subscribers. The Asia-Pacific region has more than 330 million, with the United States and Canada combining for 140 million, 100 million in the rest of the Americas, 45 million in Eastern Europe, 29 million in Africa, and 15 million in the Middle East (EMC forecast, 2001).

FACTORS TO WATCH

The dramatic rise of mobile telephony shows no sign of abating, as new technologies such as 3G promise to extend the capabilities of mobile telephony to provide a host of new services and features, including ubiquitous Internet access. With legislation and etiquette just trying to catch up as the technology leaps ahead, expect numerous efforts to govern use of mobile telephony devices, including both legislative and social forces.

One important question regarding individual use of 3G phones is how much people are willing to pay to add Internet access to their mobile telephone service. With an average monthly bill of $45.56, it is not far-fetched to project that most subscribers will be willing to pay an extra $20 to $40 per month for Internet access. The fact that most consumers have limited disposable income, however, suggests that any such increase must be accompanied by a decrease in other spending. It remains to be seen whether the money will be shifted from cellular telephone budgets, from other Internet access budgets, from entertainment budgets, or whether, faced with this choice, consumers will simply choose to forego mobile Internet access until prices fall.

The consolidation of the mobile telephone industry and the diffusion of GSM systems in the United States is steadily increasing the ability of users to use a single phone and subscription as they travel from one country to another. This trend is as much a negative one for satellite-based mobile telephony as it is positive for its terrestrial counterparts. Mobile telephony via satellite will continue to struggle until it either finds a market or dies.

One important difference remains between mobile telephony and its landline counterpart. Mobile telephones are tied to people rather than to places, and no systematic effort has yet begun to create a directory of mobile telephone users. It is remarkable that the telephone directory that is such a critical aspect of POTS (plain old telephone service) has no analogue in mobile telephony. The reasons for the difference include the economic factor (people don't want to pay for unsolicited, incoming telephone calls), as well as social factors (mobile phones are perceived as being a much more personal service).

There is little doubt that 3G services will roll out worldwide in the next five years. The "killer application" of the service, though, is yet to be determined. The main function of the mobile phone today and in the past is to communicate with other people. Whether that communication will shift to e-mail instead of voice communication remains to be seen.

BIBLIOGRAPHY

3G and UMTS frequently asked questions. (n.d.). *UMTS World News*. Retrieved May 6, 2002 from http://www.umtsworld.com/umts.faq.htm.

3G network. (n.d.). *3G Newsroom*. Retrieved May 6, 2002 from http://www.3gnewsroom.com/ html/network/index.shtml.

3G overview. (n.d.). *SmartHome Forum*. Retrieved May 6, 2002 from http://www.smarthomeforum.com/ 3g.shtml.

All eyes on Edge. (2002, April 30). *Ericcson*. Retrieved May 6, 2002 from http://www.ericsson.com/ network_operators/mobilesystems/Buildarticle.asp?Articleid=CC8F59DCF-5BE9-11D6-99C4-0030474E2F8a&SelectedStructureNodeid=23C0CE4B-3.

About Globalstar. (n.d.). *Globalstar*. Retrieved May 6, 2002 from http://www.globalstar.com/ about_globalstar.html.

AT&T Wireless introduces mMode. (2002, April 16). *AT&T Wireless*. Retrieved May 6, 2002 from http://www.attws.com/press/releases/2002_04/041602.html.

Carlson, C. (n.d.). Third gen: What's behind the ballyhoo? *3G Newsroom*. Retrieved May 6, 2002 from http://www.3gnewsroom.com/html/intro_3g/intro_into_3g.shtml.

Cell phone health: Still confused. (2001, May 22). *Wired*. Retrieved May 6, 2002 from http://www.wired.com/ news/wireless/0,1382,44023,00.html.

Changing to an i-mode world. (n.d.). *NTT DoCoMo*. Retrieved May 6, 2002 from http://www.nttdocomo.com/ html/imode01_3.html.

Chidi, G. (2002, April 16). AT&T Wireless launches mMode service. *Infoworld*. Retrieved May 5, 2002 from http://staging.infoworld.com/cgi/redesign/subject.index.wbs?year=&month=&section=&startcount=1 &topic=telecommunications.

Competitive advantages. (n.d.). *NTT DoCoMo*. Retrieved May 6, 2002 from http://investor.nttdocomo.com/ faq.cfm?page=ca.

Cellular Telecommunications & Internet Association. (2002). *CTIA's semi-annual wireless survey industry results*. Retrieved May 5, 2002 from http://www.wow-com.com/pdf/wireless_survey_2000a.pdf.

Ellen, D. (1989, October 9). The phone flushway: Uncle Sam wants you to win $20 billion. *New Republic*, 13-15.

EMC forecasts subscribers to top 1 billion by end 2001. (2001, July 18). *Cellular Online*. Retrieved May 8, 2002 from http://www.cellular.co.za/analysts/07182001-emc_forecasts_subscribers_to_top.htm.

FDA to study cell phone health risks. (2000, June 9). *USA Today*. Retrieved May 6, 2002 from http://www.usatoday.com/life/cyber/tech/review/crh206.htm.

Federal Communications Commission. (2001, March 30). *Final report: Spectrum study of the 2500-2690 MHz band. The potential for accommodating third generation mobile systems*.

Federal Communications Commission. (2002). *Broadband PCS fact sheet*. Retrieved May 8, 2002 from http://wireless.fcc.gov/pcs/bbfctsh.html.

Flynn, K. (2002, March 29). Cell phone, car may not mix. *MSNBC*. Retrieved May 6, 2002 from http://www.msnbc.com/ocal/rmn/drmn_1056727,asp.

Glasner, J. (2002, March 5). Supremes have say on spectrum. *Wired*. Retrieved May 8, 2002 from http://www.wired.com/news/wireless/0,1382,50806,00.html.

IMT-2000 Global Standard, International. (n.d.). *Mobile/Cellular Technology*. Retrieved May 7, 2002 from http://www.mobilecomms-technology.com/projects/imt2000/index.html.

Introduction to 3G. (n.d.). *3G Newsroom*. Retrieved May 6, 2002 from http://www.3gnewsroom.com/ html/intro_3g/index.shtml.

LaForge, P. (2001, April). The 3G evolution at high speed. *CDMA Development Group*. Retrieved May 6, 2002 from http://www.cdg.org/resource_center/GuestCl/laforge_april_01.asp.

Mark, R. (2002, May 8). House votes to delay spectrum auction. *Wireless News*. Retrieved May 8, 2002 from http://www.internetnews.com/wireless/article/0,,10692_1041341,00.html.

Murray, J. B. (2001). *Wireless nation: The frenzied launch of the cellular revolution in America*. Cambridge, MA: Perseus Publishing.

New York cell phone ban signed into law. (2001, June 28). *USA Today*. Retrieved May 6, 2002 from http://www.usatoday.com/life/cyber/wireless/2001-06-28-phone-ban.htm.

Nies, K. A. (2000). *Hedy Lamar: The inventor of frequency hopping, a Web technology*. Retrieved May 8, 2002 from http://www.geocities.com/vidkid_allison/h_lamarr/scigrrl.html.

Nokia FAQ. (n.d.). *Nokia*. Retrieved May 6, 2002 from http://www.nokia.com/networks/systems_and_solutions/whatis_faq/1,23786,1,00.html.

Nowlin, S. (2002, April 16). Cingular, AT&T in merger talks, report says. *Wireless News*. Retrieved May 7, 2002 from http://www.wirelessnewsfactor.com/perl/story/17287.html.

NTIA statement regarding new plan to identify spectrum for advanced wireless mobile services (3G). (2001, October 5). *National Telecommunications Industry Association*. Retrieved May 6, 2002 from http://www.nia.oc.ov/ntiahome/threeg/3gplan_100501.htm.

Our story. (n.d.). *Iridium Satellite*. Retrieved May 6, 2002 from http://www.iridium.com/corp/ir_corp-story.asp?storyid=2.

Our technology. (n.d.). *AT&T Wireless*. Retrieved May 6, 2002 from http://www.attws.com/our_company/technology/networ.jhtml.

Revolutionary 3G service. (n.d.). *NTT DoCoMo*. Retrieved May 6, 2002 from http://www.nttdocomo.com/html/imode05_1.html.

Some schools rethink cell phone ban. (2001, November 12). *CNN.com*. Retrieved May 6, 2002 from http://fyi.cnn.com/2001/fyi/teachers.ednews/11/12/schools.cellphones.ap/.

Stout, K. L. (2001, September 17). In-flight cell phone ban called into question. *CNN.com*. Retrieved May 6, 2002 from http://www.cnn.com/2001/BUSINESS/asia/09/17/hk.airbornecellular/index.html.

Third generation ("3G") wireless. (n.d.). *Federal Communications Commission*. Retrieved May 2, 2002 from http://www.fcc.gov/3G/.

What is UMTS? (n.d.). *UMTS Forum*. Retrieved May 6, 2002 from http://www.umts-forum.org/what_is_umts.html.

Why 3G. (n.d.). *National Telecommunications Industry Association*. Retrieved May 6, 2002 from http://www.ntia.doc.gov/ntiahome/threeg/3gintro.htm.

24

TELECONFERENCING

Michael R. Ogden, Ph.D.[*]

Teleconferencing has always held out the promise of increased productivity and efficiency, improved communications, enhanced business opportunities, and reduced travel expenses. But, "we humans are a 'touchy-feely' species, who, in general, prefer travel and face-to-face encounters over the more impersonal [teleconference] experience" (Kuehn, 2002). According to the Travel Industry Association, business travel rose 14% between 1994 and 1999, with nearly half of the 44 million travelers in 1998 attending a meeting, trade show, or convention. Forecasts predicted a continued increase of up to 3% per year well into the early 21st century (Seaberry, 1999).

Following the events of September 11, 2001, however, many executives accustomed to frequent cross-country trips began grappling with growing concerns over their business travel. As evidence, the National Business Travel Association (NBTA) found companies cutting business travel plans by as much as 60% (Bamnet, 2001), and airline ticket sales in the United States fell 20% to 40% in the months immediately after the attack (McLuhan, 2002).

"Lucy Johnson, senior VP of daytime television for CBS, used to fly from LA to New York every two months to meet with the East Coast staff. Now, they meet via videoconferences, phone conference calls, and a plethora of e-mails" (Ryssdal, 2001). Even as more recent NBTA surveys found evidence of a rebound in travel by the end of 2001 (Bamnet, 2001), most companies were already cutting travel budgets to varying degrees prior to September 11, replacing expensive business travel with less expensive electronic conferencing options. Quaker Chemical Corporation typically spent "...upward of $30,000 to fly participants to a single [conference] site. [V]ideoconferencing and teleconferencing

---

[*] Associate Professor of Communication, Department of Communication, Central Washington University (Ellensburg, Washington) and Affiliate Graduate Faculty, Telecommunications and Information Resource Management (TIRM) Graduate Certificate Program, University of Hawaii, Manoa (Honolulu, Hawaii).

costs are minuscule by comparison—about $2,000, ... $50 an hour for video hookups and $10 an hour for telephone connections" (Chabrow, 2001). A Yankee Group survey of small and home-based businesses indicated that nearly 68% of these businesses have broadband access (Dukart, 2002), leading industry observers to foresee increased demand for teleconferencing—video, audio, and computer-based solutions—as companies look for ways to connect with employees and customers while remaining on *terra firma* (Swanson 2001).

Definitions of what exactly constitutes "teleconferencing" tend to differ across the industry. Attempts to find a consensus in defining terms is not entirely an academic exercise. It has been the experience of many industry professionals that disputes over the feasibility of teleconferencing are often perpetuated by the fact that few of the negotiating parties know exactly what the other party believes a teleconference is (Hausman, 1991).

In the broadest since, teleconferencing can be defined as "small group communication through an electronic medium" (Johansen, et al., 1979, p. 1). However, this very broad definition has not been widely adopted. Jan Sellards, then president of the International Teleconferencing Association, defined the term in 1987 as, "the meeting between two or more locations and two or more people in those remote locations, where they have a need to share information. This does not necessarily mean a big multimedia event. The simplest form of teleconferencing is an audioconference" (Hausman, 1991, p. 246). Some practitioners prefer to call conferences using video "videoconferences," those employing mainly audio "audioconferences," and those using a range of computer-based technologies either "computer-conferences" or "Web-conferences"—oftentimes arranged on a continuum to distinguish those technologies that facilitate the "most natural" type of meetings from those that are "least natural" (see Figure 24.1). More typically, the term "teleconferencing" is used as a shorthand term to represent an array of technologies and services ranging from a three-way telephone conversation to full-motion color television to highly interactive, multipoint Web-based electronic meeting "spaces"—each varying in complexity, expense, and sense of immediacy. Be aware, however, that, as in any emerging technology, vocabularies are not precise across the entire range of users.

While seeking a definition for teleconferencing, it is also important to distinguish between the two different "modes" of teleconferencing: *broadcast* and *conversational*. In broadcast teleconferences, the intention is to reach a large and dispersed audience with the same message at the same time while providing a limited opportunity to interact with the originator. Typically, this takes the form of a "one-to-many" video broadcast with interaction facilitated via audience telephone "call-ins." These events (e.g., "state of the organization" addresses, employee relations conferences, new product kick-offs, etc.) require expensive transmission equipment, careful planning, sophisticated production techniques, and smooth coordination among the sites to be successful (Stowe, 1992).

On the other hand, conversational teleconferences usually link only a few sites together in an "each-to-all" configuration with a limited number of individuals per site. Although prior arrangement is usually necessary, such meetings are often more spontaneous, relatively inexpensive, and simple to conduct. They seldom require anything more complicated than exchanging e-mail messages, synchronizing meeting times (especially if across time zones), and/or making a phone call to a colleague. Conversational teleconferencing supports numerous kinds of business activities from management and administrative meetings (e.g., project, budget, staff, etc.), to marketing, sales, finance, and human resources (Stowe, 1992).

Figure 24.1
Teleconferencing Continuum

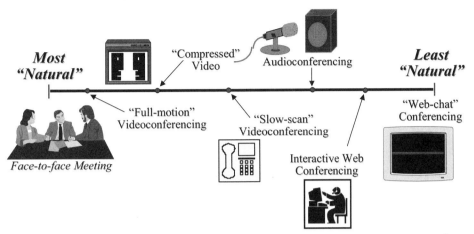

Source: M. Ogden

Another way to categorize teleconferencing applications is according to whether the facilities are used *in-house* as part of the normal routine, or *ad hoc*, only occasionally as the need arises. Once the exclusive domain of Fortune 500 firms, in-house or institutional teleconferencing—dedicated, room-based, and/or roll-about audio- and/or videoconferencing systems—are becoming more commonplace across a wide spectrum of organizations (e.g., schools, churches, hospitals, etc.). For example, Quaker Chemical offices have maintained PictureTel Corporation (now owned by Polycom) videoconferencing systems for years (Chabrow, 2001). Alyeska Pipeline Service Company, headquartered in Anchorage with business units in Fairbanks and Valdez and with seven active pump stations, uses their dedicated, in-house videoconferencing network to battle isolation, save travel costs, and improve training for its geographically-dispersed and weather-challenged workforce (Rosen, 2001).

Ad hoc teleconferencing, more commonly referred to now as "conferencing on demand," can run the gamut from large-scale, one-off videoconferences, to much smaller-scale, desktop conferencing. Many of the major hotel chains (e.g., Holiday Inn, Hilton, Marriott, Sheraton, Hyatt, etc.) first began offering teleconferencing services in the 1980s to attract corporations and professional associations (Singleton, 1983), and several business copy centers such as Kinko's have developed their own teleconferencing networks tailored for occasional use by small-businesses and/or individuals. For the much smaller-scale solutions, there is no need to schedule special rooms or equipment, "just push the Conference button on your touchtone phone, switch on the … camera on top of your PC, and you're ready to meet with others in your organization or outside it" (Weiland, 1996, p. 61).

As a case in point, take the Los Angeles law firm of Paul, Hastings, Janofsky & Walker LLP, which had scheduled a mid-September, face-to-face meeting in New York City for all of the managers from its seven U.S. offices. Because of the uncertainty of travel after the events of September 11, no one could travel. So the firm set up an online meeting through Latitude Communication Inc.'s MeetingPlace that provided an integrated voice and Web conference solution with only one day's notice. "The meeting wasn't perfect. Some managers were in hotel rooms and had to choose between participating through [an audio] conference or through the Web, and others had slow modem connections. However, the six-hour meeting took place" (Chabrow, 2001).

ʸ For most people, the key benefit of teleconferencing is that it eliminates the need for travel and enhances communications because many people can share information directly and simultaneously. Businesses are implementing teleconferencing strategies to reduce non-telecommunications costs, while improving productivity. Educational institutions are using teleconferencing for cost-effective distance learning programs, while government agencies are using teleconferencing for crisis management as well as daily information exchange (Muller, 1998). Because more people can participate in teleconferences than can affordably travel to face-to-face meetings, more input into problem-solving as well as new channels of communication can be opened, resulting in broader support for tough decisions.

On the downside, broader participation can also bring to the surface underlying conflicts within an organization, aggravating differences instead of cultivating unity. Because of limitations unique to each of the respective teleconferencing options, electronic meetings tend to be more orderly and focused. But there is a thin line between orderly agendas and narrow, repressive ones. Finally, many believe that the long-term growth of the teleconferencing industry will contribute real human value only to the extent that it improves the productivity of its users beyond the conservation of time and material resources expended in travel. Given the ease and flexibility of today's teleconferencing options, the most difficult challenge will be to hold more meaningful and productive meetings instead of just more meetings.

BACKGROUND

The basic technology for teleconferencing has existed for years. In its simplest form, teleconferencing has been around since the invention of the telephone—allowing people for the first time to converse in "real time" even though they were separated physically (Singleton, 1983). Today, we understand teleconferencing to be more involved than a simple telephone conversation. Some contend that teleconferencing has its roots in comic strips and science fiction. Certainly, comic strips such as *Buck Rogers*, popular in the early 1900s, made it easy to imagine people of the future communicating with projected images, or *Dick Tracy* calling instant meetings through a communication device on his wrist. The videophone in Stanley Kubrick's *2001: A Space Odyssey* furthered the notion of interactive video communications as a common feature of our near future. Likewise, *Star Trek*'s holodeck popularized the notion of fully-immersive virtual environments for diversion or business as accepted conventions of our far future.

The teleconferencing gear that invaded science fiction was pioneered by AT&T in the 1950s and 1960s. These early efforts were impressive for their time, but were also considered little more than novelties and failed to catch on with consumers (Noll, 1992). Today, these visions have been brought to life in three alternatives to face-to-face meetings easily classified by their broad medium of application: audio, video, and computer-mediated teleconferencing.

AUDIO

Audioconferencing relies only on the spoken word, with occasional extra capacity for faxing documents or "slow-scan" image transmissions (see discussion under *Video* below). Historically, audioconferencing began with the familiar "conference call," generally set up by an operator working with the local telephone company (Stowe, 1992). Today, an audioconference can be implemented in a variety of ways. For a basic telephone conference involving a limited number of participants between two sites, a telephone set with either a three-way calling feature on the line, or a conferencing feature supported by the PBX or key system is all that is required. If participants are expected to be in the same room at each location, a speakerphone can be used. Adding additional sites to an audioconference requires an electronic device called a *bridge* to provide the connection (Singleton, 1983)—simply plugging several telephone lines together will not yield satisfactory results. Audio bridges can accommodate a number of different types of local and wide area network (LAN and WAN) interfaces and are capable of linking several hundred participants in a single call, while simultaneously balancing the volume levels (allowing everyone to hear each other as though they were talking one-on-one) and reducing noise (echo, feedback, clipping, dropout, attenuation, and artifacts) (Muller, 1998).

A bridged audioconference can be implemented in several ways: "dial-out," "prearranged," or "meet-me." In the dial-out mode, the operator (or conference originator) places a call to each participant at a prearranged number and then connects each one into the conference. Prearranged audioconferences are dialed automatically or users dial a predefined code from a touch-tone telephone. In either case, the information needed to set up the conference is stored in a scheduler controlling the bridge. In the meet-me mode, participants are required to call into a bridge at a prearranged time to begin the conference. Regardless of the mode of implementation, the more participants brought into an audioconference through the bridge, the more free extensions are required (Muller, 1998).

VIDEO

Since AT&T demonstrated the Picturephone at the 1964 World's Fair, corporate America has had an on-again, off-again fascination with video communications (Borthick, 2002). The assumption behind videoconferencing, usually unquestioned, has been that the closer the medium can come to simulating face-to-face communications, the better (Johansen, et al., 1979). Engineers have thus struggled to make video images more lifelike in size and quality. As a result, full-motion videoconferencing offers glamour, but has yet to catch on as a routine business tool. Also, videoconferencing has traditionally had a technical complexity that tended to limit its scale and scope, despite periodic engineering breakthroughs, steady price/performance improvements, and gradual market growth.

Today, the technology has progressed to the point where—whether a room-based, roll-about, or desktop system—videoconferencing has become crucial to the new, post-September 11 business reality (Kontzer 2001). Recent data reports show that there are only a few large players that continue to dominate the videoconferencing hardware market. In 2001, Polycom (after acquiring PictureTel) continues to dominate the market with 61% market share (down from 65%), with Tandberg (16%) and Sony (11%) showing solid gains since 2000. Whereas VTel slipped to 3%, the most noticeable change came at the expense of smaller companies that saw their collective market share reduced by almost half (to 9%) between 2000 and 2001 (Big boys, 2002).

Videoconferencing systems are frequently grouped into two main types of end-point systems: group systems and personal systems. Prices for room-based group videoconferencing systems such as Polycom's ViewStation FX—designed for boardrooms, large conference rooms, or classrooms (Polycom, 2002)—can range from $15,000 to over $100,000, depending on options and/or equipment add-ons. Roll-about group videoconferencing systems—once considered viable solutions for organizations that did not have extensive need for dedicated videoconferencing facilities—are on the wane as the range of options available for set-top and personal videoconferencing systems increase in popularity. Typical roll-about systems consist of a television monitor, a video codec, a single camera, and microphone (the last three items are usually packaged as an integrated unit) on a mobile cart and range in price from $4,000 to $9,000 depending on options. Newer systems such as Polycom's iPower line of videoconferencing devices (a PictureTel heritage product)—although not technically a "roll-about" solution—are filling the niche for small conference rooms, executive suites, and professional office environments that were previously the reserve of roll-about systems (Polycom, 2002). Set-top videoconferencing systems, such as Sprint's turnkey IP-based service (Rendleman, 2001) or Polycom's ViaVideo, are becoming increasingly popular because they allow organizations to leverage existing assets. Videoconferencing becomes just another application running on the user's computer desktop. Fully-equipped desktop videoconferencing systems are available in the $1,200 to $5,000 per unit range, depending on options (Muller, 1998).

Slow-scan or freeze-frame videophones, once popular in the 1980s for more impulsive, one-on-one communications between sites similarly equipped, have all but disappeared from the business videoconferencing radar screen. Videophone units include a small screen (typically black and white), a built-in camera, video codec, audio system, and a keypad. The handset lets the unit work as an ordinary telephone as well as a videoconferencing system. Mitsubishi's VisiTel and Luna Picturephone systems are typical examples of consumer-grade freeze-frame videophones. By using special lenses, the users can send each other close-up images of themselves, photographs, graphics, or even drawings. The transmission takes about five seconds; during this time, neither party can talk as the device requires the full telephone bandwidth to send the picture. Commercial-quality slow-scan devices, such as the transceivers from Colorado Video, greatly increase the size and improve the resolution of the image, while allowing for two-way voice conversation to continue while the image is being sent (Stowe, 1992).

Videoconferencing between more than two locations requires a multipoint control unit (MCU). An MCU is a switch that acts as a video "bridge" connecting the signals among all locations and enabling participants to see one another, converse, and/or view the same graphics (Muller, 1998). The MCU also provides the means to control the videoconference in terms of who is seeing what at any given time. Most MCUs include voice-activated switching, presentation or lecture mode, and moderator control (Muller, 1998). Likewise, because of the large amount of information contained in an uncompressed full-motion video signal (about 90 million bits per second), two-way videoconferencing that even remotely approaches broadcast television quality would require incredible bandwidth at equally great cost. Therefore, all but the most expensive full-motion videoconferencing options utilize video codec compression/decompression technology to take advantage of the fact that not all of the 90 million bits of a video signal are really necessary to reconstruct a "watchable" image. In fact, the majority of the information in a typical videoconferencing image is redundant; most of the image remains exactly the same except for the speaker's head movements and occasional gestures.

Video compression techniques take advantage of this and other factors to greatly reduce the number of bits that must be transmitted, and thus reduce the bandwidth requirements (Stowe, 1992). A popular transmission rate for higher-end videoconferencing technology is about 1.5 Mb/s (Megabits per second). This is only about one-sixtieth of the original information in the video image, but it produces acceptable pictures for most purposes. More important, this rate corresponds to the telephone T1 rate so that, with the proper equipment, such videoconferences can be transmitted by telephone circuits in and among most major cities (Stowe, 1992). Other videoconferencing systems have been developed that run at varying rates up to 384 Kb/s (kilobits per second), while smaller roll-about and/or desktop systems typically run at 128 Kb/s. However, at the lower data rates, some design tradeoffs are made in the relationship between quality per frame and frames per second. "Some vendors have engineered their products to maintain a constant frame rate by sacrificing clarity when there is a high motion component to the image. This compromise presents a problem, since there are no established units of measurement for how clear an image appears, or whether the video is smooth or choppy" (Finger, 1998).

COMPUTER-MEDIATED

Murray Turoff, developed one of the first computer conferencing systems in 1971 for the Office of Emergency Preparedness (EMISARI, a management information system) to deal with the wage-price freeze (Lucky, 1991). Since then, the capacity of two or more personal computers to interconnect, send, receive, store, and display digitized imagery has expanded rapidly—thanks in large part to rapid developments in computer hardware and the spread of the Internet. At the most basic level, computer conferencing is a written form of a conference call. Participants in a computer conference could communicate via a simple conferencing application such as Internet Relay Chat (IRC) or AOL's Instant Messenger, both of which provide a simple text-based chat function.

Another form of computer conferencing takes advantage of the ability to "time-shift" a presentation via Webcasting. In a Webcast, the originator tries to anticipate the viewer's questions and concerns during the recording of a presentation, then makes the finished product available to anyone who wants to download it at a time more convenient to the viewer—sort of an asynchronous broadcast on the Web. But to make a computer-based meeting more interactive, most users want to do either real-time audio or video (or both), while sharing documents or a common workspace. Enter Web conferencing, perhaps one of the fastest growing sectors in teleconferencing. Web conferencing allows users to use either software or a service to show presentations or work on the same program or application in real time simultaneously, all while being linked on a shared telephone line (Lafferty, 2002).

Choosing the right teleconferencing technology for the right purpose is the subject of great concern for many business executives. Social psychologists have pointed out that "in terms of the immediacy that they can afford, media can be ordered from the most immediate to the least: face-to-face, [videoconferencing], picturephone, telephone, [below this, synchronous and asynchronous computerized conferencing] … the choice of media in regard to intimacy should be related to the nature of the task, with the least immediate or intimate mode preferable for unpleasant tasks" (Hiltz & Turoff, 1993, p. 118). This would suggest that, for the less intimate task, the most immediate medium (face-to-face) would lead to favorable outcomes. In this same vein, a slightly more intimate task would require a medium of intermediate immediacy (videoconferencing), while those tasks that are highly intimate, perhaps embarrassing, personal, or charged with potential conflict would benefit from using a medium of lowest immediacy (audioconferencing or even computer conferencing).

Interestingly, though, Hiltz and Turoff (1993)—reporting the results of a 1977 study—indicated that "numerous carefully conducted experiments … found all vocal media to be very similar in effectiveness [including face-to-face, audio, and video]. However, … people *perceived* audio to be less satisfactory than video" and both to be less satisfactory than face-to-face (emphasis added, p. 121). Little recent experimentation has been done in this area, and it appears that there is a great deal of room for further research.

RECENT DEVELOPMENTS

The combination of the September 11 terror attacks and the 2001 economic downturn have markedly curtailed business travel. Opinions vary about whether this phenomenon will be temporary, but the current situation is leading many corporate managers to take another look at teleconferencing and collaboration tools (Krapf, 2002). In the aftermath of September 11, stocks for companies in the teleconferencing and Web businesses fared better than most. Stock prices for companies such as Polycom, ACT Teleconferencing, and WebEx Communications surged as analysts noted many companies were turning to Web, voice, and video communications to conduct conferences instead of flying to meetings.

Additionally, businesses that provided such services saw their product usage increase by as much as 50% (Schaffler & Wolfe, 2001), as a newly reinforced aversion to business travel, along with advances in teleconferencing technologies and services, began to bring teleconferencing capabilities to the masses at reasonable costs (Lafferty 2002). Leading vendors say the teleconferencing industry was growing steadily at about 30% prior to the attacks of September 11. Since then, the industry's growth has reached 40%, and all indications point to a steady rise (Vinas, 2002). Jay Williams, senior vice president and CTO for Concours Group global consultants in Kingwood, Texas, believes that "the sole focus of collaborative technologies [today] is not to replicate the live experience of a meeting, that's been a fallacy; that we are going to be able to recreate our casual meeting environment online. We are going to create something new, and we are going to create something that works" (Vinas, 2002, p. 30).

Many business people still maintain that face-to-face meetings will always offer the best level of interactivity. But they are also beefing up their Web sites for actual end users "…using animation for demonstrations of products as well as using streaming video for testimonials" (Vinas, 2002, p. 32). The key point in the use of teleconferencing is ease of use and ease of access. The other obstacles are cultural—many people are uncomfortable speaking to a camera and very uncomfortable knowing they are being recorded. Paul Saffo, a director of the Institute for the Future in Menlo Park (California), believes this is normal. "It takes about 30 years for new technolog[ies] to be absorbed into the mainstream. Products for videoconferencing [and by extension, Web conferencing]… invented 15 to 20 years ago, are just now beginning to come into their own" (Weiland, 1996, p. 63). By Saffo's standard, American society is probably entering its second stage of acceptance of teleconferencing technology—perhaps accelerated by the events of September 11—and, while this transformation is underway, the industry may seem in a constant state of flux and somewhat confusing. The most confusing, and yet promising of the newest teleconferencing developments are not in the hardware or software, but in the international standards that make it all work.

STANDARDS

Certain international standards for teleconferencing have been worked out by the International Telecommunications Union (ITU). These standards have set the foundation for network transmission technologies and vendor interoperability in the teleconferencing industry, especially videoconferencing. Prior to the establishment of these standards, vendors employed a range of proprietary algorithms and packaged hardware and software so that systems from different vendors could not communicate with each other. By establishing worldwide teleconferencing standards, the ITU helps ensure that teleconferencing technologies can "talk" to each other regardless of brand. Likewise, as new innovations are developed, the new systems remain "backward compatible" with existing installed systems. The most important standards for the implementation and adoption of teleconferencing are under the umbrella standard of H.320 that defines the operating modes and transmission speeds for videoconferencing system codecs, including the procedures for call setup, call teardown, and conference control (Muller, 1998). All videoconferencing codecs and MCUs that comply with H.320 are interoperable with those of different manufacturers. Other important standards for the implementation and adoption of teleconferencing are:

> ➤ *H.322*—A standard for LAN-based videoconferencing with guaranteed bandwidth (ITU, 1996).

> ➤ *H.323*—A standard for LAN-based videoconferencing with non-guaranteed bandwidth such as LAN and WAN networks using packet switched, Internet protocol (IP) (ITU, 2000). This standard supports the move by many desktop teleconferencing systems to sophisticated bridging systems and IP multicasting software that reduces desktop clutter and frees up valuable bandwidth on the network. Microsoft's NetMeeting, Polycom's WebOffice, and one of the biggest names in Web conferencing, WebEx, all utilize H.323 IP videoconferencing specifications to deliver their services to the desktop. Likewise, many popular Web browsers have H.323-compliant videoconferencing capabilities embedded into their latest versions (Lafferty, 2002).

> ➤ *H.324*—A standard developed to facilitate low bit-rate desktop teleconferencing via standard telephone lines (ITU, 1998). Many computer manufacturers are now including teleconferencing capabilities as part of their standard personal computer package to take advantage of the growing home teleconferencing and videophone markets brought about by the growth in telecommuting.

As these standards become more widely implemented, and as more bandwidth is added to the Internet, the quality of Internet videoconferencing will likely improve dramatically.

CURRENT STATUS

It is ironic that some of the simplest ideas can be put into practice only when a very complex level of development has been reached in a related field. As technologies have expanded, morphed, and matured, teleconferencing options have transformed and expanded as well. As discussed earlier, teleconferencing is no longer a simple conference call, nor is it exclusively the domain of complex two-way video hook-ups. Roopam Jain, a strategic analysis for Frost & Sullivan, noted that videoconferencing has reached a new level of accessibility despite still having issues with quality, reliability, and overall cost (Lafferty, 2002). Still, Jain is optimistic, "videoconferencing systems worldwide are expected to grow from $819.9 million in 2001, to $1.55 billion by 2005. At the same time, ... videoconferencing services in the United States are expected to grow from $1.68 billion in 2001 to $2.54 billion by 2005" (Lafferty, 2002).

As the cost of codecs and other end-user equipment has fallen, making videoconferencing much more attractive than ever before, network access has grown to represent by far the largest portion of operational costs. The greatest expense for most companies using today's videoconferencing systems "is the transport rate or cost-per-minute, not monthly access, installation, or one-time, initial equipment investment. This cost-per-minute varies based on carrier and on the type of terminating and originating access" (Earon, 1998). Prices are, however, often lower than commonly perceived: "So-called full-bandwidth, ISDN-based calls [384 Kb/s] cost between $0.50 to $1.20 per minute, and IP-based calls can be virtually free, depending on the type and design of the corporate network and application of the video call" (Wisehart, 2002, p. 8).

Because of this, there is a gradual shift toward IP networks. "Devices including ISDN [Integrated Services Digital Network] switches and H.323/H.324 gateways provide dial plan support, allow users to take advantage of competitive tariffs from multiple carriers, and enable users to access the least expensive type of bandwidth as is needed" (Earon, 1998). According to Jain, IP will ultimately bring cost efficiencies and network reliability that ISDN has not been able to deliver and will eventually bring videoconferencing into the mainstream (Lafferty, 2002). Anticipating the coming broadband/IP wave, Polycom, one of the most recognized names in videoconferencing, has begun to equip all of their ViewStations with Ethernet jacks and has even brought out their own line of IP-only conferencing systems (Krapf, 2002).

Paul Berberian, co-founder, president, and CEO of audio and data conferencing provider Raindance Communications, notes that, "for every boardroom-type videoconference, there are thousands of audio conferences" (Borthick, 2002). Audioconferencing over IP networks (voice over IP or VoIP) has also garnered some attention lately as Internet telephony software makes it possible for users to engage in long-distance conversations between virtually any location in the world without regard for per-minute usage charges. In most cases, all that is needed is an Internet-connected computer equipped with telephony software, a sound card, microphone, and speakers or a headset (Muller, 1998). However, because the voice of each party is compressed and packetized, there may be significant delays that result in noticeable gaps in a person's speech. In the future, some industry analysts contend VoIP's use of the session initiation protocol (SIP) signaling protocols could provide increased reliability and enable new audioconferencing features such as breakout sessions, sub-conferencing sessions, and real-time failover without losing connectivity (Borthick, 2002).

The most recent Web conferencing market figures, reported by strategic analyst David Alexander from the consulting firm Frost & Sullivan, indicated that the market for Web conferencing equipment totaled $22.3 million in 1999 (Lafferty, 2002). These figures grew by 177% to approximately $62 million in 2000 and were forecast to increase again by about 65% in 2001. Although Alexander had yet to compute the 2001 figures as of this writing, market growth in the last two quarters of 2001 indicates that the Web conferencing market could already be in the $120 million to $150 million neighborhood (Lafferty, 2002).

Wainhouse Research, an industry market research firm, expects Web conferencing to grow at 47% annually, Alexander expects to see the Web conferencing market to grow by more than 170% in the next two years (Porter, 2002). "Web conferencing—where participants share a voice connection that's synched with a Web connection that allows, among other things, real-time PowerPoint presentations, file sharing, and collaboration—is quickly becoming a multimillion dollar service that shows no signs of slowing down" (Lafferty, 2002).

Lewis Ward, a senior analyst with Collaborative Strategies, LLC, a San Francisco-based management consulting firm, predicts that, by 2005, more than 79 million people and more than 94,000 companies and organizations will be using Web-based conferencing systems worldwide (Ward 2002). This is good news for companies such WebEx, which describes itself as "an interactive communications infrastructure provider" (WebEx, 2002), with a global network consisting of 10 data centers around the world and privately leased bandwidth (Lafferty, 2002). On top of this network rides a software platform that allows people to connect computer-to-computer to share files and applications in real time (see Figure 24.2). According to Subrah Iyar, CEO of WebEx Communications, the company enjoyed dramatic growth every quarter during its first three years of operation, resulting in an over 300% increase in revenue and usage rising by nearly 50% (Schaffler & Wolfe, 2001). WebEx now claims to be the largest provider of Web conferencing services in Europe, Asia, and North America and is viewed as the primary competitive service to Polycom's recently rolled-out WebOffice application (Krapf, 2002).

Other companies have also joined the Web conferencing bandwagon. Raindance, a Colorado-based Web conferencing service, offers "reservationless" conferencing through its Web Conferencing Pro service (Lafferty, 2002). Tom Whitten, COO at Light & Power Communications, Ltd. in Troy (New York), states that his 25-year-old, full-service communications company got into Web conferencing about three years ago and has seen a dramatic rise in service—a service that has received more attention since September 11 (Lafferty, 2002).

FACTORS TO WATCH

Over the past few years, a lot of consolidation has occurred in the multimedia conferencing market. Growing industry interest in "rich media communications and real-time conferencing" has been borne out by Collaborative Strategies senior analyst Lewis Ward's research showing that the fastest-growing type of teleconferences are those in which conferees can talk with each other while viewing the same Web-based data (Borthick, 2002; Ward, 2002). Ward predicts that, by 2005, "the ongoing globalization and decentralization of corporations, coupled with the wider availability of easy-to-use conferencing and integration tools will keep the three elements of the real-time collaboration market—audio, video and data/Web conferencing—growing at a healthy rate, both separately and in combination" (Ward, 2002, p. 53). The results of a recent survey of *VideoSystems* magazine subscribers conducted by Primedia Business magazines and Media, Inc. indicate that, by 2005, most respondents expect 42% of their work to be done via Internet/streaming technologies (How will, 2002). No doubt, virtual communication and collaboration will arguably be the lifeblood of business in the 21st century.

Figure 24.2
WebEx Screen Shots

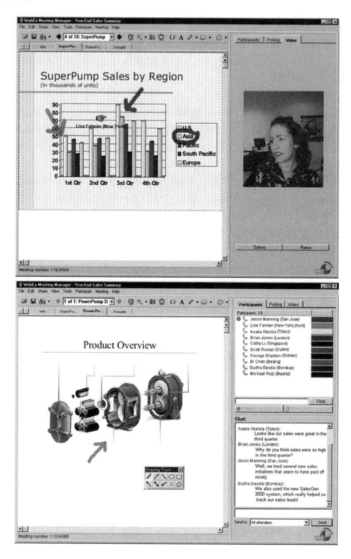

Source: WebEx

Still, industry keynoters rarely leave full-motion video off their list of future "killer" applications. Primarily, this is because many users of today's videoconferencing technology frequently state that, "While the experience wasn't as good as meeting in person, ... the technology was so strong [they] could see gestures, facial expressions, everything" (Goodridge, 2001). Others contend that videoconferencing is better for conducting meetings that need to produce decisions, whereas Web conferencing is the preferred tool for a working session or an informal presentation to a group of people (Porter, 2002). Videoconferencing would thus seem to be a better substitute for face-to-face meetings than audio-only conferencing or those mediated by computer-based technology. However, many argue that videoconferences have a different "flavor" than face-to-face meetings; they are typically more formal, there is little opportunity for side conversations, and the technology itself is fairly intrusive (Lucky, 1991). Furthermore, videoconference participants have difficulty making eye contact due to camera placement limitations and often complain of a lack of shared workspace for collaborative brainstorming (Ditlea, 2000).

Lately, the most attractive thing about videoconferencing is its high-bandwidth requirement—a good use for overbuilt, underutilized fiber networks and a reason to believe in broadband access (Borthick, 2002). Although the near future does not seem likely to promise the replacement of face-to-face meetings with electronic togetherness—video, Web, or otherwise—new kinds of computer-mediated interactions among people are nevertheless likely to be facilitated where participants can share an aural as well as visual virtual meeting space (Lucky, 1991).

What is the next step in pursuit of a more realistic virtual meeting? Some would say the answer is "tele-immersion." As one of the principal applications anticipated for Internet 2—tomorrow's much faster, next-generation Internet—"tele-immersion visually replicates, in real time and in three dimensions, slabs of space surrounding remote participants in a cybermeeting. The result is a "shared, simulated environment that makes it appear as if everyone is in the same room" (Ditlea, 2000, p. 28). Demonstrated for the first time in May 2000, researchers at Advanced Network and Services in Armonk (New York), University of Pennsylvania in Philadelphia, and the University of North Carolina at Chapel Hill tested a three-dimensional virtual meeting room where participants could see and interact with their life-sized colleagues viewed through "windows"—each participant physically located hundreds of miles away from the others (Ditlea, 2000).

Although described as being somewhat crude, requiring users to wear awkward goggles and head tracking devices (not to mention requiring nine separate video cameras per participating location), the test provided vindication for two years of collaborative research between the participants and afforded a brief glimpse of what might lie beyond today's videoconference room. Jaron Lanier, chief scientist for the project and one of the primary scientists who helped invent and popularize virtual reality in the 1980s and 1990s, believes the test demonstrated viewpoint-independent, real-time scene sensing and reconstruction (Ditlea, 2000). Despite suffering from video glitches, Lanier states that the ultimate tele-immersion experience is meant to be seamless (Ditlea, 2000). As discussed in Chapter 15, scientists continue to work on autostereo screens (for three-dimensional views without bulky headgear) and advances in haptics (for tactile simulations). As broadband access continues to increase, fully-immersive virtual conferencing may become a reality, but today, it is still years away.

BIBLIOGRAPHY

Bamnet, S. (2001, December). Rethinking telecom architectures. *Business Communications Review*. Retrieved March 24, 2002 from http://www.bcr.com/bcrmag/2001/12/p16.asp.
Big boys dominate videoconferencing. (2002, April). *VideoSystems, 28* (4), 12.

Borthick, S. (2002, March). Video: Nice but not necessary? *Business Communications Review*. Retrieved March 24, 2002 from http://www.bcr.com/bcrmag/2002/03/p10.asp.

Chabrow, E. (2001, November 5). Technology brings far-flung colleagues together. *InformationWeek.com*. Retrieved March 21, 2002 from http://www.informationweek.com/story/IWK20011102S0014.

Ditlea, S. (2000, September/October). Meeting the future: Tele-immersion makes virtual conferencing more real. *Technology Review, 103* (5), 28.

Dukart, J. (2002, January). The broadband age cometh. *Utility Business*, 27. Retrieved January 18, 2002 from http://industryclick.com/magazine.asp?magazineid=11&siteid=30 &releaseid=9786.

Earon, S. (1998, November). The economics of deploying videoconferencing. *Business Communications Review*. Retrieved March 24, 2002 from http://www.bcr.com/bcrmag/1998/11/p43.asp.

Finger, R. (1998, June). Measuring quality in videoconferencing systems. *Business Communications Review*. Retrieved March 24, 2002 from http://www.bcr.com/bcrmag/1998/06/p51.asp.

Goodridge, E. (2001, October 22). Virtual meetings yield real results. *InformationWeek.com*. Retrieved March 21, 2002 from http://www.informationweek.com/story/IWK20011018S0082.

Hausman, C. (1991). *Institutional video: Planning, budgeting, production, and evaluation*. Belmont, CA: Wadsworth Publishing Company.

Hiltz, S., & Turoff, M. (1993). *The network nation: Human communication via computer*. Cambridge, MA: The MIT Press.

How will you display your work? (2002, April). *VideoSystems, 28* (4), 13.

International Telecommunications Union. (1996, March). *Visual telephone systems and terminal equipment for local area networks which provide a guaranteed quality of service (H.322)*. Retrieved April 14, 2002 from http://www.itu.int/rec/recommendation.asp? type=folders&lang=e&parent=T-REC-h.322.

International Telecommunications Union. (1998, February). *Terminal for low bit-rate multimedia communications (H.324)*. Retrieved April 14, 2002 from http://www.itu.int/rec/recommendation.asp? type=folders&lang=e &parent=T-REC-h.324.

International Telecommunications Union. (2000, November). *Packet-based multimedia communications systems (H.323)*. Retrieved April 14, 2002 from http://www.itu.int/rec/recommendation.asp? type=folders&lang=e &parent=T-REC-h.323.

Johansen, R., Vallee, J., & Spangler, K. (1979). *Electronic meetings: Technical alternatives and social choices*. Reading, MA: Addison-Wesley.

Kontzer, T. (2001, October 12). Video-to-desktop set to emerge as killer app. *InformationWeek.com*. Retrieved March 21, 2002 from http://www.informationweek.com/story/IWK20011012S0032.

Krapf, E. (2002, January). Web conferencing from Polycom. *Business Communications Review*. Retrieved March 24, 2002 from http://www.bcr.com/bcrmag/2002/01/p62.asp.

Kuehn, R. (2002, January). 2002: Year of the conundrum. *Business Communications Review*. Retrieved March 24, 2002 from http://www.bcr.com/ bcrmag/2002/01/p66.asp.

Lafferty, M. (2002, February). Convergence: Teleconferencing takes-off! *Communications Engineering & Design*. Retrieved February 8, 2002 from http://www.cedmagazine.com/ced/2002/0202/02a.htm.

Lucky, R. (1991). In a very short time. In D. Leebaert (Ed.), *Technology 2001: The future of computing and communications*. Cambridge, MA: The MIT Press.

McLuhan, R. (2002, January 17). Webcasts bolster access to events—Conferences are turning to technology to widen their reach. *Marketing*, 27.

Muller, N. (1998). *Desktop encyclopedia of telecommunications*. New York: McGraw-Hill.

Noll, A. (1992). Anatomy of a failure: Picturephone revisited. *Telecommunications Policy*, 307-317.

Polycom. (2002). *Corporate Website*. Retrieved April 22, 2002 from http://www.polycom.com/naindex.html.

Porter, S. (2002, April). The new conferencing: A meeting of minds ... and media. *VideoSystems, 28* (4), 43-50.

Rendleman, J. (2001, June 25). Sprint offers IP-based videoconferencing. *InformationWeek, 843*, 91.

Rosen, E. (2001, April 16). Extreme videoconferencing. *InformationWeek.com*. Retrieved March 21, 2002 from http://www.informationweek.com/833/oovideo.htm

Ryssdal, K. (Anchor), & Gray, C. (Reporter). (2001, October 15). Some executives using teleconferencing instead of flying across country for meetings [radio program]. *Marketplace Morning Report,* Minnesota Public Radio (6:50 A.M. ET), Syndicated.

Schaffler, R., & Wolfe, C. (2001, September 21). Teleconferencing, web-based meetings on the rise. *CNNfn Market Call* (09:30 A.M. EST), Transcript #092108cb.105, Federal Document Clearing House, Inc.

Seaberry, J. (1999, December 6). Business travel continues to increase. *The Dallas Morning News.* Retrieved February 26, 2002 from http://www.dallasnews.com/.

Singleton, L. (1983). *Telecommunications in the information age.* Cambridge, MA: Ballinger.

Stowe, R. (1992). Teleconferencing. In A. Richardson (Ed.), *Corporate and organizational video.* New York: McGraw-Hill.

Swanson, S. (2001, September 27). Travel fears fuel Web conferencing. *InformationWeek.com.* Retrieved March 21, 2002 from http://www.informationweek.com/story/IWK20010927S0017.

Vinas, T. (2002, February). Meetings makeover. *Industry Week, 251* (2), 29-35.

Ward, L. (2002, March). The rise of rich media and real-time conferencing. *Business Communications Review.* Retrieved April 13, 2002 from http://www.bcr.com/bcrmag/2002/03/p53.asp.

WebEx. (2002). *Corporate Website.* Retrieved April 14, 2002 from http://www.webex.com.

Weiland, R. (1996). 2001: A meetings odyssey. In, R. Kling (Ed.). *Computerization and controversy: Value conflicts and social choices*, 2nd edition. San Diego, CA: Academic Press.

Wisehart, C. (2002, April). Can you see me now? *VideoSystems, 28* (4), 8.

V

CONCLUSIONS

RETROSPECTIVE: 10 YEARS OF COMMUNICATION TECHNOLOGIES

August E. Grant, Ph.D.[*]

In the 10 short years since the first volume of the *Communication Technology Update* was published, the telecommunications landscape has changed dramatically. A few technologies, such as digital video compression, that were seen as a distant force in 1992 have diffused more rapidly then expected, but the progress in other technologies, such as interactive television and HDTV, has lagged all but the most pessimistic predictions.

The purpose of this chapter is to take a look back at each of the specific technologies discussed in this book to provide some perspective on the pace of change in the telecommunications landscape. Along the way, we will examine a few technologies that were included in previous editions of this book, but are not included in this edition.

---

[*] Senior Consultant, Focus 25 Research & Consulting and Visiting Associate Professor, College of Mass Communications and Information Studies, University of South Carolina (Columbia, South Carolina)

ELECTRONIC MASS MEDIA

CABLE TELEVISION

The most important developments in cable television in 1992 were the imminent introduction of digital video compression and re-regulation of the U.S. cable television industry. The average cable system offered 50 channels, and gross revenues were just under $17 billion (Chen & Stout, 1992). Digital video compression was seen as a means of expanding channel capacity so that cable operators could offer video on demand and multiplexed popular channels. In the regulatory arena, Congress was poised to pass the Cable Television and Consumer Protection Act of 1992, which regulated most cable television rates, mandated that local broadcasters be given a choice of "must-carry" or retransmission consent, and required cable television programmers to make their services available to cable's competitors, including DBS and MMDS (Dunatchik, 1993).

Ten years later, digital cable is commonplace, with many cable systems offering hundreds of channels of programming. Virtually every premium service is multiplexed, and cable television revenues have risen to $48 billion. The only trend reported in 1992 that has not been sustained is the emergence of local cable television channels serving a single city or geographic area.

PREMIUM CABLE

The most important factor reported in the first edition of this book regarding premium cable services was the fact that all but one of the major premium cable services had lost subscribers in 1991. The pay cable industry was concerned over competition from pay-per-view and video rentals and the impact of the 1991 recession on subscribership (Morgan, 1992).

The 1992 chapter correctly predicted that premium cable services such as HBO would focus on multiplexing and on providing original programming that viewers could not find elsewhere. A separate chapter addressed the prospects of pay-per-view, predicting that the outcome of the NBC *Triplecast* (which was to provide pay-per-view access to the 1992 Summer Olympics) would foretell the future of pay-per-view (Faske, 1992). The *Triplecast* was a failure, and PPV remains a service that has yet to live up to most predictions.

INTERACTIVE TELEVISION

Few technologies have changed as little since 1992 as interactive television. The 1992 chapter on ITV explored a number of different trials and technologies to deliver interactivity to television viewers, indicating that most of the then-current ventures were losing money, and that consumers had yet to demonstrate that they wanted interactivity (Koester, 1992).

Ten years later, most of the companies have changed, but the outcome remains the same. In 1992, the driving technology for ITV was a set of frequencies set aside by the FCC for interactive television services; in 2002, the same statements are being made about the Internet. The only broadly successful form of interactive television today is the same one that existed in 1992: television shopping.

DBS

On the other hand, few technologies discussed in the first edition of *Communication Technology Update* have experienced the degree of success earned by DBS. In 1992, the viability of satellite-delivered program services was (literally) up in the air, as the DBS frequencies allocated by the FCC were yet to be developed. One key observation in the chapter was that access to cable television channels by DBS services would be critical to the survival of the medium (Spease, 1992).

That access was guaranteed by the Cable Television Consumer Protection and Competition Act of 1992, which opened the window of opportunity for then-unknown companies such as DirecTV and EchoStar. The leading company to watch in 1992 was Skypix, which is long gone, but DBS has developed to provide a significant challenge to cable television.

DIGITAL TELEVISION

The concept of digital television was brand new in 1992, with six systems (including five digital systems) competing to be chosen as the high-definition television (HDTV) standard in the United States. Although the merger of digital competitors to create the "grand alliance" system was not foreseen, the chapter correctly predicted that HDTV would first be available by 1998 with a 15-year transition from NTSC to HDTV (Van Tassel, 1992).

Today's digital television environment is different in two important ways from that envisioned in 1992: the FCC has mandated an accelerated transition to digital television, but consumer adoption is taking longer than anticipated.

STREAMING MEDIA

Streaming video was barely mentioned in the 1992 edition of the *Communication Technology Update*. The only reference was the to first version of Apple's QuickTime, which was described as a "sequence of multimedia standards designed to ease the integration and implementation of multimedia data" (Taxon, 1992, p. 57). In short, QuickTime was a technology that allowed video to be displayed on a computer. The development of the Internet extended the capabilities for transmitting multimedia information, leading to the development of streaming media technologies in the latter 1990s.

The first mention of video over a data connection occurred in the 1994 edition, which provided an extensive discussion of "video dialtone," delivery of video programming by the telephone company over ordinary telephone wires (Goldsmith, 1994). (Video dialtone was the forerunner of DSL, which has become one of the primary technologies for delivering video from the Internet to the home.)

RADIO

In 1992, digital radio had been proven in the laboratory, with the key debate being whether it would be provided via satellite or as a terrestrial broadcast service (Grant, 1992). The first chapter on this technology did not envision that both technologies would ultimately be available, nor did it anticipate the creation of terrestrial digital radio that uses the existing AM and FM bands (iBiquity's IBOC system, discussed in Chapter 9).

One aspect of radio that is no longer a factor is AM stereo. In 1992, AM stereo was still seen as one source of economic salvation for AM radio; the other was improved frequency response. Ten years later, many AM stations continue to broadcast in stereo, but listeners are, by and large, oblivious to the technology. AM radio has experienced a resurrection due to the popularity of talk, news, and sports programming, but it continues to lag behind FM in audience size, revenue, and respect.

OTHER ELECTRONIC MASS MEDIA

A few other electronic mass media have failed to find a market since 1992. Low-power television stations were seen in 1992 as a major new force in television broadcasting. Although thousands of LPTV stations are on the air in the United States, the technology is considered relatively unimportant because most viewers are limited to channels provided by their local cable television company, and most LPTV channels are not carried by cable television systems.

The 1992 chapter also discussed two potential competitors to cable television: multichannel, multipoint distribution service (MMDS) and satellite master antenna television (SMATV). MMDS uses microwave frequencies to beam up to 33 analog channels or up to 200 digital channels within a local coverage area. A few MMDS systems continue to battle cable television and DBS for market share, but most MMDS operators are now looking at new uses for the spectrum, including delivering high-speed Internet services. In the meantime, SMATV, which has traditionally provided cable-television-type services to hotels, apartment complexes, etc., has seen a migration from low power, C-band technology to high-power, DBS technology.

COMPUTERS AND CONSUMER ELECTRONICS

PERSONAL COMPUTERS

Few technologies have achieved as many technical advances with as little change in patterns of usage as the personal computer. Today's computers have more than 20 times the processing power and memory as their 1992 counterparts, but the most common uses remain the same: word processing, spreadsheets, and electronic mail.

The penetration of home computers has grown significantly, from about 25% of U.S. households in 1992 to almost 60% today. This growth in home computer use has been driven primarily by lower prices and the desire for Internet access (including e-mail) at home. Perhaps more important, the most recent research indicates a gradual disappearance of the "digital divide," as low-income households are becoming equally likely as high-income houses to have computers, Internet access, and the benefits that flow from these technologies.

ELECTRONIC GAMES

One of the most difficult technologies to analyze in this type of retrospective is electronic games. As with computers, today's game consoles are generations more advanced than their 1992 counterparts, with advances in speed, graphics, memory, and user interaction. On the other hand, the form and function of the core device remains unchanged—these systems consist of set-top boxes that allow a person to play games that are displayed on a television screen.

Ten years ago, a group of technologies presented a more sophisticated promise. Philips had introduced an interactive video technology called CD-I (for compact disc-interactive) that was designed to integrate all of the capabilities of video games, multimedia computer programs, and video (with limited storage capacity) in a single format. CD-I struggled along with other formats such as DV-I (digital video-interactive) for consumer attention, but these technologies ultimately lost out to the traditional video game consoles (Taxon, 1992).

Several new video game systems in 2002 use DVDs to store individual games, with the consoles also serving as DVD players. The fact that these devices offer the same integration of games and home video as offered by the earlier market failures suggests that we should watch these consoles carefully to see how their fate differs from their predecessors.

INTERNET

The Internet was more than 20 years old in 1992, but the word "Internet" did not appear in the book. A chapter on electronic messaging explored the interconnection of computer systems using "gateways" (Mason, 1992), and a chapter on videotext explored the nascent systems that produced "electronic newspapers," public databases, and fledgling services such as Compuserve and Prodigy.

The early view of each of these services was that each was a closed system, with limited interconnections that provided only for transmission of e-mail across systems. In the intervening years, of course, the Internet emerged as the network connecting this disparate system, with the Internet's TCP/IP protocol becoming the standard for data transmission across most networks, enabling the development of the connectivity we enjoy today through the Internet.

INTERNET COMMERCE

In 1992, the concept of electronic commerce discussed in the first edition of the *Communication Technology Update* was limited to television shopping. At the time, television shopping was just overcoming the aftereffects of a market bust, with all but a handful of the 40-some shopping channels going out of business or being bought by the remaining services. Stock prices were dramatically inflated in the mid-1980s, as investors fought for a piece of the "gold rush." Many investors lost some or all of their investment when the stock prices of television shopping companies boomeranged to their initial offering prices as the decade ended (Kim, 1992).

The core business strategy of those shopping networks remained sound, with gross revenues for the television shopping industry growing from $2 billion in 1992 to more than $4 billion in 2001. The parallels with Internet commerce and the dot.com boom and bust are remarkable, with the primary difference due to the sheer scale of the dot.com events. The lesson is that electronic commerce will likely emerge as a sound and profitable business, not dominating other forms of commerce, but carving out its own niche.

OFFICE TECHNOLOGIES

One of the weaknesses of the first edition of the *Communication Technology Update* was that it included too many chapters with overlapping information. This was especially true in the case of office technologies, with separate chapters on desktop publishing, fax machines, electronic messaging, and voice messaging.

Each of these technologies was quite revolutionary in 1992. In the intervening 10 years, they have become taken for granted. One measure of our reliance upon communication technologies in business is the fact that it is almost inconceivable that a business would not have a fax machine, voice mail, electronic mail, and word processing capabilities (with today's typical word processing programs providing much the same capabilities as high-end desktop publishing systems did 10 years ago).

VIRTUAL AND AUGMENTED REALITY

Virtual reality demonstrated exceptional promise for entertainment applications in 1992 (Caillouet, 1992), but the technology has not yet proven itself as a viable entertainment technology. Advances in computer processing and the input/output devices required in VR have increased the capabilities of VR systems, but large-scale applications in entertainment seem as far away today as they were in 1992.

VR has emerged, however, as an important tool in architecture, medicine, data visualization, and training. In each case, the value of the VR application offsets the comparatively high cost of the technology; such a tradeoff has not yet emerged for entertainment applications.

One substantial similarity stands out between the discussion of VR in the first edition and the current edition: Both chapters address the inclusion of VR in movies and science fiction. Clearly, VR is one technology that is easier to simulate than to create.

HOME VIDEO

The VCR was a well-diffused technology in 1992, already in three out of four U.S. homes. The primary issue addressed in the discussion of home video in that chapter was the imminent introduction of HDTV and the need for a home recording device that would record HDTV signals. The chapter also discussed the recent "victory" of VHS over Betamax and the Supreme Court decision legalizing home taping. The chapter also detailed prospective VCR functions such as an on-screen program guide and a beeping feature to help locate a lost remote control (Schwarz, 1992).

A separate chapter detailed developments in videodiscs, the 12-inch "laserdisc" format that was the precursor to today's DVD format. A videodisc recorder costing "only $30,000" had just been developed, and sales of videodisc players were growing modestly, with just under one million players sold in 1991 (Johnson, 1992).

Ten years later, HDTV has not progressed to the point that there is a market for high-definition video recorders, but, as predicted 10 years ago, that need is just around the corner. In the meantime, the 12-inch laserdisc format has disappeared, DVD has begun to challenge VHS as the primary distribution format for home video, and a battle may be brewing between PVRs and recordable DVD units as replacements for the VCR.

DIGITAL AUDIO

In 1992, the latest developments in digital audio were the introduction of digital audiotape and the digital compact cassette, both of which demonstrated the imminent move from analog to digital audio. Although neither format earned any success in the consumer market, compression, sampling, and recording technologies developed for both formats have played a major role in today's digital audio formats.

One primary concern with both technologies was the capability of each to make perfect copies of recordings, thereby robbing the recording industry of revenue. The two solutions explored in 1992—the incorporation of anti-copying circuitry in the form of the serial copy management system (SCMS) and payment of royalties on recorders and blank media—remain primary elements of the latest audio formats.

TELEPHONY AND SATELLITE TECHNOLOGIES

LOCAL AND LONG DISTANCE TELEPHONY

Plain old telephone service (POTS) may have experienced less change over the past 10 years than any technology discussed in this book. In 1992, the aftershocks from the 1984 breakup of AT&T were beginning to die down, and most prognosticators predicted that competitive companies would soon challenge the incumbent local exchange carriers (ILECs) in the same manner that competitive long distance companies had begun to compete with AT&T.

Not only has competition in the local telephone market failed to materialize, but the seven "baby Bells" that were created by the AT&T breakup have been reduced to four through mergers, further reducing the likelihood of these companies entering each others' markets.

BROADBAND NETWORKS

In contrast to developments in POTS, broadband connectivity to the home has achieved a measure of success that was not expected in 1992, with everyday consumers today enjoying speeds that would be the envy of those connecting at the then-blazing speed of 9,600 b/s. In the 1992 edition of the *Communication Technology Update*, two chapters dealt with broadband networks. The first explored ISDN, a relatively slow (144 Kb/s) and expensive technology, that was projected to be available to half of the United States by 1994 (Tenouri, 1992) (availability never reached those levels). The second explored fiber optics, which promised, at some undetermined point in the future, to deliver high-speed entertainment programming directly to the home.

Although neither ISDN nor fiber optics has achieved significant penetration to the home, the services promised by both technologies are being delivered through ordinary telephone lines using DSL technology and through cable television networks using a blended system of fiber optics and coaxial cable known as hybrid fiber/coax. The media for delivering broadband have changed, but the allure of fast data connectivity to the home is more attractive than ever.

HOME NETWORKS

Home networks were not mentioned in the first six editions of the *Communication Technology Update*. They attained prominence in the late 1990s as a popular adjunct to broadband connectivity to the home. In retrospect, increases in computer penetration and the fact that most computer purchases were by homes that already owned one or more computers should have suggested the impending need to interconnect these computers. (Preparing this chapter causes this editor to wonder which important technologies may have been left out of *this* edition of the *Update*.)

SATELLITES

The technology for geosynchronous satellites has changed little since 1992, but these high-flying satellites have since been supplemented by a new generation of satellites that orbit at lower altitudes, allowing a new generation of communication devices utilizing satellites. Point-to-point satellite communications using VSATs (very small aperture terminals) has been supplanted, to some degree, by high-speed terrestrial networks and Internet connectivity. Satellites remain a mainstay of distribution in the television industry, however, with a strong DBS industry that has emerged since 1992 to offer a challenge to cable television.

DISTANCE LEARNING

The use of technology for distance learning was almost as common in 1992 as it is today. The primary change in this technology is the manner in which information is distributed. In 1992, satellites offered the most common means of delivering educational programs to remote areas; today, the Internet has emerged to provide an alternate medium for distance learning.

Video compression techniques were pioneered by distance learning proponents seeking more efficient use of expensive satellite time. The resulting video compression standards are used extensively in all forms of distance education, from Internet streaming to traditional satellite networks.

WIRELESS TELEPHONY

Today's cellular telephone industry includes two technologies explored separately in 1992: cellular telephony and personal communication systems (PCS). The former consists of the original cellular telephone service introduced in the 1980s, while the latter consists of a new set of technologies and new spectrum allocated to mobile telephony in the 1990s.

Two features of the cellular telephone industry stand out. The first is the speed with which the FCC and industry acted to create a means of allocating, utilizing, and marketing PCS spectrum. The second is the fact that, rather than become a separate service, PCS has become an extension of cellular telephone service. Advances in PCS technology have led to compatible changes in cellular telephone systems, from analog to digital, and adding numerous features has been easily afforded by the digital technology.

It is likely that most people cannot tell you whether their mobile telephone uses traditional cellular technology or PCS. It is equally likely that most people who have mobile telephones today would not have expected back in 1992 that they would possess (and make such extensive use of) the technology. The imminent introduction of mobile Internet access on the next generation of mobile telephones may be slated for the same fate.

TELECONFERENCING

The 1992 vision of teleconferencing technology to cut costs and increase efficiency has remained consistent for the intervening 10 years. Despite significant reductions in the cost of teleconferencing equipment, however, use of the technology has grown only moderately since 1992. (In 1992, the least expensive videoconferencing setup cost $20,000; comparable systems are available today for less than $5,000.)

The technologies used in videoconferencing have changed significantly in the interim, with video compression reducing the cost and complexity of the interconnections among videoconferencing sites. The development of standards has also facilitated connectivity through local area networks and the Internet, greatly expanding access to connection points for videoconferencing.

It is important to note, however, that no edition of the *Communication Technology Update* has viewed audioconferencing or videoconferencing as a replacement for all travel and meetings. Rather, the market for these technologies is limited by the type of meeting that is appropriate for teleconferencing (for example, routine meetings where the participants already know each other). As a consequence, teleconferencing should continue to enjoy steady but slow growth over the next decade.

VIDEOPHONES

The introduction of a new generation of videophones in 1992 prompted inclusion of a chapter on this topic. This chapter was careful to differentiate the capabilities of the then current videophone technology from the challenges that would be faced in making the videophone a commonplace technology (Bolmarcich, 1992).

That generation of videophones failed. It is likely that a new generation of videophones will be introduced before 2005; that generation will fail as well (just as the ones introduced in the 1960s, 1970s, and 1980s failed, too). Two obstacles must be overcome for videophones to become a success: People must be comfortable with being seen during a video conversation, and a "critical mass" of users must exist so that a prospective user has someone to call when he or she acquires a videophone. To date, these two issues have been more important than technology in deciding the fate of each generation of videophones; there is no reason to believe that the next generation will be any different.

CONCLUSIONS

The goal of this chapter was to provide a context for the developments discussed throughout this book. There is no way that a single chapter can provide a complete analysis of each technology; rather, specific information was chosen about each technology discussed in order to help you identify patterns that will predict the fate of technologies that are yet to be produced.

Perhaps the biggest surprise in 10 years is that the *Communication Technology Update* has become as popular as it has. The *Update* itself has changed little since the second edition, but the technology used to produce the update has changed dramatically. For the earliest editions, as much time and effort was spent translating data formats, waiting for delivery of floppy disks, and creating simple graphics as was spent editing and preparing the text for publication. Advances in the communication technologies discussed in this book have caused a dramatic change in the manner in which the book is produced, allowing more time and effort to be spent on content with less focus on mechanical processes.

The other major change occurred in 1996 when we switched from a one-year publication cycle to a two-year cycle, using a companion Web page to provide updates on every chapter to cover the intervening year. One thing hasn't changed: The goal of the *Update* is to provide the most comprehensive, most up-to-date source of information on communication technologies.

On a personal note, the eight editions I've edited over the past 10 years have been a labor of love for me. There is no question that each edition has improved over the previous ones, thanks largely to the efforts of the contributors and input from readers and faculty who use the book in their courses.

I am looking forward to doing a 20-year retrospective in 2012. I hope you will enjoy that edition as much as this one.

BIBLIOGRAPHY

Bolmarcich, T. (1992). Videophones. In A. E. Grant & L. Sung (Eds.). *Communication technology update.* Austin, TX: Technology Futures, Inc.

Caillouet, C. (1992). Virtual reality. In A. E. Grant & L. Sung (Eds.). *Communication technology update.* Austin, TX: Technology Futures, Inc.

Chen, F., & Stout, J. (1992). Cable television. In A. E. Grant & L. Sung (Eds.). *Communication technology ipdate.* Austin, TX: Technology Futures, Inc.

Dunatchik, C. L. (1993). Cable television. In A. E. Grant & K. T. Wilkinson (Eds.). *Communication technology update: 1993-1994.* Austin, TX: Technology Futures, Inc.

Faske, R. (1992). Pay-per-view cable television. In A. E. Grant & L. Sung (Eds.). *Communication technology update.* Austin, TX: Technology Futures, Inc.

Goldsmith, L. (1994). Video dialtone. In A. E. Grant (Ed.). *Communication technology update,* 3rd edition. Boston: Focal Press.

Grant, A. E. (1992). Digital audio broadcasting. In A. E. Grant & L. Sung (Eds.). *Communication technology update.* Austin, TX: Technology Futures, Inc.

Johnson, V. (1992). Videodiscs. In A. E. Grant & L. Sung (Eds.). *Communication technology update.* Austin, TX: Technology Futures, Inc.

Kim, H. (1992). TV shopping. In A. E. Grant & L. Sung (Eds.). *Communication technology update*. Austin, TX: Technology Futures, Inc.

Koester, S. (1992). Interactive television. In A. E. Grant & L. Sung (Eds.). *Communication technology update*. Austin, TX: Technology Futures, Inc.

Mason, D. (1992). Electronic messaging. In A. E. Grant & L. Sung (Eds.). *Communication technology update*. Austin, TX: Technology Futures, Inc.

Morgan, L. A. (1992). Pay cable television. In A. E. Grant & L. Sung (Eds.). *Communication technology update*. Austin, TX: Technology Futures, Inc.

Schwarz, G. (1992). Videocassette recorders. In A. E. Grant & L. Sung (Eds.). *Communication technology update*. Austin, TX: Technology Futures, Inc.

Spease, M. (1992). Direct broadcast satellites. In A. E. Grant & L. Sung (Eds.). *Communication technology update*. Austin, TX: Technology Futures, Inc.

Taxon, M. (1992). Digital video-interactive. In A. E. Grant & L. Sung (Eds.). *Communication technology update*. Austin, TX: Technology Futures, Inc.

Tenouri, M. (1992). Integrated services digital network. In A. E. Grant & L. Sung (Eds.). *Communication technology update*. Austin, TX: Technology Futures, Inc.

Van Tassel, J. (1992). High-definition television. In A. E. Grant & L. Sung (Eds.). *Communication technology update*. Austin, TX: Technology Futures, Inc.

26

CONCLUSION

Jennifer H. Meadows, Ph.D.[*]

This is the 10th anniversary of the *Communication Technology Update*. As the previous chapter demonstrates, it is both surprising and illuminating to review the technologies discussed in this book over the years. It is also useful to review the technologies covered in this edition of the book because several factors reach across all chapters.

The conversion from analog to digital continues. While some technologies such as audio have almost completed this conversion, others, such as digital television, seem to be struggling for wide consumer adoption. The case of digital television is a conundrum because consumers will not buy the receivers until the prices come down and there is sufficient programming to warrant purchasing a set. On the other hand, broadcasters and program producers are resistant to providing digital broadcasts and programming because the audience just isn't there yet. The federal government has required broadcasters to complete the conversion but, as each deadline approaches, factors pile up which force extensions to be granted. Radio broadcasting, on the other hand, is just beginning to engage in the conversion to digital. While recording media and station technology for audio production are digital, terrestrial radio is still transmitted in an analog form. Digital audio broadcasting, while available in Europe and Canada, is still in the earliest stages of adoption in the United States.

One of the major factors driving the conversion of many technologies to digital is the Internet. The packet switched technology that is the basis of the Internet is increasingly being adapted to other technologies as a means to send data and to make software communicate seamlessly with the Internet and other communication technologies. For example, 3G wireless and H.323 for videoconferencing both employ packet switched networks to transmit data. For wireless telephony, this means that users

---

[*] Associate Professor, Department of Communication Design, California State University, Chico (Chico, California).

can receive not only phone calls but also video and audio in file formats such as MP3, as well as surf the Web quickly and easily using a phone or other personal communication device (such as a PDA).

Other technologies such as PVRs, televisions, and DBS receivers are increasingly Internet enabled. Home networking technologies allow users integrate the technologies in a single network in the home, enabling for more efficient use of resources such as storage and broadband access.

All was not roses, though, for Internet-related companies since 2000. The "dot.com bust" hit the United States hard in 2000, with the impact continuing into 2002 as companies and investors flush with new ideas and unlimited money in the early days find only limited opportunities today. Investors are more cautious now. It was said that, in the early days of the dot.com explosion, if you had an idea and a PowerPoint presentation, you could get funded. Those days are over; investors lost big as technology stocks fell dramatically. For example, high profile e-commerce businesses such as grocery service Web Van and pet store Pets.com went out of business. Even giant powerhouses are losing. Consider that computer manufacturer Gateway lost market share and over $600 million in the last quarter of 2001 and the first quarter of 2002 (Fried, 2002).

Despite the losses, the pace of technology development shows no sign of slowing down. Moore's law appears to be in full effect in 2002, and it looks like, at this point, there is no end in sight. For example, in 2000, PVRs could store up to 40 hours of television programming. In 2002 a ReplayTV unit can store 320 hours of programming (ReplayTV, n.d.). Personal computers have CPUs running at speeds above 2.4 GHz with 80 gigabytes of storage—all while prices have remained stable or dropped for consumers.

As storage becomes cheaper and processing speed increases, it becomes difficult to predict the kinds of new technologies that will come in the future. Consumer camcorders that record onto DVD would have sounded incredible in 2000 but, in 2002, that camcorder is a reality (Summary of, 2002). While it is impossible to ignore the fast pace of communication technologies, it is important to note that technology has no real future unless users adopt it. It is easy to become enamored with new technologies and their possibilities, but it is equally important to understand and study users and the motivating and inhibiting factors that drive them to either adopt or reject technologies. For example, Sega Dreamcast may have been an excellent and advanced gaming system but users did not adopt it, choosing instead to adopt another game system, the Playstation 2. The reasons for the rejection of a technology are often not inherent in the hardware itself. For example, the demise of Dreamcast may have been due to marketing elements associated with the Playstation 2, which was backward-compatible with Playstation 1 games and also carried strong brand loyalty with users.

Finally, it would be remiss on the part of the author to fail to mention the tremendous impact the terrorist attacks of September 11, 2001 had on communication technology. Almost every chapter in this book discusses, in some way, the effects this tragic and horrible event had on America's use of communication technologies. For example, many Americans are spending more time in the home as evidenced by the increases in home entertainment hardware and software purchases since September 11. Businesses are employing more telecommuting and videoconferencing technology to cut back on travel, which not only saves money but also alleviates concerns about aviation safety. An increased concern over terrorism in response to the attacks led to anti-terrorism legislation, which deals not only with security but also with privacy. The federal government, for example, now has more authority to monitor Internet use and e-mail with the USA Patriot Act.

The events of September 11 highlight, in the most extreme manner, the notion that the world is unpredictable. Communication technologies are in a constant state of change. In fact, by the time this book is published, there will have been significant developments in every area. Please visit the Communication Technology Website at http://www.tfi.com/ctu/redev to read updates on the technologies discussed in the book.

BIBLIOGRAPHY

Fried, I. (2002, April 18). Gateway reports loss as expected. *News.com*. Retrieved on April 28, 2002 at http://www.news.com.com/2100-1040-886410.html.
ReplayTV 4000. (n.d.). Retrieved on April 28, 2002 at http://www.sonicblue.com/video/replaytv/default.asp.
Summary of DVD-CAM FAQ. (2002, March). Hitachi Home Electronics.

GLOSSARY

2-wire line. The set of two copper wires used to connect a telephone customer with a switching office, loosely wrapped around each other to minimize interference from other twisted pairs in the same bundle. Synonymous with twisted pair.

802.11. An IEEE specification for 1 Mb/s and 2 Mb/s wireless LANs.

802.11a. An IEEE specification for 54 Mb/s high rate wireless LANs in the 5 Ghz band that uses orthogonal frequency division multiplexing.

802.11b. An IEEE specification for 11 Mb/s high rate wireless LANs in the 2.4 Ghz band that uses direct sequence spread spectrum technology. (Also known as 802.11HR—for high rate.)

802.3. An IEEE specification for SCMA/CD based Ethernet networks.

802.5. An IEEE specification for token ring networks.

A

Adaptive transform acoustic coding (ATRAC). A method of digital compression of audio signals used in the MD (minidisc) format. ATRAC ignores sounds out of the range of human hearing to eliminate about 80% of the data in a digital audio signal.

Addressability. The ability of a cable system to individually control its converter boxes, allowing the cable operator to enable or disable reception of channels for individual customers instantaneously from the home office.

ADSL (*asymmetrical digital subscriber line*). A system of compression and transmission that allows broadband signals up to 6 Mb/s to be carried over twisted pair copper wire for relatively short distances.

Advanced television (ATV). Television technologies that offer improvement in existing television systems.

Agent. Any being in a virtual environment.

Algorithm. A specific formula used to modify a signal. For example, the key to a digital compression system is the algorithm that eliminates redundancy.

AM (*amplitude modulation*). A method of superimposing a signal on a carrier wave in which the strength (amplitude) of the carrier wave is continuously varied. AM radio and the video portion of NTSC TV signals use amplitude modulation.

American National Standards Institute (ANSI). An official body within the United States delegated with the responsibility of defining standards.

American Standard Code for Information Interchange (ASCII). Assigns specific letters, numbers, and control codes to the 256 different combinations of 0s and 1s in a byte.

Analog. A continuously varying signal or wave. As with all waves, analog waves are susceptible to interference, which can change the character of the wave.

ANSI. See *American National Standards Institute.*

ASCII. See *American Standard Code for Information Interchange.*

Aspect ratio. In visual media, the ratio of the screen width to height. Ordinary television has an aspect ratio of 4:3, while high-definition television is "wider" with an aspect ratio of 16:9 (or 5.33:3).

Asynchronous. Occurring at different times. For example, electronic mail is asynchronous communication because it does not require the sender and receiver to be connected at the same time.

ATM (*asynchronous transfer mode*). A method of data transport whereby fixed length packets are sent over a switched network. Speeds of up to 2 Gb/s can be achieved, making it suitable for carrying voice, video, and data.

Audio on demand. A type of media that delivers sound programs in their entirety whenever a listener requests the delivery.

Augmented reality. The superimposition of virtual objects on physical reality. For example, a technician could use augmented reality to display a three-dimensional image of a schematic diagram on a piece of equipment to facilitate repairs.

Available bit rate (ABR). An ATM service type in which the ATM network makes a "best effort" to meet the transmitter's bandwidth requirements. ABR differs from other best-effort service types by employing a congestion feedback mechanism that allows the ATM network to notify the transmitters that they should reduce their rate of data transmission until the congestion decreases. Thus, ABR offers a qualitative guarantee that the transmitter's data can get to the intended receiver without experiencing unwanted cell loss.

Avatar. An animated character representing a person in a virtual environment.

AVI (*audio-video interleaved*). Microsoft's video driver for Windows.

B

Backbone. The part of a communications network that handles the major traffic using the highest-speed, and often longest, paths in the network.

Bandwidth. A measure of capacity of communications media. Greater bandwidth allows communication of more information in a given period of time.

Basic rate interface ISDN (BRI-ISDN). The basic rate ISDN interface provides two 64-Kb/s channels (called B channels) to carry voice or data and one 16-Kb/s signaling channel (the D channel) for call information.

BISDN. See *Broadband integrated services digital network*.

Bit. A single unit of data, either a one or a zero, used in digital data communications. When discussing digital data, a small "b" refers to bits, and a capital "B" refers to bytes.

Bluetooth. A specification for short-range wireless technology created by a consortium of computer and communication companies that make up the Bluetooth Special Interest Group.

Bridge. A type of switch used in telephone and other networks that connects three or more users simultaneously.

Broadband. An adjective used to describe large-capacity networks that are able to carry several services at the same time, such as data, voice, and video.

Broadband integrated services digital network (BISDN). A second-generation ISDN technology that uses fiber optics for a network that can transmit data at speeds of 155 Mb/s and higher.

Buy rate. The percentage of subscribers purchasing a pay-per-view program divided by the total number of subscribers who can receive the program.

Byte. A compilation of bits, seven bits in accordance with ASCII standards and eight bits in accordance with EBCDIC standards.

C

Carrier. An electromagnetic wave or alternating current modulated to carry signals in radio, telephonic, or telegraphic transmission.

CATV (community antenna television). One of the first names for local cable television service, derived from the common antenna used to serve all subscribers.

C-band. Low-frequency (1 GHz to 10 GHz) microwave communication. Used for both terrestrial and satellite communication. C-band satellites use relatively low power and require relatively large receiving dishes.

CCD (charge coupled device). A solid-state camera pickup device that converts an optical image into an electrical signal.

CCITT (International Telegraph and Telephone Consultative Committee). CCITT is the former name of the international regulatory body that defines international telecommunications and data communications standards. It has been renamed the Telecommunications Standards Sector of the International Telecommunications Union.

CDDI (copper data distributed interface). A subset of the FDDI standard targeted toward copper wiring.

CD-I (compact disc-interactive). A proprietary standard created by Philips for interactive video presentations including games and education.

CDMA (code division multiple access). A spread spectrum cellular telephone technology, which digitally modulates signals from all channels in a broad spectrum.

CD-ROM (compact disc-read only memory). The use of compact discs to store text, data, and other digitized information instead of (or in addition to) audio. One CD-ROM can store up to 700 megabytes of data.

CD-RW (compact disc-rewritable). A special type of compact disc that allows a user to record and erase data, allowing the disc to be used repeatedly.

Cell. The area served by a single cellular telephone antenna. An area is typically divided into numerous cells so that the same frequencies can be used for many simultaneous calls without interference.

Central office (CO). A telephone company facility that handles the switching of telephone calls on the public switched telephone network (PSTN) for a small regional area.

Central processing unit (CPU). The "brains" of a computer, which uses a stored program to manipulate information.

Circuit-switched network. A type of network whereby a continuous link is established between a source and a receiver. Circuit switching is used for voice and video to ensure that individual parts of a signal are received in the correct order by the destination site.

CLEC. See *Competitive local exchange carrier.*

CO. See *Central office.*

Coaxial cable. A type of "pipe" for electronic signals. An inner conductor is surrounded by a neutral material, which is then covered by a metal "shield" that prevents the signal from escaping the cable.

CODEC (COmpression/DECompression). A device used to compress and decompress digital video signals.

COFDM (coded orthogonal frequency division multiplexing). A flexible protocol for advanced television signals that allows simultaneous transmission of multiple signals at the same time.

Common carrier. A business, including telephone companies and railroads, which is required to provide service to any paying customer on a first-come/first-served basis.

Competitive local exchange carrier (CLEC). An American term for a telephone company that was created after the Telecommunications Act of 1996 made it legal for companies to compete with the ILECs. Contrast with *ILEC.*

Compression. The process of reducing the amount of information necessary to transmit a specific audio, video, or data signal.

Constant bit rate (CBR). A data transmission that can be represented by a non-varying, or continuous, stream of bits or cell payloads. Applications such as voice circuits generate CBR traffic patterns. CBR is an ATM service type in which the ATM network guarantees to meet the transmitter's bandwidth and quality-of-service requirements.

Cookies. A file used by a Web browser to record information about a user's computer, including Websites visited, which is stored on the computer's hard drive.

Core network. The combination of telephone switching offices and transmission plant connecting switching offices together. In the U.S. local exchange network, core networks are linked by several competing interexchange networks; in the rest of the world, the core network extends to national boundaries.

CPE. See *Customer premises equipment.*

CPU. See *Central processing unit.*

Crosstalk. Interference from an adjacent channel.

Customer premises equipment (CPE). Any piece of equipment in a communication system that resides within the home or office. Examples include modems, television set-top boxes, telephones, and televisions.

Cyberspace. The artificial worlds created within computer programs.

D

DBS. See *Direct broadcast satellite.*

DCC (digital compact cassette). A digital audio format resembling an audiocassette tape. DCC was introduced by Philips (the creator of the common audio "compact cassette" format) in 1992, but never established a foothold in the market.

Dedicated connection. A communications link that operates constantly.

Desktop publishing (DTP). The process of producing printed materials using a desktop computer and laser printer. Commonly-used peripherals in DTP include scanners, modems, and color printers.

Dial-up connection. A data communication link that is established when the communication equipment dials a phone number and negotiates a connection with the equipment on the other end of the link.

Digital audio broadcasting (DAB). Radio broadcasting that uses digital signals instead of analog to provide improved sound quality.

Digital audiotape (DAT). An audio recording format that stores digital information on 4mm tape.

Digital signal. A signal that takes on only two values, off or on, typically represented by a "0" or "1." Digital signals require less power but (typically) more bandwidth than analog, and copies of digital signals are exactly like the original.

Digital subscriber line (DSL). A data communications technology that transmits information over the copper wires that make up the local loop of the public switched telephone network (see Local loop). It bypasses the circuit-switched lines that make up that network and yields much faster data transmission rates than analog modem technologies. Common varieties of DSL include ADSL, HDSL, IDSL, SDSL, and RADSL. The generic term xDSL is used to represent all forms of DSL.

Digital subscriber line access multiplexer (DSLAM). A device found in telephone company central offices that takes a number of DSL subscriber lines and concentrates them onto a single ATM line.

Digital video compression. The process of eliminating redundancy or reducing the level of detail in a video signal in order to reduce the amount of information that must be transmitted or stored.

Direct broadcast satellites (DBS). High-powered satellites designed to beam television signals directly to viewers with special receiving equipment.

Discrete multitone modulation (DMT). A method of transmitting data on copper phone wires that divides the available frequency range into 256 subchannels or tones, and which is used for some types of DSL.

Domain name system (DNS). The protocol used for assigning addresses for specific computers and computer accounts on the Internet.

Downlink. Any antenna designed to receive a signal from a communication satellite.

DSL. See *Digital subscriber line*.

DSLAM. See *Digital subscriber line access multiplexer*.

DVD (digital video or versatile disc). A plastic disc similar in size to a compact disc that has the capacity to store up to 20 times as much information as a CD and uses both sides of the disc.

DVD-RAM. A type of DVD that can be used to record and play back digital data (including audio and video signals).

DVD-ROM. A type of DVD that is designed to store computer programs and data for playback only.

DVI (digital video interactive). An interactive video system that uses a computer to record and/or play back compressed video, text, and audio.

E

E-1. The European analogue to T1, a standard, high-capacity telephone circuit capable of transmitting approximately 2 Mb/s or the equivalent of 30 voice channels.

Echo cancellation. The elimination of reflected signals ("echoes") in a two-way transmission created by some types of telephone equipment; used in data transmission to improve the bandwidth of the line.

Electromagnetic spectrum. The set of electromagnetic frequencies that includes radio waves, microwave, infrared, visible light, ultraviolet rays, and gamma rays. Communication is possible through the electromagnetic spectrum by radiation and reception of radio waves at a specific frequency.

E-mail. Electronic mail or textual messages sent and received through electronic means.

Extranet. A special type of Intranet that allows selected users outside an organization to access information on a company's Intranet.

F

FDDI-I (fiber data distributed interface I). A standard for 100 Mb/s LANs using fiber optics as the network medium linking devices.

FDDI-II (fiber data distributed interface II). Designed to accommodate the same speeds as FDDI-I, but over a twisted-pair copper cable.

Federal Communications Commission (FCC). The U.S. federal government organization responsible for the regulation of broadcasting, cable, telephony, satellites, and other communications media.

Fiber-in-the-loop (FITL). The deployment of fiber optic cable in the local loop, which is the area between the telephone company's central office and the subscriber.

Fiber optics. Thin strands of ultrapure glass or plastic that can be used to carry light waves from one location to another.

Fiber-to-the-cabinet (FTTCab). Network architecture where an optical fiber connects the telephone switch to a street-side cabinet. The signal is converted to feed the subscriber over a twisted copper pair.

Fiber-to-the-curb (FTTC). The deployment of fiber optic cable from a central office to a platform serving numerous homes. The home is linked to this platform with coaxial cable or twisted pair (copper wire). Each fiber carries signals for more than one residence, lowering the cost of installing the network versus fiber-to-the-home.

Fiber-to-the-home (FTTH). The deployment of fiber optic cable from a central office to an individual home. This is the most expensive broadband network design, with every home needing a separate fiber optic cable to link it with the central office.

Fixed satellite services (FSS). The use of geosynchronous satellites to relay information to and from two or more fixed points on the earth's surface.

FM (frequency modulation). A method of superimposing a signal on a carrier wave in which the frequency on the carrier wave is continuously varied. FM radio and the audio portion of an NTSC television signal use frequency modulation.

Footprint. The coverage area of a satellite signal, which can be focused to cover specific geographical areas.

Frame. One complete still image that makes up a part of a video signal.

Frame relay. A high-speed packet switching protocol used in wide area networks (WANs), often to connect local area networks (LANs) to each other, with a maximum bandwidth of 44.725 Mb/s.

Frequency. The number of oscillations in an alternating current that occur within one second, measured in Hertz (Hz).

Frequency division multiplexing (FDM). The transmission of multiple signals simultaneously over a single transmission path by dividing the available bandwidth into multiple channels that each cover a different range of frequencies.

FTP (file transfer protocol). An Internet application that allows a user to download and upload programs, documents, and pictures to and from databases virtually anywhere on the Internet.

FTTC. See *Fiber-to-the-curb.*

FTTH. See *Fiber-to-the-home.*

Full-motion video. The projection of 20 or more frames (or still images) per second to give the eye the perception of movement. Broadcast video in the United States uses 30 frames per second, and most film technologies use 24 frames per second.

Fuzzy logic. A method of design that allows a device to undergo a gradual transition from on to off, instead of the traditional protocol of all-or-nothing.

G

G.dmt. A kind of asymmetric DSL technology, based on DMT modulation, offering up to 8 Mb/s downstream bandwidth, 1.544 Mb/s upstream bandwidth. "G.dmt" is actually a nickname for the standard officially known as ITU-T Recommendation G.992.1. (See *International Telecommunications Union.*)

G.lite [pronounced "G-dot-light"]. A kind of asymmetric DSL technology, based on DMT modulation, that offers up to 1.5 Mb/s downstream bandwidth, 384 Kb/s upstream, does not usually require a splitter, and is easier to install than other types of DSL. "G.lite" is a nickname for the standard officially known as G.992.2. (See *International Telecommunications Union.*)

Geosynchronous orbit (GEO, also known as geostationary orbit). A satellite orbit directly above the equator at 22,300 miles. At that distance, a satellite orbits at a speed that matches the revolution of the earth so that, from the earth, the satellite appears to remain in a fixed position.

Gigabyte. 1,000,000,000 bytes or 1,000 megabytes (see *Byte*).

Global positioning system (GPS). Satellite-based services that allow a receiver to determine its location within a few meters anywhere on earth.

Gopher. An early Internet application that assisted users in finding information on and accessing the resources of remote computers.

Graphical user interface (GUI). A computer operating system that is based on icons and visual relationships rather than text. Windows and the Macintosh computer use GUIs because they are more user friendly.

Groupware. A set of computer software applications that facilitates intraorganizational communication, allowing multiple users to access and change files, send and receive e-mail, and keep track of progress on group projects.

GSM (Group standard mobile). A type of digital cellular telephony used in Europe that uses time-division multiplexing to carry multiple signals in a single frequency.

GUI. See *Graphical user interface.*

H

Hardware. The physical equipment related to a technology.

HDSL. See *High bit-rate digital subscriber line.*

Hertz. See *Frequency.*

HFC (hybrid fiber/coax). A type of network that includes a fiber optic "backbone" to connect individual nodes and coaxial cable to distribute signals from an optical network interface to the individual users (up to 500 or more) within each node.

High bit-rate digital subscriber line (HDSL). A symmetric DSL technology that provides a maximum bandwidth of 1.5 Mb/s in each direction over two phone lines, or 2 Mb/s over three phone lines.

High bit-rate digital subscriber line II (HDSL II). A descendant of HDSL that offers the same performance over a single phone line.

High-definition television (HDTV). Any television system that provides a significant improvement in existing television systems. Most HDTV systems offer more than 1,000 scan lines, in a wider aspect ratio, with superior color and sound fidelity.

Hologram. A three-dimensional photographic image made by a reflected laser beam of light on a photographic film.

Home networking. Connecting the different electronic devices in a household by way of a local area network (LAN).

HTML. See *Hypertext markup language.*

http (hypertext transfer protocol). The first part of an address (URL) of a site on the Internet, signifying a document written in hypertext markup language (HTML).

Hybrid fiber/coax (HFC). A type of network that includes coaxial cables to distribute signals to a group of individual locations (typically 500 or more) and a fiber optic backbone to connect these groups.

Hypertext. Documents or other information with embedded links that enable a reader to access tangential information at programmed points in the text.

Hypertext markup language (HTML). The computer language used to create hypertext documents, allowing connections from one document or Internet page to numerous others.

Hz. An abbreviation for Hertz. See *Frequency*.

I

ILEC. See *Incumbent local exchange carrier*.

Image stabilizer. A feature in camcorders that lessens the shakiness of the picture either optically or digitally.

Incumbent local exchange carrier (ILEC). A large telephone company that has been providing local telephone service in the United States since the divestiture of the AT&T telephone monopoly in 1982.

Institute of Electrical & Electronics Engineers (IEEE). A membership organization comprised of engineers, scientists, and students that sets standards for computers and communications.

Instructional Television Fixed Service (ITFS). A microwave television service designed to provide closed-circuit educational programming. Underutilization of these frequencies led wireless cable (MMDS) operators to obtain FCC approval to lease these channels to deliver television programming to subscribers.

Interactive TV. A television system in which the user interacts with the program in such a manner that the program sequence will change for each user.

Interexchange carrier. Any company that provides interLATA (long distance) telephone service.

Interlaced scanning. The process of displaying an image using two scans of a screen, with the first providing all the even-numbered lines and the second providing the odd-numbered lines.

International Organization of Standardization (ISO). Develops, coordinates, and promulgates international standards that facilitate world trade.

Internet Engineering Task Force (IETF). The standards organization that standardizes most Internet communication protocols, including Internet protocol (IP) and hypertext transfer protocol (HTTP).

Internet service provider (ISP). An organization offering and providing Internet access to the public using computer servers connected directly to the Internet.

Intranet. A network serving a single organization or site that is modeled after the Internet, allowing users access to almost any information available on the network. Unlike the Internet, Intranets are typically limited to one organization or one site, with little or no access to outside users.

IP (Internet protocol). The standard for adding "address" information to data packets to facilitate the transmission of these packets over the Internet.

ISDN (Integrated Services Digital Network). A planned hierarchy of digital switching and transmission systems synchronized to transmit all signals in digital form, offering greatly increased capacity over analog networks.

ISDN digital subscriber line (IDSL). A type of DSL that uses ISDN transmission technology to deliver data at 128 Kb/s into an IDSL "modem bank" connected to a router.

ISO. See *International Organization of Standardization.*

ISP. See *Internet service provider.*

ITU (International Telecommunications Union). A U.N. organization that coordinates use of the spectrum and creation of technical standards for communication equipment.

J

JPEG (Joint Photographic Experts Group). A committee formed by the ISO to create a digital compression standard for still images. Also refers to the digital compression standard for still images created by this group.

K

Killer app. Short for "killer application," this is a function of a new technology that is so strongly desired by users that it results in adoption of the technology.

Kilobit. One thousand bits (see *Bit*).

Kilobyte. 1,000 bytes (see *Byte*).

Ku-band. A set of microwave frequencies (12 GHz to 14 GHz) used exclusively for satellite communication. Compared to C-band, the higher frequencies produce shorter waves and require smaller receiving dishes.

L

LAN. See *Local area network.*

Laser. From the acronym for "light amplification by stimulated emission of radiation." A laser usually consists of a light-amplifying medium placed between two mirrors. Light not perfectly aligned with the mirrors escapes out the sides, but light perfectly aligned will be amplified. One mirror is made partially transparent. The result is an amplified beam of light that emerges through the partially transparent mirror.

Last mile. See *Local loop.*

LATA (local access transport area). The geographical areas defining local telephone service. Any call within a LATA is handled by the local telephone company, but calls between LATAs must be handled by long distance companies, even if the same local telephone company provides service in both LATAs.

LEO (low earth orbit). A satellite orbit between 400 and 800 miles above the earth's surface. The close proximity of the satellite reduces the power needed to reach the satellite, but the fact that these satellites complete an entire orbit in a few hours means that a large number of satellites must be used in a LEO satellite system in order to have one overhead at all times.

Liner notes. The printed material that accompanies a CD or record album, including authors, identification of musicians, lyrics, pictures, and commentary.

LMDS (local multipoint distribution service). A new form of wireless technology similar to MMDS that uses frequencies above 28 GHz.

Local area network (LAN). A network connecting a number of computers to each other or to a central server so that computers can share programs and files.

Local exchange carrier (LEC). A local telephone company.

Local loop. The copper lines between a customer's premises and a telephone company's central office (see *Central office*).

M

Mb/s. Megabits per second.

Megabit. One million bits.

Megabyte. 1,000,000 bytes or 1,000 kilobytes (see *Byte*).

Microcell. The area, typically a few hundred yards across, served by a single transmitter in a PCS network. The use of microcells allows the reuse of the same frequencies many times in an area, allowing more simultaneous users.

MIDI (musical instrument digital interface). An international standard for representing music in digital form. Music can be directly input from a computer keypad and stored to disc or RAM, then played back through a connected instrument. Con-versely, a song can be played by the performer on an instrument interfaced with the computer.

MIPS (millions of instructions per second). This is a common measure of the speed of a computer processor.

MMDS (multichannel multipoint distribution systems). A service similar to cable television that uses microwaves to distribute the signals instead of coaxial cable. MMDS is therefore better suited to sparsely-populated areas than cable.

Mobile satellite services (MSS). The use of satellites to provide navigation services and to connect vehicles and remote regions with other mobile or stationary units.

Modem (MOdulator/DEModulator). Enables transmission of a digital signal, such as that generated by a computer, over an analog network, such as the telephone network.

Monochromatic. Light or other radiation with one single frequency or wavelength. Since no light is perfectly monochromatic, the term is used loosely to describe any light of a single color over a very narrow band of wavelengths.

MPEG (Moving Picture Experts Group). A committee formed by the ISO to set standards for digital compression of full-motion video. Also stands for the digital compression standard created by the committee that produces VHS-quality video.

MPEG-1. An international standard for the digital compression of VHS-quality, full-motion video.

MPEG-2. An international standard for the digital compression of broadcast-quality, full-motion video.

MTSO (mobile telephone switching office). The "heart" of a cellular telephone network, containing switching equipment and computers to manage the use of cellular frequencies and connect cellular telephone users to the landline network.

Multicast. The transmission of information over the Internet to two or more users at the same time.

Multimedia. The combination of video, audio, and text in a single platform or presentation.

Multiple system operator (MSO). A cable company that owns and operates many local cable systems.

Multiplexing. Transmitting several messages or signals simultaneously over the same circuit or frequency.

Must-carry. A set of rules requiring cable operators to carry all local broadcast television stations.

N

Nanometer. One billionth of a meter. Did you know that "nano" comes from the Greek word "dwarf?"

Narrowband. A designation of bandwidth less than 56 Kb/s.

Narrowband ISDN. Same as ISDN.

National information infrastructure (NII). A Clinton administration initiative to support the private sector construction and maintenance of a "seamless web" of communication networks, computers, databases, and consumer electronics that will put vast amounts of information at users' fingertips.

Near video on demand (NVOD). A pay-per-view service offering movies on up to eight channels, each with staggered start times so that a viewer can watch one at any time. NVOD is much more flexible than PPV, and costs far less than VOD to implement.

Network access provider (NAP). Another name for a provider of networked telephone and associated services, usually in the United States.

Network service provider (NSP). A high-level Internet provider that offers high-speed backbone services.

Newbies. New users of an interactive technology, usually identified because they have not yet learned the etiquette of communication in a system.

Nodes. Routers or switches on a broadband network that provide a possible link from point A to point B across a network.

N-ISDN. Narrowband ISDN. See *ISDN.*

NTSC (National Television Standards Committee). The group responsible for setting the U.S. standards for color television in the 1950s.

O

OCR (optical character recognition). Refers to computer programs that can convert images of text to text that can be edited.

Octet. A byte, more specifically, an eight-bit byte. The origins of the octet trace back to when different networks had different byte sizes. Octet was coined to identify the eight-bit byte size.

Operating system. The program embedded in most computers that controls the manner in which data are read, processed, and stored.

Optical carrier 3 (OC3). A fiber optic line carrying 155 Mb/s, a U.S. designation generally recognized throughout the telecommunications community worldwide.

Optical network unit (ONU). A form of access node that converts optical signals transmitted via fiber to electrical signals that can be transmitted via coaxial cable or twisted pair copper wiring to individual subscribers. See *Hybrid fiber/coax.*

Overlay. The process of combining a graphic with an existing video image.

P

Packet switched networks. A network that allows a message to be broken into small "packets" of data that are sent separately by a source to the destination. The packets may travel different paths and arrive at different times, with the destination site reassembling them into the original message. Packet switching is used in most computer networks because it allows a very large amount of information to be transmitted through a limited bandwidth.

Passive optical network (PON). A fiber-based transmission network with no active electronics.

PCN (personal communications network). Similar to PCS, but incorporating a wider variety of applications including voice, data, and facsimile.

PCS (personal communications services). A new category of digital cellular telephone service which uses much smaller service areas (microcells) than ordinary cellular telephony.

Peripheral. An external device that increases the capabilities of a communications system.

Personal digital assistants (PDAs). Extremely small computers (usually about half the size of a notebook) designed to facilitate communication and organization. A typical PDA accepts input from a special pen instead of a keyboard, and includes appointment and memo applications. Some PDAs also include fax software and a cellular telephone modem to allow faxing of messages almost anywhere.

Photovoltaic cells. A device that converts light energy to electricity.

Pixel. The smallest element of a computer display. The more pixels in a display, the greater the resolution.

Point of presence (POP). The physical point of connection between a data network and a telephone network.

Point-to-multipoint service. A communication technology designed for broadcast communication, where one sender simultaneously sends a message to an unlimited number of receivers.

Point-to-point service. A communication technology designed for closed-circuit communication between two points such as in a telephone circuit.

POTS (plain old telephone service). An acronym identifying the traditional function of a telephone network to allow voice communication between two people across a distance.

POTS splitter. A device that uses filters to separate voice from data signals when they are to be carried on the same phone line. Required for several types of DSL service.

PPV (pay-per-view). A television service in which the subscriber is billed for individual programs or events.

Primary-rate interface ISDN (PRI-ISDN). The primary rate ISDN interface provides twenty-three 64-Kb/s channels (called B channels) to carry voice or data and one 16-Kb/s signaling channel (the D channel) for call information.

Progressive scanning. A video display system that sequentially scans all the lines in a video display.

PTT. A government organization that offers telecommunications services within a country. (The initials refer to the antecedents of modern communication: the post [mail], telephone, and telegraph.)

Q

QuickTime. A computer video playback system that enables a computer to automatically adjust video frame rates and image resolution so that sound and motion are synchronized during playback.

R

RBOC (regional Bell operating company). One of the seven local telephone companies formed upon the divestiture of AT&T in 1984. The original seven were Bell Atlantic, NYNEX, Pacific Telesis, BellSouth, SBC Corporation, U S WEST, and Ameritech.

Retransmission consent. The right of a television station to prohibit retransmission of its signal by a cable company. Under the 1992 Cable Act, U.S. television stations may choose between must-carry and retransmission consent.

RF (radio frequency). Electromagnetic carrier waves upon which audio, video, or data signals can be superimposed for transmission.

Roaming. Movement of a wireless node between two microcells.

Router. The central switching device in a computer network that directs and controls the flow of data through the network.

S

SCMS (Serial Copy Management System). A method of protecting media content from piracy that allows copies to be made of a specific piece of content, but will not allow copies of copies.

SCSI (small computer system interface) [pronounced "scuzzy"]. A type of interface between computers and peripherals that allows faster communication than most other interface standards.

SDTV (standard-definition television). Digital television transmissions that deliver approximately the same resolution and aspect ratio of traditional television broadcasts, but do so in a fraction of the bandwidth through the use of digital video compression.

SMPTE (Society of Motion Picture and Television Engineers). The industry group responsible for setting technical standards in most areas of film and television production.

Software. The messages transmitted or processed through a communications medium. This term also refers to the instructions (programs) written for programmable computers.

SONET (Synchronous Optical Network). A standard for data transfer over fiber optic networks used in the United States that can be used with a wide range of packet- and circuit-switched technologies.

Spot beam. A satellite signal targeted at a small area, or footprint. By concentrating the signal in a smaller area, the signal strength increases in the reception area.

Symmetric digital subscriber line (SDSL). A DSL technology that provides a maximum bandwidth of 1.5 Mb/s using one phone line, with a downstream transmission rate that equals the upstream transmission rate and allows use of POTS service on the same phone line. SDSL also refers to single-line digital subscriber line.

Synchronous transmission. The transmission of data at a fixed rate, based on a master clock, between the transmitter and receiver.

T

T1. A standard for physical wire cabling used in networks. A T-1 line has the bandwidth of 1.54 Mb/s.

T3. A standard for physical wire cabling that has the bandwidth of 44.75 Mb/s.

TCP/IP (transmission control protocol/Internet protocol). A method of packet-switched data transmission used on the Internet. The protocol specifies the manner in which a signal is divided into parts, as well as the manner in which "address" information is added to each packet to ensure that it reaches its destination and can be reassembled into the original message.

TDMA (time division multiple access). A cellular telephone technology that sends several digital signals over a single channel by assigning each signal a periodic slice of time on the channel. Different TDMA technologies include North America's Interim Standard (IS) 54, Europe's global system for mobile communications, and a version developed by InterDigital Corporation. These systems differ in circuits per channel, timing, and channel width.

Telecommuting. The practice of using telecommunications technologies to facilitate work at a site away from the traditional office location and environment.

Teleconference. Interactive, electronic communication among three or more people at two or more sites. Includes audio-only, audio and graphics, and videoconferencing.

Teleport. A site containing multiple satellite uplinks and downlinks, along with microwave, fiber optic, and other technologies to facilitate the distribution of satellite signals.

Terabyte. 1,000,000,000,000 bytes or 1,000 gigabytes (see *Byte*).

Time division multiplexing (TDM). The method of multiplexing where each device on the network is provided with a set amount of link time.

Transponder. The part of a satellite that receives an incoming signal from an uplink and retransmits it on a different frequency to a downlink.

TVRO (television receive only). A satellite dish used to receive television signals from a satellite.

Twisted pair. The set of two copper wires used to connect a telephone customer with a switching office. The bandwidth of twisted pair is extremely small compared with coaxial cable or fiber optics.

U

UHF (ultra high frequency). Television channels numbered 14 through 83.

Universal ADSL Working Group (UAWG). An organization composed of leading personal computer industry, networking, and telecommunications companies with the goal of creating an interoperable, consumer-friendly ADSL standard entitled the G.992.2 standard, and commonly referred to as the G.lite standard.

Universal serial bus (USB). A computer interface with a maximum bandwidth of 1.5 Mb/s used for connecting computer peripherals such as printers, keyboards, and scanners.

Universal service provider (USP). A company that sells access to phone, data, and entertainment services and networks.

Universal service. In telecommunications policy, the principle that an interactive telecommunications service must be available to everyone within a community in order to increase the utility and value of the network for all users.

Uplink. An antenna that transmits a signal to a satellite for relay back to earth.

URL (uniform resource locator). An "address" for a specific page on the Internet. Every page has a URL that specifies its server and file name.

V

Variable bit rate (VBR). Data transmission that can be represented by an irregular grouping of bits or cell payloads followed by unused bits or cell payloads. Most applications other than voice circuits generate VBR traffic patterns.

VDSL (Very high bit-rate digital subscriber line). An asymmetric DSL that delivers from 13 Mb/s to 52 Mb/s downstream bandwidth and 1.5 Mb/s to 2.3 Mb/s upstream.

Vertical blanking interval (VBI). In an NTSC television signal, the portion of the signal that is not displayed on a television receiver. Some of the lines in the VBI contain "sync" information that is used to identify the beginning of a new picture. Some of the blank lines in the VBI can be used to carry data such as closed captions.

Vertical integration. The ownership of more than one function of production or distribution by a single company, so that the company, in effect, becomes its own customer.

VHF (very high frequency). Television channels numbered 2 through 13.

Videoconference. Interactive audio/visual communication among three or more people at two or more sites.

Video on demand (VOD). A pay-per-view television service in which a viewer can order a program from a menu and have it delivered instantly to the television set, typically with the ability to pause, rewind, etc.

Videophone. A telephone that provides both sound (audio) and picture (video).

Videotext (also known as videotex). An interactive computer system using text and/or graphics that allows access to a central computer using a terminal or personal computer to engage in data retrieval, communication, transactions, and/or games.

Virtual reality (VR). A cluster of interactive technologies that gives users a compelling sense of being inside a circumambient environment created by a computer.

VRML (virtual reality markup language). A computer language that provides a three-dimensional environment for traditional Internet browsers, resulting in a simple form of virtual reality available over the Internet.

VSAT (very small aperture terminal). A satellite system that uses relatively small satellite dishes to send and receive one- or two-way data, voice, or even video signals.

W

Wide area network (WAN). A network that interconnects geographically distributed computers or LANs.

Wireless cable. See *MMDS.*

Wireless node. A user computer with a wireless network interface card.

WORM (Write once, read many). A technique that allows recording of information on a medium only once, with unlimited playback.

X

X.25 data protocol. A packet switching standard developed in the mid-1970s for transmission of data over twisted-pair copper wire.

xDSL. See *Digital subscriber line.*